Autism and the Brain

Tatyana B. Glezerman

Autism and the Brain

Neurophenomenological Interpretation

Springer

Tatyana B. Glezerman
Independent Practitioner
New York, NY, USA

ISBN 978-1-4614-4111-3 ISBN 978-1-4614-4112-0 (eBook)
DOI 10.1007/978-1-4614-4112-0
Springer New York Heidelberg Dordrecht London

Library of Congress Control Number: 2012941499

Printed on acid-free paper

Springer is part of Springer Science+Business Media (www.springer.com)

For Ilya, Oliver and Abigail and in honor of my parents, Boris and Maria Altschuler

Preface

"Autism and the Brain" is an interdisciplinary study that provides neurophenomenological analysis of autism. It hypothesizes the brain networks involved in autism allowing a new understanding of why symptoms of autism go together and how they are connected to one another in a complex pattern that repeats itself in each autistic child.

In this book, introspections of high functioning autistic individuals (HFA), products of autistic savants (as a means of inferring their subjective experience), and behavioral descriptions of low functioning autistic individuals (LFA) are analyzed in connection with the function of various cortical and subcortical regions (and the networks they belong to).

The specific function of each particular region (network) is considered within the conceptual frame of human brain three-dimensional differentiation: *hierarchical* (low order–higher order function); *Left–Right Brain* (analytic–holistic information processing); *Posterior–Anterior Brain* (spatial–temporal organization of information).

Considering general principles of brain organization, together with the specific functions of the separate regions while explaining symptoms of autism, this work strives to overcome reductionism of modern neuroscience and arrive to a new cooperation linking phenomenology, psychopathology, and neuroscience.

The conducted neurophenomenological analysis gives rise to a new hypothesis of the brain pattern that underlies clinical manifestation of autism. The pattern includes the primary, fundamental brain abnormality, and secondary changes resulting from reorganization of brain functional systems in response to the primary deficit.

The book tracks the fundamental problem to part of the left hemisphere which is responsible for awareness of personal identity, self as *one* and apart from others. In this connection, the primary problem of autism is the problem of consciousness, of the self, not of cognition. Self as an agent is missing or underdeveloped.

This book offers a unique and original approach. It gives a panoramic view of autistic brain's functioning. It defines the brain networks that explain the diversity of the clinical manifestation of autism: from the low functioning autists to high functioning autists to autistic savants. On the other hand, it delineates autistic clinical-brain model—what is necessary and sufficient for diagnosis. This is important as at present we witness unjustified broadening of autistic-spectrum-disorder boundaries.

New York, NY, USA Tatyana B. Glezerman

Acknowledgments

I am grateful to Dr. Natalia Novinskaya for critical comments and important suggestions.

I am indebted to Dr. Amanda Moberg Wilson for her invaluable help with editing and preparation of the manuscript. Her sense of responsibility and devotion cannot be exaggerated.

Contents

Chapter 1
Introduction

1.1 Prelude

The 6-year-old boy ran into my office and immediately, without making eye contact and without showing any interest in the examiner, rushed for the small objects on the table and began quickly but carefully carrying them to another slightly smaller table on an adjoining wall, one by one. Finished, he carried them back to the table, one by one, in exactly the same order.

Absorbed in this activity, the boy paid no attention to anything else, squealing with satisfaction from time to time. He pushed my hands away and tenaciously continued moving the objects from place to place. Periodically, always with the same inflection, he repeated, "Where is mommy, mommy is coming, mommy is in the car." (His mother had remained in the waiting room.) Otherwise, he did not speak and would not follow instructions or answer any questions.

Next, I showed the boy pictures and asked him to name the objects pictured. His responses were excellent. I gave the boy two more tests, quite challenging, of visual recognition and verbal naming: distinguishing the figures superimposed on one another (Fig. 1.1a) and distinguishing a figure from the background (Fig.1.1b). He performed flawlessly on the first test (Fig. 1.1a) and failed completely on the second (Fig. 1.1b).

Both tests are designed to diagnose difficulties in visual recognition of objects caused by lesions in the specific brain areas that process complex visual and verbal information. People who have such problems might fail both tests. Normal people should do fine on both tests. Yet the 6-year-old boy, an autism patient, excelled on one test and failed the other. Had it happened by chance? Or might it reflect something intrinsic to autism?[1]

[1] I will return to this example in Chap. 10.

T.B. Glezerman, *Autism and the Brain: Neurophenomenological Interpretation*,
DOI 10.1007/978-1-4614-4112-0_1, © Springer Science+Business Media New York 2013

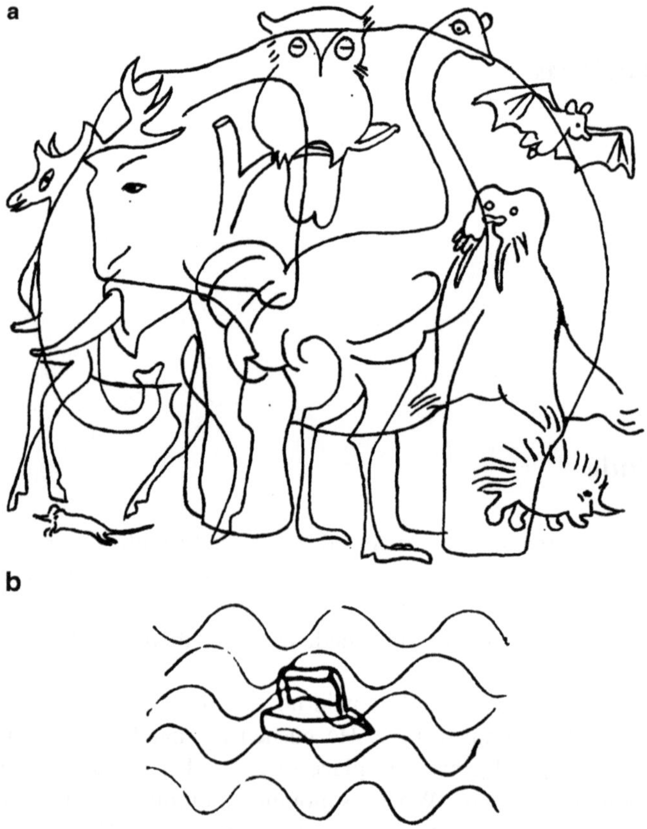

Fig. 1.1 (**a**) Poppelreuter overlapping figure test and (**b**) figure-ground test

It is often possible to predict the site of a lesion in the brain from a specific clinical pattern (neuropsychological syndrome). There are various neuropsychological syndromes associated with specific (focal) brain dysfunctions. None of these resemble autism.

Autism, however, is a highly distinctive and permanent clinical pattern. Autistic individuals differ from one another, not only by degree of intellectual ability, but also by their special interests—which are often extraordinarily varied and original. Even so, a "pattern of autism" is superimposed on all the individual variations. "Once one has properly recognized an autistic individual one can spot such children instantly. They are recognizable from small details, for instance, the way they enter the consulting room at their first visit, their behavior in the first few moments and the first words they utter" (Asperger, translated by U. Frith, 1991, p. 68).

The mystery of autism began with its discovery. Leo Kanner, a psychiatrist at the John Hopkins Clinic in Baltimore, and Hans Asperger, a pediatrician with the University Paediatric Clinic in Vienna, independently but almost exactly contemporaneously in 1943–1944[2] described a puzzling clinical entity which they both—by remarkable coincidence—called "autism."[3]

Leo Kanner and Hans Asperger had never met, and during World War II communication between the United States and Austria was hardly possible.

Both men distinguished the core abnormality, which may account for the symptoms and disabilities in autism: a fundamental abnormality of affective contact—the children's inability, from the beginning of life, to relate themselves to people and situations in the ordinary way.

The pattern of autism as described by Kanner and Asperger:

1. There is a profound lack of affective contact with other people accompanied by vacant facial expression, lack of eye contact, and a monotonous, robot-like voice, or peculiar intonation inappropriate to the situation.

   From the moment he set foot on the ward he stood out from the rest of the group, and this did not change. He remained an outsider and never took much notice of the world around him (Asperger, 1991, p. 42).

   He paid no attention to persons around him. When taken into a room, he completely disregarded the people and instantly went for objects, preferably those that could be spun. Commands or actions that could not possibly be disregarded were resented as unwelcome intrusions. But he was never angry at the interfering person. He angrily shoved away the *hand* that was in his way or the *foot* that stepped on one of his blocks, at one time referring to the foot on the block as 'umbrella'. Once the obstacle was removed, he forgot the whole affair (Kanner, 1943, p. 220).

   He had few facial expressions and gestures... He talked slowly and in a deadpan way without much modulation. He never looked at his interlocutor while talking. His gaze was far away (Asperger, 1991, p. 52).

2. Language development in the so-called low functioning autists (LFA) differs significantly from that of high functioning autists (HFA). LFA may lack spontaneous speech; some of them remain mute. LFA may produce parrot-like repetitions of heard word combinations (echolalia), or may repeat words and phrases out of context, both the words and their intonation never varying (speech stereotypes). HFA may have quite sophisticated speech. Neither LFA nor HFA use language for communication, i.e., to convey meaning to others.

   Upon entering the room, he instantly went after objects...he got hold of a pair of scissors and patiently and skillfully cut a sheet of paper into small bits, singing the phrase 'Cutting paper,' many times. He helped himself to a toy engine, ran around the room holding it up high and singing over and over again, 'The engine is flying'. While these utterances, made always with the same inflection, were clearly connected with his actions, he ejaculated others that could not be linked up with immediate situations. These are a few examples:

[2] Kanner published his article in 1943; Asperger submitted his work as his doctoral thesis also in 1943, and it was published in 1944.

[3] Moreover, they both borrowed this title from Bleuler's term *autism*, which designates the core deficit in schizophrenia—loss of contact with reality. In schizophrenia, autism is accompanied by autistic thinking. Both Kanner and Asperger indicated that the entity described by them is *not* characterized by autistic thinking.

'The people in the hotel'; 'Did you hurt your leg?' 'Candy is all gone, candy is empty' (Kanner, 1943, p. 227).

His speech...conveyed the impression bordering on caricature... He spoke incessantly, regardless of the questions he was being asked. Everything he did was accompanied by elaborate explanations. He constantly justified why he did something in a particular way. He had to tell others at once whatever it was that captured his attention, whether or not the remark was relevant to the situation (Asperger, 1991, pp. 60–61).

His 'conversation' consisted of repetitive questions of an obsessive nature: 'How many days in a week, years in a century, hours in a day, hours in half a day, weeks in a century, centuries in half of a millennium', etc. (Kanner, 1943, p. 222).

3. There is a marked limitation of spontaneous activity. Rather, the autist's pattern of activities—from movement to action to complex behavior—is as stereotypically repetitious as his verbal utterances, with an obsessive desire for the preservation of sameness.

He ran around in circles emitting phrases in an ecstatic-like fashion. He took a small blanket and kept shaking it, delightedly shouting, "Ee! Ee!" he could continue in this manner for a long time and showed great irritation when he was interfered with. All these and many other things were not only repetitions but also recurred day after day with almost photographic sameness (Kanner, 1943, p. 228).

He wandered about smiling, making stereotyped movements with his fingers, crossing them about in the air... He spun with great pleasure anything he could seize upon to spin... He arranged beads, sticks, or blocks in groups of different series of colors... Most of his actions were repetitions carried out in exactly the same way in which they had been performed originally. If he spun a block, he must always start with the same face uppermost. When he threaded buttons, he arranged them in a certain sequence that had no pattern to it but happened to be the order used by the father when he first had shown them to Donald (Kanner, 1943, p. 219).

1.2 Current Approach to Autism: Emphasizing "Trees" and Losing "The Forest"?

Autism[4]—an enigmatic and devastating disorder of childhood that persists throughout life—has attracted a vast amount of research in recent years, with a focus on brain imaging and genetic studies. There is an abundance of new data and facts, but autism continues to elude understanding.

It is well established at present that autism is a neurodevelopmental disorder, and it is the most strongly genetic condition among the neurodevelopmental conditions

[4] In this book, the term *autism* includes low functioning autism (LFA), high functioning autism (HFA) and Asperger syndrome (AS). The term AS is not used by itself, it is included into the category of HFA. I did not use the umbrella term Autistic Spectrum Disorder (ASD) common in the recent literature because it is still not uniformly defined. For example, some authors include in it Pervasive Developmental Disorder Not Otherwise Specified (PDD NOS) (Akshoomoff, 2006), while others distinguish four subgroups within ASD: Asperger syndrome, high, medium, and low functioning autism (Baron-Cohen et al., 2002), etc.

(Bailey et al., 1995). A number of genetic variants have been studied that have demonstrated association with autism. However, the majority of findings have not been widely replicated (Baron-Cohen & Belmonte, 2005). There is as yet no confirmed genetic finding in autism that has been explained in terms of its specific relation to brain development (Belmonte et al., 2004).

Prevalence estimates of autism have increased significantly from 4/10,000 before 1980 to 1/150. No gene abnormality alone can explain such an increase in prevalence over the last 20 years. Despite a number of theories, the actual reasons for this increase are still unclear (Theoharides, Kempuraj, & Redwood, 2009).

Structural neuroimaging and morphological studies in autism have revealed abnormalities in various cortical and subcortical brain regions. However, these results have not been always replicated. Moreover, brain abnormalities were small, nonspecific, and took many different forms not consistent from subject to subject or from report to report. The brain changes reported to be associated with autism might as well be considered as overlapping with the cytoarchitectural aberrations existing in the general population.

Interesting data about abnormal brain growth in autism, together with the abnormalities in white matter, led to the hypothesis that a disorder of neural connectivity might be central in the brain pathogenesis of autism (Herbert et al., 2003; Piven et al., 1995; Just, Cherkassky, Keller, & Minshew, 2004; Welchew et al., 2005). There is still no well-accepted account of an underlying brain dysfunction (disconnectivity pattern) shared by all persons with autism.

Most intriguing are the findings in autism obtained from functional neuroimaging studies: in people with autism versus persons not affected by this disorder, different cortical areas are activated while performing the same cognitive tasks.

Cognitive tasks are designed to explore the specific function of a particular cortical area, and this area is usually activated in normal people during fMRI task performance. Autistic persons' task performance might be similar to the control group, but they do not activate the target area. Instead, they activate cortical areas that are not task-specific. For example, autists were able to perform adequately on face recognition tasks but did not activate the fusiform facial area (FFA), the face-specific area in the brain (Schultz et al., 2000), and on a facial expression recognition task, autists did not activate the amygdala, an area specific for this task (Baron-Cohen et al., 2000; Critchley et al., 2000a; 2000b). During word processing in autistic people, increased activation was observed in the right inferior temporal area, instead of its left hemispheric counterpart, which was the target area (Muller et al., 1999). Shape processing in autism activated the cortical area specific in the norm for object recognition (Ring et al., 1999). Each new research study about an abnormality in the neuroanatomy of a particular cortical function in autism seems to claim that this particular difference might be *the* underlying brain mechanism of autism. However, the meaning of these findings in the context of the functional organization of the autistic brain as a whole remains unclear. Also, practically every cortical function explored in autism had neuroanatomy that differed from the norm. As this is the case (especially considering autists are able to adequately perform the corresponding cognitive tasks), one should doubt the current approach to research in autism.

The reductionistic approach to the brain pathogenesis of autism is closely related to the behavioral definition of autism.

Current classification of psychiatric disorders (including autism) is strictly behavioral (DSM-IV, American Psychiatric Association). This behavioral classification of psychiatric disorders is not medical, but rather a statistical model. The medical model's *syndrome* is useful for describing medical conditions showing frequently co-occurring symptoms but having unknown etiology. Identifying a *common factor* between individuals showing variation in the features of a syndrome is an important step toward identifying a cause (Gerrard & Rugg, 2009). The medical model, thus, imposes a search for pathogenesis (and etiology). The behavioral or statistical classification model is a contrasting philosophy under which behavioral manifestations of disease are viewed not as symptoms but as criteria. To make a diagnosis under the behavioral model means to find the minimum number of criteria necessary for statistical (viz., mathematical) significance. With this "behavioral" approach, the problem of autism seemed to lose its mystery but in reality became even more confusing.

In the DSM-IV's behavioral classification, the three symptom "domains"—social impairment, communication impairment, and stereotypical, repetitive behavior—are considered independent and equal, separate entities. This behavioral definition, therefore, does not imply that the three domains characterizing autism must have a common underlying brain mechanism (pathogenesis). Correspondingly, having not found significant statistical correlation (quantitative expression) between different symptom domains in autism, researchers have concluded that different domains may have independent genetic etiology (Buxbaum et al., 2001; Dworzynski, Happe, Bolton, & Ronald, 2009; Ronald et al., 2006; Ronald, Happe, Price, Baron-Cohen, & Plomin, 2006).

Penn (2006) notes: "Although a number of neuropathological abnormalities have been linked to autism, most research has a correlational as opposed to causational focus…many research findings in autism "exist in a vacuum" (Belmonte et al., 2004, p. 658), with limited links between structural abnormalities, functional abnormalities, and behavioral correlates" (p. 72). Not surprisingly, there is some disappointment and confusion in the field of autism research as illustrated by the following statement, which combines the extreme of the reductionistic viewpoint with the opposite—mind-brain dualism. "Autism's surface symptoms might…be understood as the developmental reaction of a normal human mind to abnormal neural hardware" (Belmonte et al., p. 649). Even the well-esteemed expert in autism, Uta Frith, is disappointed in autism brain research. In her statement (that follows) we can still see a disjointed philosophy of behaviorism through her claim that mind and brain can be understood separately from each other. "Dare I say that the really interesting facts about autism are not about the brain and not about genes? They are about the mind. I firmly believe that even if we did know everything about the causes of autism, we would still not understand autism. We need to know what is like to be autistic [I agree with the last sentence here.]" (Frith, 2008, p. 65).

I believe the first thread to grasp in the work of disentangling the mystery of autism is the pattern of autism. Expressing itself in each autistic child with the "chased reiteration" of a coin, the pattern of autism gives a powerful impression of wholeness.

The pattern of autistic behavior induces a compelling and intriguing thought: there must be a particular brain pattern that is producing with such regularity this *co-occurrence* of seemingly unrelated symptoms (clinical pattern).

Autism varies in severity and can occur at all levels of intellectual ability. What does the classical description of infantile autism—a person with severe mental impairment—have in common with a highly intelligent, successful individual like the outstanding animal scientist Temple Grandin, who is herself autistic? Both extremes have in common the set of core autistic features, the "pattern of autism." In severe cases of autism, there is a significant mental handicap in which the absence of communicative speech, lack of spontaneous behavior, and sometimes non-responsiveness result in a very low measured IQ or even impossible-to-define IQ. But children with autism have a normal or attractive appearance, in stark contrast to many other forms of mental handicap.

Language deficit and low IQ are no more common in families of autistic individuals than in the general population. Therefore, low IQ and language deficit do not represent genetic vulnerabilities to autism—that is, they do not predispose a person to autism—but must instead, where present in autistic individuals, be considered manifestations or results of the autism. Many high functioning autistic individuals exhibited severe impairments, including mutism, in early childhood. The tantalizing question of which children will become high functioning, and which will remain low functioning autists (LFA), cannot be answered at an early age. Do high functioning autists have a different potential than LFA, or do they achieve high functioning via more adequate training? If HFA actually do have different potential (i.e., the ability to achieve better compensation for their autistic condition), then what exactly is that potential; how can we diagnose it at an early age; and what is the optimal environment and training for its expression?

1.3 Clinical Neurophenomenological Approach to Autism

In examining psychiatric patients, there are two types of symptoms (1) objective and (2) subjective. Objective symptoms can be directly observed, described, and quantified.

Subjective symptoms have to be grasped by transferring oneself, so to speak, into another individual's psyche, that is, by empathy. They can only become an inner reality for the observer by his participating in the other person's experiences (Jaspers, 1968, 1997).[5]

[5] The behavioral approach uses mathematical methods to determine the statistical significance (or lack thereof) of separately observed features of a particular clinical group (study population). There is nothing wrong with such detailed descriptions. Indeed they are very important and give invaluable information. What is wrong, or at least incomplete, is treating the observation and mathematical description of behaviors as a goal. Behavioral manifestations may or may not adequately reflect altered subjective experience. Jaspers (1968, 1997) warned that a claim to be concerned with objective data *only* has a natural consequence: it will be psychiatry without psyche—elimination of everything that can be called mental or psychic.

Jaspers outlined three kinds of methods by which we can carry out a phenomenological analysis and determine what patients really experience (1) immersing oneself, so to speak, in their gestures, behavior, and expressive movements; (2) exploration, by direct questioning of patients and by means of the accounts which they themselves, under our guidance, give of their own experiences; and (3) written self-descriptions. Jaspers indicated that in all these instances we are pursuing phenomenology in so far as we are oriented towards subjective experience and not towards objective manifestations, which are the means, not the object, of our investigation (Jaspers, 1968, 1997).

Autists are unable to emotionally relate to other people. This characteristic inability of autism might be an inability to experience "something" or an experiencing of real-world events differently than "normal" persons. Autism, then (like other psychiatric disorders such as schizophrenia) may be considered a disorder of subjective experience.[6]

The question is: how can we empathize with the autistic child if we are struck by the lack of replies and the impossibility of making contact with the child? Here we refer again to the founder of psychiatric phenomenology, Karl Jaspers. He distinguished three groups of pathological phenomena. The first consists of phenomena known to us all from our own experience. Next, there are phenomena to be understood as exaggerations, diminutions, or combinations of phenomena that we ourselves experience. The third group of pathological phenomena is distinguished from the two previous groups by their complete inaccessibility to any empathic understanding. We perceive them individually, "not through any positive understanding of them but through the shock which the course of our comprehension receives in the face of the incomprehensible" (Jaspers, 1968, p. 1318). Autism fits into this third group of pathological phenomena.[7]

We must pay tribute to the great vision and clinical insight of Kanner and Asperger, who discovered the unique syndrome—the pattern of autism. They meticulously described behavioral manifestations (objective symptoms) of autism, but I believe they grasped the "gestalt," the "pattern of autism" by empathy.

---

[6] Autism and schizophrenia are very different disorders. Strikingly, as in autism, neuroimaging and morphological studies of schizophrenia have similarly confusing results: findings are perplexingly heterogenous, "global" but variable, not consistent and non-specific, overlapping with those in the normal population. This rings the bell that we are dealing with artifacts. Standard neuropsychological tests and in particular neurocognitive tasks that are used in fMRI studies are based on our knowledge about localization of cortical functions. These tests are designed to diagnose focal brain lesions. However, we know *now* that neither schizophrenia nor autism are focal brain disorders. Therefore, a standard neuropsychological approach is not adequate to autism.

[7] It is interesting to note that no other neurodevelopmental condition gets "assigned" with so many "unscientific" epithets such as "mysterious," "enigmatic," "puzzling," "fascinating," "paradoxical," as does autism. This is in contradiction to the rather "boring" behavioral cliché it represents, the lack of imagination described in autistic children—their literal, "unstrange" mind. It seems to me that this paradox reflects *"the shock which the course of our comprehension receives in the face of the incomprehensible"* in Jaspers' terms.

This book is offering a new and different approach—to look at autism from a neurophenomenological perspective. Self-reports of high functioning autists and the products of autistic savants (as a means of inferring their subjective experience) will be analyzed in relation to different hierarchical levels in the brain and function of concrete brain networks.

1.3.1 *The Three-Dimensional Brain Organization as Conceptual Framework for Neurophenomenological Analysis of Autism*

The conceptual framework my work relies upon is a derivative of the pioneering Russian neurophysiologist Nikolai Bernstein's ideas on cerebral organization of movements (1947/1990). The transmutation of Bernstein's discoveries regarding physical movement to subjective experience and construction of self has been detailed in previous works (Glezerman, 1986; Glezerman & Balkoski, 1999), but I will outline it again here, more briefly, starting with a summary of Bernstein's model. According to Bernstein, "construction" of movement in the brain can be represented as a "multistory building" composed of hierarchically overlaid stories of different phylogenetic age, each corresponding to a certain functional level. Each functional level is considered a relatively autonomous system comprised of two parts operating concurrently: posterior brain regions associated with afferentation and anterior brain regions associated with efferent systems. The mode of operation of the posterior brain is spatial organization of information, whereas the mode of operation of the anterior brain provides temporal organization of information. Each of Bernstein's level has its own unique spatial and temporal organization of information. I will often refer in this work to the *space* and *time* of a particular level, but as Bernstein aptly noted, it is "subjective space" and "subjective time." At each level, afferentation is connected with *space*, while movement is connected with *time*. Afferentaion gives information about the field of space where movement is performed, which is a "model" or the spatial configuration of action—Bernstein called it the semantic structure of movement. The anterior brain unfolds these spatial images in time, creating the *dynamic scheme*, or representation of action. In any complex movement, each level has its specific contribution (its own "movement"), however, Bernstein emphasized, the highest level participating in the movement's formation is the leading one—conscious and voluntary, directly responding to the action's task. Levels lower than the leading one participate in the formation of movement in an assimilated way, as a background.

Thus, Bernstein's model is two-dimensional: it incorporates the interactive functioning of hierarchical (vertical) and intrahemispheric (horizontal) dimensions of brain differentiation. Having found Bernstein's comprehensive system to provide a unifying framework for understanding brain functional organization, it was expanded to include the *symbolic* level that subserves higher cortical functions (Glezerman, 1986). Table 1.1 demonstrates the interplay between the hierarchical

Table 1.1 Hierarchical and intrahemispheric brain organization [Based on Bernstein (1947, 1990); Glezerman (1986); Glezerman and Balkoski (1999)]

	Intrahemispheric dimension					
	Posterior brain			Anterior brain		
Hierarchical dimension (function level)	Anatomic base	Major modality	"Space"	Anatomic base	Type of movement	"Time"
Level 1: Basic regulation of rhythmical, homeostatic processes of the organism; muscle tone	Cerebellum	Proprioceptive Vestibular	Body position against gravity	Midbrain (substantia nigra and nucleus ruber group)	Muscular tone	Simple rhythm
	Hypothalamus	Visceral Interoceptive	Internal milieu of the organism			Circadian rhythm
Level 2: Thalamic	Thalamus	Proprioceptive tactile	Spatial coordinates of one's own body	Pallidum	Synergetic movements	Complex rhythm, individual pattern
Level 3: Sensory-motor	Primary sensory cortex	Visual Auditory Proprioceptive Tactile Vestibular	External spatial field	Striatum Primary motor cortex	Goal-directed movements in external space	Moment, speed, duration; temporal sequence
Level 4: Gnostic-praxic	Secondary (associative) sensory cortex	Visual Auditory Proprioceptive Tactile	Object, topological schema	Premotor cortex	Object action (praxis)	Semantic sequence
Level 5: Symbolic	Tertiary (temporal–parietal) cortex	Supramodal	Concept, individual symbol	Prefronal cortex	Thinking, Language, goal-directed behavior	Semantic sequence

and the intrahemispheric dimensions. It shows the five functional levels of different phylogenetic age.

The first level (cerebellum [and hypothalamus]–mid brain) is concerned with regulation of the most basic rhythmic, homeostatic processes of the organism. The second (thalamic-pallidum) level defines internal space derived from spatial coordinates of one's body and is subserved mostly by proprioceptive modality. Level three (striato-cortical) is called the sensory-motor and concerned with the external spatial field; visual, auditory, vestibular, and proprioceptive/tactile modalities provide its afferentation. Next is the gnostic-praxic (cortical) level, concerned with object action; visual, auditory, tactile, and proprioceptive modalities inform this level. Level five (cortical) is the symbolic level; it is supramodal, and its work is operating with symbols.

1.3.1.1 Adding the Third Dimension

Movements, actions and, on a broader scale, behavior are the domain of the left hemisphere (LH). Thus, Bernstein's model deals for the most part with LH functioning. His model has been expanded to include the third dimension of human brain differentiation—interhemispheric (Glezerman, 1986; Glezerman and Balkoski, 1999). Interhemispheric dimension refers to the different modes of information processing in the LH (analytic) versus the right hemisphere (RH)—holistic. The principle of LH processing consists of the breaking down of processed information (analysis) and successive sifting of the resultant variants, leading to synthesis (a secondary model of the world). It is supposed that LH representations do not exist in a whole, integrated form but rather are put together as certain combinations of discrete units (features). The RH, however, appears to process information in a holistic manner. Here, the emerging image of the whole is not dissected and parceled: the integral image is represented in its non-repeatable uniqueness.

What follows is a more detailed description of the function levels incorporating interrelatedness of the three brain dimensions (hierarchical, intrahemispheric, and interhemispheric) beginning with thalamic level.

Bernstein described the thalamic level as the level of synergy and pattern. Afferentation (sensory synthesis) of this level consists exclusively of bodily "feelings"—proprioceptive, tactile, and deep tactile (pressure sense, vibration). Bernstein built his concept of the thalamic level on the fact that it has no afferentation from distant modalities (visual and auditory) or from vestibular. Bernstein emphasized that the major component in afferentation at this level is the proprioceptive (kinesthetic) sense.[8] "If any sensory synthesis in the brain can be called proprioceptive, it is the synthesis of the thalamic level. Afferentation of this level is the afferentation of one's own body, proprioception *par excellence.* Body is (for this level)

[8] The synonymous terms "proprioceptive" and "kinesthetic" will herein be used interchangeably.

both the initial coordinate system to which proprioception is related and the final goal of the sensory feedback—movement" (Bernstein, 1947, p. 66).

This *afferentation-space* thus described is the cornerstone for understanding thalamic level movements. Neither lower nor higher levels' effector systems, Bernstein indicated, receive such exhaustive and detailed information about body motor apparatus, as does the globus pallidum from the thalamus. Such comprehensive data allow highly coordinated and simultaneous contractions of numerous muscle groups: a "synergetic chorus" (hence Bernstein's naming it the level of synergy). All complex movements for which the leading level is higher than the thalamic one (i.e., walking, running, swimming, working with tools, dancing, playing tennis, etc.), wholly depend on this level as a coordinational base.

What then are the movements of the thalamic level per se? Table 1.1 shows that *space* of the thalamic level is the spatial coordinate system of one's body. Correspondingly, movements led by this level are directed towards one's own body—with no regard to the external world. Thalamic level movements are ends unto themselves, corrected only along the line of their own harmony and organization. In essence, they are "simply" alternations in body position. Bernstein called the thalamic level movement a propriomotor function or reaction. He also emphasized that the main sensory-proprioceptive input for the thalamic level comes from the extremities: "body at the thalamic level is the locomotor machine equipped with four extremities-engines."[9] Our extremities are pendulums prone to oscillation. Moreover, they are multilinked pendulums, with each extremity consisting of several links (pendulums) connected by joints. Each link has its own oscillation frequency. Proprioceptive feedback (for example, muscle sense of stretching) from the extremities-pendulums, informs on the position of each pendulum at a particular moment. This information is integrated according to the body's spatial coordinates, allowing for reigning in the varied rhythmical patterns of the separate pendulums into one resulting rhythm (Bernstein, 1947, 1990). The united rhythm is individual-specific and reflects the unique configuration of one's body and its extremities-pendulums. Furthermore, according to Bernstein, one's unique whole-body rhythm represents the temporal organization of this level's movement (*time* of this level). Due to the construction of the human body's extremities, level-of-synergy-movements have a significant tendency to be cyclical, rhythmical, and repetitive. However complicated the rhythm may be, it is always exactly reproduced in that individual. For this striking sameness of the thalamic level's rhythmical movement, Bernstein called it the level of cliché, or pattern. Thus a unique signature can be recognized in all kinds of complex motions and behaviors of a particular individual—in gestures, mimicry, handwriting, dancing, playing music or sport, as well as individual voice and speech characteristics—due to a consistent underlying rhythm.

In the LH, kinesthetic sense from the extremities provides afferentation for movements, giving feedback about body positions and allowing sensory corrections. It was proposed (Glezerman, 1986; Glezerman & Balkoski, 1999) that the RH takes

[9] Contrast this to the cerebellum–midbrain level where the proprioceptive–vestibular input comes from the trunk and head, constructing the body as a weight in the gravitation field.

Table 1.2 The sensory-motor level in the brain: anterior/posterior and left/right dimensions

Left hempisphere (Bernstein, 1947, 1990)	Right hemisphere (Glezerman & Balkoski, 1999)
Posterior brain (afferentation/space)	
One aperiodic space (external spatial field where movement is unfolded)	Multiple spaces—VSSs
External spatial field is homogenous, indifferent to what fills it and devoid of its own features	VSS is a global entity, including visual picture with its objects and action, and emotion at the moment it was perceived. It is experiential sense of events
External spatial field is a continuous space	While each VSS is continuous in itself, these multiple spaces are separate, they do not get intermixed or crossed
Anterior brain (movement/time)	
Time is the sequence of moments of "real" environmental time according to which movement is unfolded. Time is linear and is characterized by moment, speed, and duration	Moment is internalized as an intimate part of VSS—moment of experience of the external world. Time is experienced through movement-continuously changing impressions of the external world

the same kinesthetic input and creates a holistic image: one's own body as *space*, one and the same with the kinesthetic sense from the extremities that makes the boundaries of this space. Likewise, while in the LH, individual-specific rhythmical pattern of the thalamic level provides a temporal "frame" for its cyclical movements, in the RH, such rhythm is an indivisible part of the body-space gestalt.

Next in the hierarchy is the sensory-motor level. *Space* of the sensory-motor level, according to Bernstein, is the external spatial field. It is aperiodic and homogenous. Afferentation is provided by the distant modalities, mostly visual and auditory, giving sensory feedback about the outside space where the movement is to be performed. The sensory-motor level's *time* is linear, "the arrow of time." Unfolded in this "objective" environmental time, movements performed at this level are aperiodic, with a beginning and end-point destination. Movements are organized by the *time* of the sensory-motor level: moments, speed, and duration. The sensory-motor level thus is the level of possession of external space and overcoming its forces. Correspondingly, movements for which this level is the leading one are all goal directed movements in the external spatial field.

As an attempt to understand what the sensory-motor level of the RH (with its holistic processing) would be like, the construct of *Visual Scene-Situation*—VSS (Glezerman, 1986; Glezerman & Balkoski, 1999) was proposed. VSS stands for the immediate impression of the outside world, stored as perceived in the RH. It is a totality of experience, including visual picture, action and affects present at the moment. The moment (*time* of the sensory-motor level) is a *temporal marker* of the VSS and a product of the anterior brain, while the VSS is stored in the posterior brain. Stored VSSs remain singular, they are not divided into components parts and not connected with each other. Owing to the *temporal mark*, they are internalized as an ordered succession of moments of experience of the external world. Table 1.2 summarizes LH/RH dichotomy of *space/time* at the sensory-motor level.

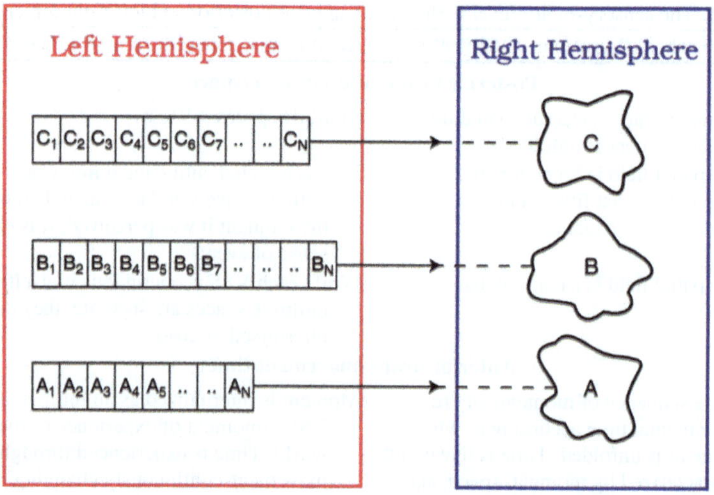

Fig. 1.2 Interhemispheric specialization: analytic versus holistic information processing. From *Language, Thought and the Brain*, (Fig. 3, p. 22), by T. Glezerman and V. Balkoski, 1999, Kluwer Academic/Plenum Publishers, copyright 1999; with kind permission from Springer Science + Business Media B.V

The sensory-motor level's external spatial field remains the most objective of all levels—"metric" and "geometric." At higher function levels, reorganization of *space* leads to loss in objectivity but gained meaning (semantic organization). Bernstein describes how *objective space* of the sensory-motor level evolves toward schematization becoming *"topological" space* at the next level: gnostic-praxic. The *space* of the gnostic-praxic level is the *object* itself. It has been noted (Glezerman & Balkoski, 1999) that this shift must involve LH processing: distinguishing the functional signs of the object with subsequent synthesis resulting in the image (Bernstein called it topological schema of the object) that presents the meaning of the object as a tool. Movements at this level are elementary behavioral acts, from more simple like putting on and buttoning a coat, sharpening a pencil, to more complex, multilinked object actions, such as cooking, driving a car, etc. Unfolding spatial schema of object action in time follows semantic sequence rather than sequence of moments of "real" environmental time, characteristic for the lower, sensory-motor level. Thus, *time* of the gnostic-praxic level is the temporal order in which meaningful manipulation with object proceeds.

Bernstein did not describe the symbolic level, but he outlined the basic direction in evolution of the function levels' space: sensory-motor level—metric space; gnostic-praxic level—topological space; symbolic level—concept.

Before we discuss the symbolic level, I will clarify more the differences between LH and RH processing. Figure 1.2 shows that the RH operates with discrete combinations of whole continuous images whereas the LH operates with continual combinations of discrete signs (Ivanov, 1978).

Evolution (reorganization) of *space* at the symbolic level in the LH can be described as transformation from visual representations to categorical representations (categorical classification of the world), or semantic space (Glezerman, 1986; Glezerman & Balkoski, 1999). In this regard, LH symbolic level is represented by discrete units (categorical signs) tied together as members of continual logical series (Fig. 1.2). *Time* of the LH symbolic level is semantic (logical) sequence. An example of a simple logical operation is the Similarity test from WAIS (Wechsler Adult Intelligence Scale) where the subject is presented with two words and asked in what way they are alike (*orange–banana, table–chair, seed–egg, poem–statue,* etc.). Performance of these tasks is based on sequential logical operation of dividing both objects into their features (and, thus, abstracting from the objects themselves) and then comparing these objects by their features.

Reorganization of *space* at the symbolic level in the RH can be understood as the formation of a symbolic system (Glezerman, 1986; Glezerman & Balkoski, 1999). VSSs are identified through the common affect/emotion, and objects within different VSSs are identified through the resemblance of their holistic form (Fig. 1.3).

As a result, a new unchangeable whole appears—a symbolic system. VSSs become multiple aspects of a continuous and indivisible whole, identical in their meaning yet at the same time unchanged from the instant they were formed. Thus, the same unit, VSS, is used for operations at the sensory-motor and symbolic level. It is assumed to be a parallel process in which the VSS is simultaneously involved in the different brain networks. At the sensory-motor level, VSSs are organized by the *temporal marks*. At the symbolic level, VSSs are organized by emotion into the symbolic system of meaning (Fig. 1.3), even if they have completely different content and do not have anything in common objectively.

The LH cognitive mechanism implies infinite number of categorical signs unfolding in one abstract space, the RH cognitive mechanism implies infinite number of discrete spaces that are continuous wholes.

1.3.2 Bernstein's Levels and Modern Neuroscience

Contemporary brain imaging, experimental neurophysiology and myeloarchitectural studies all corroborate Bernstein's theory of relatively independent hierarchical function levels in the brain. For example, fronto-subcortical circuits have been identified that originate in the frontal cortex with a sequential flow of connections to the following nodes: to the striatum, to the globus pallidum, to the substantia nigra (Alexander & Crutcher, 1990; Mega & Cummings, 1994). These vertically organized, separate nodes exactly correspond to Bernstein's effector centers at the gnostic-praxic, sensory-motor, thalamic, and cerebellum–midbrain levels (see Table 1.1).

Furthermore, neurophysiological studies in monkeys performing motor tasks reveal that motor processing proceeds concurrently, i.e., in parallel at the cortical, striatal, and pallidal stages of the circuit (Alexander & Crutcher, 1990). Such studies elegantly illustrate function levels' autonomy, with each level providing its own

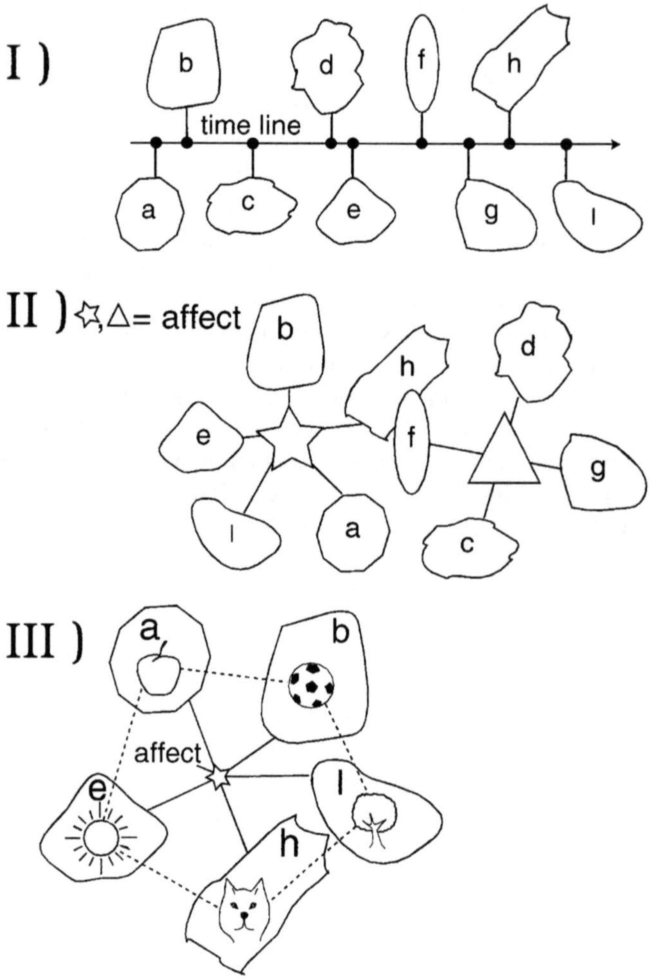

Fig. 1.3 Steps in right hemispheric cognition. (*I*) Situational Thought; (*II*) Symbolic-Situational; and (*III*) Symbolic-Object-Thought. From *Language, Thought and the Brain*, (Fig. 11, p. 53), by T. Glezerman and V. Balkoski, Kluwer Academic/Plenum Publishers, copyright 1999; with kind permission from Springer Science + Business Media B.V

unique contribution to movement. Indeed, the phylogenetic development of the frontal lobe represents "sequential unfolding" of higher-order cortical areas from the central sulcus to the frontal pole, in the transition from the motor cortex to the prefrontal cortex (PC). The PC constitutes nearly one-third of totality of the neocortex in the human brain. Evolutionary growth of the PC parallels that of temporal and parietal regions (Fuster, 2002). Paleoneurological data also suggest that two main foci of intense growth in anthropogenesis (temporal–parietal–occipital and frontal region) coincide in time with the formation of the superior longitudinal fascicle, connecting these brain regions in modern man (Kochetkova, 1973).

Neuroanatomical studies showed that each sector of the sensory association area is connected with a frontal lobe region that has basically similar architectonic features. This would imply that each sensory association sector, from first-order, through second-order, to third-order, may have developed in parallel with a specific frontal lobe region, and with that region may constitute a functional subsystem within the cerebral cortex (Pandya & Yeterian, 1985).

All the above is in full correspondence with Bernstein's concept of *anterior–posterior* brain cooperation at each function level. Recent data also point beyond the motor loop of such circuit, proving "stratification" at the higher levels, including the symbolic function level. There must have been coordination in the evolution of information processing in the posterior brain—with *units* specific for each hemisphere (LH signs/RH symbols)—and evolution of *operating with them* in the anterior brain system.

The PC is functionally heterogeneous, playing a critical role in the organization of behavioral, linguistic, and cognitive actions. However, the principle of *anterior/posterior* brain dichotomy remains the same: temporal versus spatial organization. The common denominator in the frontal lobe's functions is temporal organization (representation of action): from temporal organization of movements to temporal organization of behavior to temporal organization of (operations with) symbols, that is thinking. As Fuster (2002) aptly summarizes, "cognitive functions of the adult human prefrontal cortex are viewed as culminations of temporal integration in language and intellectual performance" (p. 374).

In contrast to Bernstein's framework, current neuropsychological, physiological, and neuroimaging research defines the cardinal function of the PC in the term *working memory*.[10]

Working memory has been defined as "the temporary retention of an item of information, e.g., a sensory cue for the solution of a problem or a mental operation" [Baddeley, 1986, cited from Fuster (2002), p. 379]. I agree with Fuster's (2002) statement that "working memory is memory for the short term, rather than short-term memory. It is attention focused on an internal representation" (p. 379). The PC does not store information, but it has access to and can wield for its purposes any information stored in the posterior brain. Like a librarian who has labels for each shelved book and can retrieve any as needed, the frontal lobe takes necessary information from the archives of the posterior brain, bringing it "on-line" to unfold in time and "read" (operate with) it. During any such operation by the PC, there will be a substantial activation in the posterior counterpart as well. Indeed, during visual memory task in monkeys there is concurrent and sustained activation of cell assemblies in the PC and in the inferotemporal, visual memory cortex (Fuster, 2002). Fuster suggests that working memory is based on the reverberation of activity between the executive network of the prefrontal cortex (PC) and the sensory networks of the posterior cortex.

[10] The term "working memory" is misleading since the frontal lobe does not store information. Information is stored where it is spatially organized, i.e., in the posterior brain. Furthermore, "working memory" is in reality not a term, but a metaphor, given that "working" and "memory" are not compatible semantically.

Such was the essence of Bernstein's general postulate about cooperation of anterior–posterior brain. My reservation regarding the parsimonious term working memory instead of temporal organization as the basic principle of the PC's function is that the former is vague and removes the hierarchy of function levels. Furthermore, the PC does so much more than bring attention to and access "memory." It operates with the highly processed information stored in the posterior brain and brings forth *dynamic schemes* from the posterior brain's *static schemes.* "Executive function" is another catchy term used broadly to define the PC's function. A more adequate description would be *execution of function.* Although "working memory" and "executive function" both speak to the frontal lobe's operating with information (in case of working memory, summoning highly processed information from the posterior brain and, in case of executive function, creation of inner programming, i.e., representation of action), they miss the importance of joint activity of the PC with a corresponding posterior brain area. Moreover, these terms remove from consideration the richness of understanding inherent in Bernstein's hierarchy of the function levels in the brain. The result of such sidestepping is that both a "touch each third button in the row" task and solving a complex mathematical problem fall undifferentiated under the rubric of "executive functions."

Why do we need Bernstein's theory today, what is so innovative and powerful about it?

Bernstein's theory emphasizes spatial and temporal frameworks as the organizing principles for the processing of external and internal information. It is not afferentation and movement themselves but specific spatial and temporal frameworks that define each function level. *Space* and *time* are not only organizing factors for movements and behavior, but they are also constructs at the very base of our subjective experience. Thus, Bernstein's *space/time* framework in the brain is a useful tool to conduct a neurophenomenological analysis of autism.

1.3.3 Patterns of Connectivity in the Brain as Widely Distributed Networks

Within the three dimensional organization (lower order–higher order function, anterior–posterior, and left–right dimensions), functional systems in the brain are organized as widely distributed networks (Mesulam, 1990). Each area participating in a network gives its own specific contribution. The networks are called "widely distributed" because their "players" (separate brain areas) can be remotely located from each other.

The cortical areas involved in networks are the so-called cytoarchitectural fields.[11] Brodmann (1909) subdivided the cortex into approximately 50 cytoarchitectonically

[11] The following parameters are usually used to distinguish one cytoarchitectural field from another: the layering in depth of the cortical sheet and, within each layer, cell characteristics—cell type ratio, cell size and number, cell density, cell spatial distribution, and their connections.

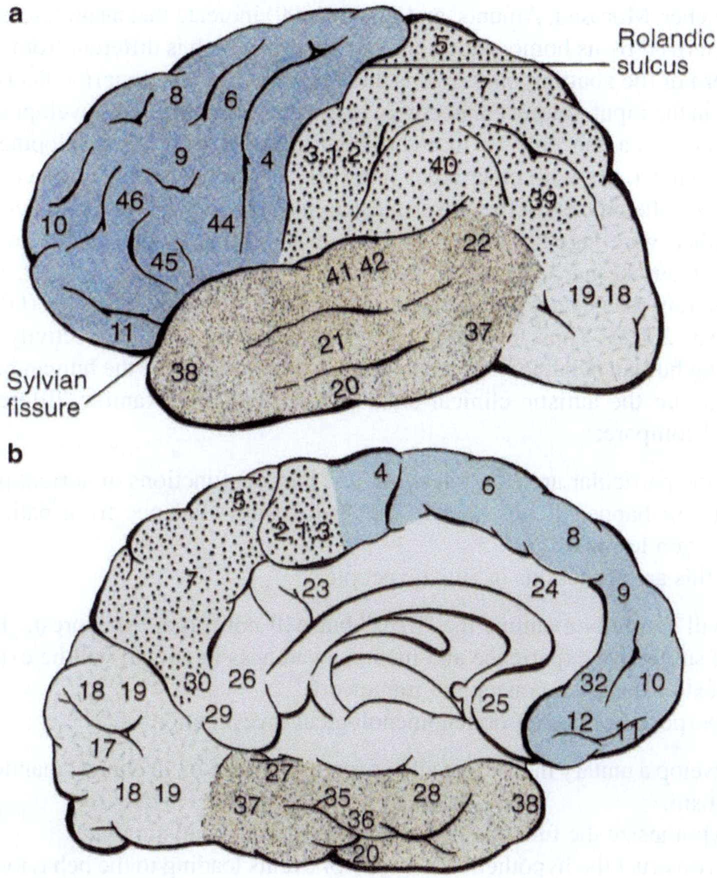

Fig. 1.4 Brodmann's cytoarchitectural map of the human brain (**a**) Lateral view and (**b**) medial view. Modified Fig. 3A and B from *Neuroscience and Medicine*, Conn, P. Michael (Ed.), Humana Press (Part of Springer Science + Business), 2008; with kind permission from Springer Science + Business Media B. V. and Dr. Daniel Tranel

distinct areas (see Fig. 1.4), and this topographical map of the human cortex is still in use (Brodmann ares—BA's).

It is agreed that cytoarchitectural division is not merely "parcellating" the cortex according to cellular physiognomy, but reflects a structural–functional association of cortical areas and their phylogenetic development (Pandya & Yeterian, 1985). For example, it has already been mentioned above that in the networks of the anterior–posterior brain, a particular area in the anterior brain is connected with a posterior area of similar stage in architectonic differentiation.

The principle of vertical hierarchy is manifested in distinguishing the primary, secondary, and tertiary cytoarchitectonic fields that differ in their phylogenetic age and the level of function they subserve.

Schleicher, Morosan, Amunts, and Zilles (2009) indicate that an architectural area is characterized by its homogenous laminar pattern, which is different from the laminar pattern of the abutting ones.[12] This distinction in laminar pattern reflects the differences in the input–output connectivity between areas. Thus, the development of an architectural area can only be thought about together with the development of its specific connectivity with other areas (network). In ontogenesis, brain networks are developed in the process of the correlated activity of the areas involved ("neurons that fire together, wire together"), triggered by sensory stimulation specific for a given network (Courchesne, Chisum, & Townsend, 1994). This development, however, proceeds from the genetic program controlling *area*lization of the cortex (O'Leary & Nakagawa, 2002). Thus, overall, cortical networks present connectivity patterns underlying human behavior that are "fixed" in the evolution of the human brain.

To explore the autistic clinical-brain pattern, we will examine different brain areas and compare:

- How the particular area (and network it is part of) functions in normal people?
- What will happen if this area is damaged (using evidence from patients with focal brain lesions)?
- How this area functions in autistic people?

We will carefully examine the "trees" but will not forget the "forest." Different levels of subjective experience and mental awareness in autism will be explored in the context of the brain conceptual framework.

The purpose of this neurophenomenological investigation is:

1. To develop a unitary theory that allows for explanation of all clinical manifestations of autism.
2. To hypothesize the fundamental abnormality of the autistic brain.
3. To reconstruct the hypothetical cascade of events leading to the behavioral manifestations of autism.

[12] For example, the border between the primary visual cortex BA17 and the secondary visual cortex BA18 is characterized by an abrupt change in lamina IV, which is broad and subdivided into sublaminae in BA17 and a thinner and less well differentiated lamina IV in BA18 (Schleicher et al., 2009).

Chapter 2
How Autistic Persons Understand Words (Cerebral Organization of Word Meaning and Autism)

Autists do not speak or when they speak, they repeat the same phrase many times meaninglessly. They say things that look nonsensical and irrelevant to others, because their utterances do not seem to have any connection to the situation in which they are voiced. Autists are unable to understand metaphors, irony, lies, and humor.[1]

Kanner and Asperger were the first to describe autism as a severe disorder of spontaneous speech. Yet it was Kanner who revealed metaphorical language in children with autism! And it was Asperger who emphasized that autistic children have a special creative attitude to language, a spontaneous way with words and can produce novel but particularly apt expressions.

To approach this paradox, we will discuss what areas are involved in word comprehension and these areas' specific functions that underlie internal representation of word meaning. After this discussion an attempt will be made at determining whether the autistic brain processes word meaning differently, and, if so, in which way cerebral organization of word meaning is different in people with autism. Would these differences account for the above contradictions about the speech of people with autism?

2.1 The *What* System and Its Major Player, BA37

Seemingly inseparable aspects of a word, sound and meaning, are processed and stored in two different areas of the cerebral cortex. This fact is demonstrated by the phenomenon of *double dissociation*: damage to the left superior–posterior area within the temporal lobe produces disorder of word sound, but not word meaning, whereas a lesion in the inferior–posterior area in the temporal lobe (bordering the occipital lobe) affects word meaning, but leaves word sound intact. In the historical development of the human brain, these two areas originated from different roots,

[1] Multiple studies demonstrate this phenomenon.

T.B. Glezerman, *Autism and the Brain: Neurophenomenological Interpretation*, DOI 10.1007/978-1-4614-4112-0_2, © Springer Science+Business Media New York 2013

auditory versus visual cortex. Thus, the region concerned with word meaning represents an evolutionary extension of the visual cortex. In the cortical map, this inferior–posterior area of the temporal lobe is designated BA37. BA37 is located between the visual occipital and auditory temporal areas. Anatomist Blinkov (1955) compared the brains of different species and established that, in the development of the human brain, the cortical zones, in which auditory and visual pathways terminate, move apart from each other, and the area between them enlarges.

Studying the primate visual system, Mishkin, Ungerleider, and Macko (1983) distinguished two cortical pathways: ventral and dorsal. The ventral visual pathway, which interconnects the occipital striate, prestriate, and inferior temporal areas, plays a crucial role in the visual identification of objects—the so-called *what* system. In the inferior temporal area, *form* is processed for the purpose of identifying the visual stimulus and assigning it with meaning (Mishkin et al., 1983). The inferior temporal region, considered a continuation of the visual system in primates, is homologous to BA37 in man. Using the terminology of the human cortical map, the visual ventral pathway can be delineated as a sequential flow of connections from the primary BA17, through the secondary BA18 and BA19, culminating in BA37. The tertiary cytoarchitectural field of BA37 projects to the tertiary areas within the prefrontal (BA46) and orbitofrontal (BA11) cortex.

BA37 is heterogeneous in structure. Its peripheral parts are transitional, retaining features similar to the bordering auditory and visual areas. Only the central, historically youngest part, the "nucleus" of BA37, is unique and specific to the human brain (Blinkov, 1938, 1955; Blinkov & Glezer, 1968). Considering the structural heterogeneity of BA37, containing both new and older phylogenetic formations, it has been proposed that the visual recognition of an object, a generic modality-specific function of the gnostic–praxic level, is connected with the posterior part of BA37,[2] whereas the human-specific "nucleus" of the field is connected with linguistic functions of the supramodal symbolic level (Glezerman, 1986; Glezerman & Balkoski, 1999). Recently, this hypothesis was supported by studies involving the new technique of transcranial magnetic stimulation (TMS). TMS can be applied to a small area in the brain, producing a transient and reversible "virtual lesion" in normal subjects and as such can support ideas about a given area being necessary for a particular type of processing. TMS applied over the posterior part of the left BA37 severely disrupted naming of everyday objects presented in pictures. The disorder induced by TMS was highly selective (Stewart, Meyer, Frith, & Rothwell, 2001).

Moore and Price's (1999) imaging studies suggest the semantic subregion[3] of the left BA37 is positioned approximately 4 cm anterior to the subregion identified by the TMS study as important for object recognition in the left hemisphere.

---

[2] This part of BA37 borders the occipital cortex; it is usually characterized as a transitional area and is phylogenetically older than the nucleus.

[3] The "semantic subregion" corresponds to what in this book is referred to as the linguistic functions of the symbolic level in BA37.

2.2 Object Recognition, Topological Scheme of the Object, and Empirical Component of Word Meaning

Looking at a table, one does not just see its physical appearance. One knows what a table is for and the category of objects to which it belongs—one knows its *meaning*. The *whatness* of the table is grasped at once, but in fact this requires LH *sequential* processing in the left posterior–anterior brain system (from BA37 to prefrontal cortex), with parallel (simultaneous) processing in lower and higher functional levels.

At the sensory-motor level, concrete and concrete-situational signs are distinguished. Signs at this level are the "objective" features of *things*,[4] such as shape, size, color, and the like, but *things* are bound to their context. For example, the synthetic LH image of a cup at the sensory-motor level would be a metric and geometric image with the exact physical parameters of the cup, as experienced in the RH singular VSS. In other words, each cup is seen and recognized by its physical features in each particular situation, and no generalization is made regarding all cups as being the *same* things.

At the gnostic–praxic level, an object is analyzed by its functional signs. The functional signs of the object are those necessary to act with it. For a cup as an object for a particular use, neither height nor width, nor roundness nor squareness has substantial meaning. What is relevant at the gnostic–praxic level is the cup's having solid sides, an unbroken bottom, and a handle (Bernstein, 1947, 1990). By these signs, every child will recognize a cup, even if he has never met a cup with this particular size, shape, etc. The combination of functional signs, as a result of LH synthesis, makes the LH synthetic image—the topological scheme of an object, according to Bernstein. A topological scheme represents not the object itself, but the rational aspect of acting with it. In other words, the topological scheme of an object is the visual representation that implies the object's meaning as a tool. This visual image is "constructed" and stored in the left BA37.

2.2.1 Topological Scheme Is the Base for Object Action

A child's world has already been shaped by human activity and is full of things designed by people to be used in human-specific activities. Children are introduced to this world of objects by the people around them. During interaction with a child, his parents and caregivers demonstrate the shared meaning of objects by drawing

[4] At the sensory-motor level, there are not yet *objects* as such, just *things*, for the construct of an "object" arises from analysis as to the functional meaning of some "thing." Such analysis begins at the gnostic–praxic level.

the child's attention to the object and then showing how it is used. In addition, other people's actions on objects also provide a model, not necessarily intentional, for the child to imitate. One important aspect of the child's developing sense of the meaning of things is the appreciation that most objects he encounters have a single, definite function. He then learns to use objects according to their established proper function. A child is capable of learning object use because his brain has an innate potential to analyze or "extract" functional signs and construct a topological scheme, which serves as inner programming for action. The time when children start using objects predictably coincides with maturation of those areas in the brain involved in LH processing at the functional level of abstraction.

Returning now to the left BA37, this region not only creates generalized visual–"functional" images of perceived objects, but also serves as the storehouse for their representations and, hence, for their later recognition. After mastering use of a cup in one particular situation, the child will be able to perform this object action in any situation and with any type of cups.

2.2.2 *Topological Scheme of an Object Is the Basis for Communicative Gestures*

When we point to a separate object, it is distinguished in our consciousness. Although the RH stores the whole object as perceived, objects in the RH VSS are trapped in their spatial–temporal context (VSS) and not distinguishable from it. Separating the object out of the visual scene-situation is achieved only if the object is analyzed and defined by its features via LH processing. Thus, when we point to an object, we in fact are indicating its topological scheme, not the object itself, and the part of BA37 involved with topological schemes of objects must be activated (Fig. 2.1).

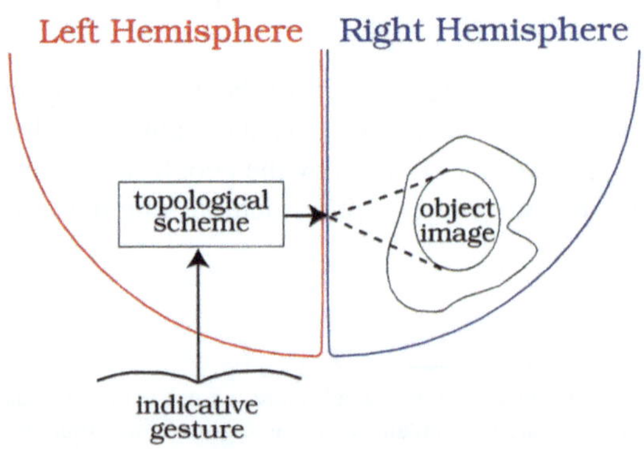

Fig. 2.1 Indicative gesture and topological scheme of an object. From *Language, Thought and the Brain*, (Fig. 7, p. 42), by T. Glezerman and V. Balkoski, 1999, Kluwer Academic/Plenum Publishers, copyright 1999; with kind permission from Springer Science + Business Media B.V

The topological scheme of an object in the LH is linked to the holistic object image in the RH. When the topological scheme is actualized, it "develops" the image of the corresponding object in the RH. Only in this indirect way, through the left BA37's connection with the right BA37, can we perceive the whole image of the object that is stored within the RH VSS (Glezerman & Balkoski, 1999).

2.2.3 Topological Scheme of an Object Is the Basis for the Empirical Component of Word Meaning

In the history of language, before articulate speech appeared, communication was performed by means of gestures. Human communication took the form of action—theatrical-like demonstration including pantomime, imitative, and indicative gestures. It is supposed that humans' first words had very broad and diffuse meaning (Blonsky, 1935; Ivanov, 1978), which reflects the RH visual-action-situation (Glezerman & Balkoski, 1999). For example, in a series of African nonwritten languages, one word, *ngu,* denoted the following: *this, I, to look, to know, nose, mouth, to drink, water, tooth, to bite, to eat, to speak, to listen, ear, hand, five, two* [Blonsky (1935) cited from Glezerman and Balkoski (1999)]. Here word meaning is diffuse and gives an image of whole visual-action-situations ("man and his activities"), however, at the same time we can also see an early sign in the emergence of separated-out objects in language by the use of *ngu* to mean "this." "This" emerged along with the first words to appear and replaced the indicating hand with sound, allowing for reference to any object, but still dependent on context or situation.

The next step in the evolution of language comprised the arrival of words that designate a *separate object.* Again, people had spoken with their hands before they spoke with words. Many languages preserve this initial reliance in language on demonstration by their having the same roots denote *hand* and *speech*, and many more languages have the same or very similar words to mean *to speak* and *to show* (Latin *dico* speak, *indico* indicate, Russian *rasskasivatj* tell, *ukasivatj* indicate, *pokasivatj* show, etc). Here then is the chain of the hypothetical historical development: indicative gesture, replacement of indicating hand by "this," and then word–name for the concrete, separate object. As noted earlier, constructing the topological scheme of an object is what allows the object to be separated out in consciousness. It can now be further postulated that the topological scheme of an object remained the common thread of what gets referred to in this evolution of language. Thus, the topological scheme is what was pointed to by gesturing in nonverbal languages, it was then what "this" referred to, and it also is what was referred to by words for separated-out objects (Glezerman, 1986; Glezerman & Balkoski, 1999). In linguistic terminology, word meaning includes an empirical component, or object reference, and a categorical component (Katznelson, 1972, 1986). We may conclude that the brain correlate for the empirical component of word meaning is the topological scheme of an object (which then refers to the whole object image in the RH).

2.3 Categorical Classification, Categorical Component of Word Meaning and Left BA37

To recapitulate, the LH constructs a visual world, and different versions of this world exist at each hierarchical level. The LH sensory-motor level's "synthetic" image of a *cup* evokes the visual and tactile sense of shape, color, and size of that particular cup from a single situation. *Cups* do not exist as such at this level, but only within the individual situation to which each cup belongs. At the gnostic–praxic level, a "synthetic" LH image of a *cup* evokes a separate object, with emphasis placed on action with the useful object. Here a *cup* exists as such, as a separate object, but without connections to other objects.

At the symbolic level, from a successive series of numerous signs and features of objects obtained during LH analysis (Fig. 1.2), one sign common to a given group of objects and differing this group from all others is distinguished. Such a sign is called a categorical one. Within each categorical group, subgroups are distinguished that possess their own distinctive signs—by a progressive distinguishing of specific categorical signs, ever more specific categories are formed from the more general. At the symbolic level, a *cup* belongs to the general category of objects and to the following subordinate categories: inanimate objects, things, man-made things (artifacts), things made for certain needs (instruments). At the categorical level, objects exist as such and also in their relation to other objects, as members of categories.

The categorical sign is an abstraction, a principle according to which signs of objects are organized into a set. For example, how can we imagine a category of living things, in particular the subcategory of animals? There must be some marker uniting such properties as "has legs," "has eyes," "can walk," "can breath," etc. These properties are not coded in the same section of BA37, and they may also have their equivalents represented in other modality-specific cortical areas. Although built upon the sensory-motor and gnostic–praxic levels, the LH symbolic (categorical) level cannot be reduced to either physical–sensory or functional–schematic (still modality-specific) representations (Glezerman, 1986; Glezerman & Balkoski, 1999). The idea that there are categorical representations in the brain independent from visual representations finds an empirical support in cases of brain-damaged patients who were intact in object recognition but had categorical deficits (Caramazza & Shelton, 1998). Also, in brain-damaged patients, category-specific deficits were observed; for example, categories of animals, plant life, fruits and vegetables, and artifacts were impaired independently from each other (Caramazza & Shelton, 1998). Anatomo-clinical correlations in patients with category-specific deficits and functional brain imaging of the different categories' processing in normal individuals revealed that the left temporal lobe is crucially involved in categorical processing, with the left inferotemporal area subserving the category of animals, while the left posterior middle and inferior temporal area is more important for tools (Damasio, Grabowski, Tranel, Hichwa, & Damasio, 1996). In addition, an important role of the left frontal lobes is indicated for both categories (Martin, Wiggs, Ungerleider, & Haxby, 1996).[5] We will discuss the role of the prefrontal cortex later.

We can see that the visual representations with which BA37 (historically visual cortex) is concerned are transformed at the symbolic level into a categorical classification of the external world.[6] Indeed, BA37 serves as a "storehouse" for categories—this categorically "packed" world (semantic memory) lies dormant until called into play by the left prefrontal cortex (Glezerman & Balkoski, 1999).

As discussed in Chap. 1, the prefrontal cortex's function can be thought of broadly as the representation of action. The left prefrontal region's action, operating with categorical signs, is what leads to their sequential hierarchical organization from more general to more specific categories. If the left prefrontal region is damaged, the hierarchy of categorical signs is lost even though the stockpile of objects' features in the posterior (temporal) region remains intact. The patients are unable to distinguish between supraordinate–subordinate categories. For example, the word "animal" may be used by such patients and it will be associated with words like "dog," "cat," "goat," etc., but these words are not united into a common category. The words "animal," "dog," "cat," "goat" are "located" in these patients at one and the same level of abstraction. "Animal" is perceived as one concrete animal at a time and not as a designation of a class.[7]

---

[5] Despite these facts, most theories of how categories are organized in the brain are reductionistic. For example, the sensory/functional theory explains category-specific deficits being a result of selective damage to noncategorically organized visual or functional semantic systems (Warrington & Shallice, 1984). Another theory claims that because members of a superordinate category share many features in common, the bundles of interrelated properties are differentially distributed in the categories of living and nonliving things (Hills et al. 1995). For example, such categorical distinction as biological versus nonbiological motion is based on the particular types of nonoverlapping features, certain kinds of "stuff" (shape, texture, color, odor, etc.) that distinguish animate and inanimate objects. According to this theory, local damage to a region of semantic space will result in impairment to those categories whose members' meaning depends on the affected semantic properties. This theory has some similarity with ours, for it considers representation of the category as a combination of interrelated properties (which is the LH processing). However, the discussed above theory remains reductionistic since it defines a category by the physical, sensory properties. Closest to my understanding of categorical representations in the brain is Caramazza and Shelton (1998) model. According to these authors, categorical knowledge has distinct brain organization within which specialized brain networks subserve different categories (specific domains). The authors even allude to the different levels of organization in the brain. Furthermore, criticizing reductionistic theories, Caramazza and Shelton (1998) argue that not only are there specific domains for categories in the brain, but that perceptual systems are sensitive to category distinctions. We can accept this last statement not literally, but in a sense of *top–down* regulation.

[6] The terms categorical classification of the external world and categorical representations used in this book are equivalents of categorical system and of semantic memory, the latter is often used in literature.

[7] Similarly, categorical deficits are observed in individuals with mental retardation (concrete thinking), except that in mental retardation the deficit is developmental.

Thus, formation of categorical classification is possible only with joint activity of the left BA37 and the left prefrontal cortex.[8] Indeed, functional brain imaging studies have consistently shown that categorization tasks[9] produce a large activation in the left hemisphere—the inferior prefrontal as well as the inferior–posterior temporal lobes (Devlin et al., 2002).

It was argued earlier that the empirical component of word meaning is connected with the gnostic–praxic level in the LH.[10] Now the categorical component of word meaning's connection to the symbolic (supramodal) level in the LH can be discussed.

In linguistics, certain hierarchical sequences of categorical signs are thought to form the categorical component of word meaning, or concept. For example, the categorical component of the word *cup* includes the following linear sequence of categorical signs: *objectness—nonanimation—thingness—artifactness—instrumentality* (meant for certain needs); the categorical component of the word *waiter* would be *objectness—animation—person—agentivity* (acting person)—*occupationness* (Katznelson, 1972).

The categorical component of word meaning is implicitly contained within the categorical system. When we say or hear a word, *only at that moment* do categorical signs, characteristic for the particular word, gather together into the categorical component of word meaning. Thus, the categorical component of word meaning is realized as a result of joint activity in the left BA37 and the left middle and superior temporal region (BA22,42,21), responsible for word sound or the phonological code of the word (Glezerman & Balkoski, 1999).[11]

Our ability to think in categories and categorical component of word meaning are not one in the same; the categorical system is broader than the categorical component of word meaning that is based upon it.[12]

The categorical component of word meaning (existing only when connected with word sound) in the history of any particular language developed concurrently with

---

[8] In the historical development of the human brain, powerful growth in two areas, corresponding to prefrontal and inferior temporal regions (BA37) in modern man, along with the formation of a massive bundle of connections between these two areas was the turning point toward brain "hominization" (Kochetkova, 1973).

[9] An example of a categorization task used in functional imaging studies is when subjects are shown a picture of three cue objects and then must decide whether a fourth (target) object belonged to the same category as the cue objects.

[10] In a previous work, two parallel lines in the phylogenesis of language were proposed (1) the articulatory praxic and empirical component of word meaning (gnostic–praxic level), and (2) the phonological code and categorical component of word meaning (symbolic level) (Glezerman, 1986; Glezerman & Balkoski, 1999).

[11] As alluded to at the beginning of this chapter, word sound by itself does not mean anything; it is literally a sound code of a word, the code that allows access to word meaning the "other side" of the same coin.

[12] Interestingly, in the history of the human brain, intensive growth in the area corresponding to modern man's BA37 is registered at earlier stages than the development of the specifically human region responsible in modern man for word sound.

word sound development. In different languages, there are variations in the number of categorical signs for equivalent concepts. This phenomenon suggests that words-concepts do not have exactly the same meaning in different languages. Categorical representations are, on the other hand, nonverbal and universal. Patients who suffer from damage to the left temporal region, responsible for word sound, cannot grasp the categorical component of word meaning, but they do not lose a categorical attitude to the external world. Presented with a classification test, these patients unite objects into categorical groups, but cannot name the categories.

To conclude, BA37 is fascinating: it is complicated and not homogeneous in historical age, with an older transitional periphery and a young, specifically human central "nucleus"; as such, BA37's generic function of object recognition is an evolutionary knot related to the development of an active attitude to the surrounding world (action with object tool), a new means of communication (language), and thinking (categorical classification). Categorical representations are "grown" from visual object perception, and the units of those representations, that is categorical signs, form the categorical component of word meaning. This evolutionary knot must be important for understanding autism, given autists do not use either objects or language in a conventional way.

2.4 Word Meaning and the Right BA37

> The hard nucleus of an uttered word is spiritually accompanied by something like a halo of evaporation from images and strong affects merged together.
>
> (Kretschmer, 1927)

> The living word does not designate the object but freely chooses, as though for lodging, this or that object meaning, the dear body. And around this thing, the word wanders freely, as the soul around the cast off but not forgotten body.
>
> (Mandelshtam, 1921)

Does the RH have any influence on *how* we understand words? The neurolinguist Moscovitch (1983) indicates:

> Hemispheric differences are large in processing syntactic or phonological properties of language; however…in processing semantic properties of language—hemispheric differences are likely to be smaller because each hemisphere contributes in its own way to the task (p. 94).

Moscovitch warns, however, "on verbal tasks, special techniques are required to free the right hemisphere from the dominance of the left to reveal its contribution to normal performance" (p. 103). To uncover RH contribution to word meaning, we must return to LH/RH differences and discuss in more detail the RH mode of information processing.

In our awake state we are subject to continuously changing impressions from the outside world. These moments of experience in the external world are perceived and directly stored by the RH as unchangeable, stationary wholes—visual scene-situations (VSS) (Glezerman, 1986; Glezerman & Balkoski, 1999). VSS was defined in Chap. 1, and here it just needs reminding that VSS includes not only visual picture but also the emotion experienced at the moment it was perceived. In short, VSS is subjectively felt. VSS is a unit for RH operations just as discrete signs of objects are units for LH operations.

Objects in the right BA37 are perceived and stored as included into their spatial–temporal–emotional context, the VSS. It is in this form that the LH receives information about objects from the RH and then interprets the VSS according to its own cognitive mechanism. The LH distinguishes the signs of objects and "remakes" them as "synthetic," separate-from-situation images. RH representations are primary, immediate experience. LH constructs are secondary and result from analyzing an object by its features, i.e., abstracting from the object itself, and then re-constructing a synthetic image as a combination of those features.

While the RH is a bearer of primary experience, the LH is always an interpreter of primary experience. Primary experience is not on the foreground of consciousness, but "covered" by LH interpretations.

2.4.1 RH Visual–Situational Level and Word Meaning

The singular VSS is a whole that is idiosyncratic for the individual, having its own unique spatial–temporal context and saturated with emotion. Singularity means that the images of one and the same object represented in different VSSs are not connected to each other. VSS as a stage in RH cognition corresponds to the sensory-motor level in the LH. At the RH situational level, the image of the *cup* is incorporated in an infinite amount of singular situations: an image of the *cup* when you were meeting with a long-lost friend, an image of the *cup* when your daughter broke it in defiance of you, etc.

This step in RH cognition was termed *visual–situational thought* (Glezerman, 1986; Glezerman & Balkoski, 1999). RH visual–situational thought can be recognized in word meaning of so-called primitive nonwritten languages. For instance, in the Australian language Aranta there is no word "leaf," instead, there are several words: "kanta"—round leaf; "ibala"—oval and fleecy leaves; "iana"—fleshy leaves. There is no word "hair" in Aranta but the following cluster of words: "panga"—long hair; "pantja"—long, trailing hair; "aratja"—straight hair standing upright. On the other hand, one word may mean more than one thing—visual situational association. For instance, in Aranta INTA means at the same time "stone" and "recumbent"; the word INKA means "foot," "footprint," and "steep" (cliff and mountain path) (examples taken from Katznelson, 1986, pp. 94–95).

2.4.2 RH Symbolic Level and Word Meaning

Having used the term *thought* in the previous section, the organizing principle for RH associative processing should now be clarified. The LH establishes connections between objects based on kinships of their properties, features, and signs. The RH is oriented toward perceiving the whole and identifies objects according to their appearance, even if their content is different. The mechanism of RH identification is simultaneous and instantaneous recognition of objects (in different situations) by the resemblance of their holistic forms. Examples of RH association by appearance would be: cat muzzle—sun—ball—apple—crown of the tree, see Fig. 1.3; ditch—plate; wall—sheet; suitcase—well. Simultaneous recognition of objects in different situations can be thought of as a form of pattern recognition. This association by holistic form in the RH roughly corresponds to the gnostic–praxic level in the LH.

Now let's return to the right BA37. In the VSS (sensory-motor level) objects are not functional objects, they are appearances devoid of content (meaning)—holistic forms. However, belonging to the VSS, they are saturated with the emotion of their situational context.[13]

Association by holistic form is also reflected in the word meaning of primitive languages. For example, in Aranta, *libala* means the oval or angular leaf but also bird's feather, bird's wing, and fin; another word designates knee, curved bone, bend of the river and earthworm (examples taken from Katznelson, 1986, p. 95).

In parallel with the described above "pattern recognition," RH processing also entails identification of VSSs by common affect/emotion giving rise to its symbolic system (described in Chap. 1). It has been called visual situational-symbolic thought (Glezerman & Balkoski, 1999).

Identification, then, is the main organizing principle of RH processing. What exactly is *identification?* Or, what does it mean to be identified?

If A and B are identified, it is not that they are equal or similar, it means that they are identical, i.e., A remains A, but at the same time it is B and vice versa. The identified entities are interchangeable. Evidence for RH identification can be found in certain cultures, where RH associations are not only externalized but embodied in societal rules and ceremonies. For example, nineteenth-century observers of a native American tribe recorded that its members claimed to be human beings and red parrots at the same time (Levy-Bruhl, 1930). It is not that they would turn into red parrots after death. They believed they were birds with red feathers in the present. It was not the name or label they gave to themselves, and it was not that they were similar to red parrots: they *were* red parrots. It was not an analogy, not an association according to some common features, it was *identification.*

Identifications of multiple VSSs (and the objects within them) by common affect, and identification of objects from different VSSs by resemblance of holistic form, mark the next stage of RH cognition—visual object-symbolic thought where the object becomes a polysemantic symbol and individual VSSs once identified get left behind (Glezerman & Balkoski, 1999).

This visual situational-symbolic and visual object-symbolic thought can also be recognized in the word meaning of primitive languages. For example, in Aranta, we see singular images from alien domains are identified by similar affect and

---

[13] We will see below how this is very important for our purposes because autists treat objects as forms (appearances), and it is form, not the meaning of the object, that holds emotion for them.

resemblance of holistic forms in word meaning: "aratja" means not only straight hair standing upright but also straight road; "pantja" means not only long, trailing hair, but also black night and deep; "alknanta" means both crimson flame and bloodthirsty man (examples are from Katznelson, 1986, p. 96).

2.5 Model of Cerebral Organization of Word Meaning

> A word has become not a seven-barreled but a thousand-barreled reed, brought to life by the breath of all centuries at once.
>
> (Mandelshtam, 1921)

Now the model for word meaning (Glezerman & Balkoski, 1999) can be summarized. Word meaning (WM) is connected with BA37 in the LH, and it has two components—empirical and categorical. "Behind" these components are RH representations connected with word, we can call them RH equivalents of WM: VSS and the objects within it, RH situational-symbolic associations, and RH object-symbolic associations. RH representations get driven away to the periphery of consciousness, parallel to the development of the language system. As a result, the degree of awareness of RH "out-of-language" content of words (different in each individual) in general is marginal.

Different components and equivalents of WM represent "layers" in language history. The most ancient are the visual–situational and the visual situational-symbolic content of words, followed by the empirical component. More "young" are the categorical component and object-symbolic equivalent. Correspondingly, the cortical representation of word meaning includes regions of different phylogenetic age (Glezerman, 1986; Glezerman & Balkoski, 1999). The empirical component is related to the peripheral area of BA37,[14] bordering with the occipital lobe, whereas the categorical component is connected with the central subfield, the "nucleus" of the left BA37. The fact that BA37 subserves both sensory-motor/gnostic–praxic as well as the symbolic level implies these layers within BA37 have different interhemispheric connections.[15] The empirical component of WM in the LH would be connected with the whole object image within the VSS, while the categorical component of WM would be related to RH symbolic images ("individual sense" of word).

[14] This hypothesis was later supported by TMS studies (see the first part of this chapter) and also by imaging studies during memorizing concrete and abstract words. There was activation in the inferotemporal area (BA37) bilaterally with LH superiority in memorizing abstract nouns. When concrete nouns were being memorized, the bordering part of the occipital region was involved as well (Goldenberg, Podreka, Steiner, & Willness, 1987). The meaning of concrete nouns includes empirical and categorical components, whereas in abstract nouns the empirical component is reduced. The difference in localization between abstract and concrete words was attributed to the empirical component being related to the peripheral, temporal–occipital part of the left BA37, and the bordering occipital BA18,19 (Glezerman & Balkoski, 1999).

[15] Cortical connectivity is stratified, meaning connections exist between regions that developed together in phylogenesis (see Chap. 1 for details).

Interhemispheric interaction at the symbolic level is a constant in the development of WM. As in the gnostic–praxic level where the topological scheme on the left "develops" the whole object image within the VSS in the right hemisphere, so at the symbolic level, the categorical component "reveals" symbolic systems on the right.

2.6 Cerebral Organization of Word Meaning and Autism

Now we can move on to analyze WM in autistic individuals. The question will be: what is the composition of WM in autism and how are different components of that WM developed. Does any component/equivalent of WM (empirical or categorical, RH situational or RH symbolic associations connected with a word) prevail in autism? Is there any pattern of WM characteristic for people with autism, independent of age and degree of severity? Is there a pattern of WM characteristic for autism notwithstanding all variations in the clinical picture?

2.6.1 RH Situational Level and Word Meaning in Autism

Below are examples of LFA's explanations of WM from Kanner's material (Kanner, 1943).

Alfred, a 9-year-old boy: *balloon*: "[It] is made out of lined rubber and has air in it and some have gas and sometimes they go up in the air and sometimes they can hold up and when they got a hole in it they'll bust up; if people squeeze they'll bust" (p. 235). We can see here a specific definition behind which are the visual scene-situations (VSSs).

John, a 6-year-old boy: *dictionary*—"'That's where you left the money'.... [O]nce his father left some money in a dictionary and asked John to tell his mother about it" (p. 239).

For this autistic boy, the object (and word meaning) is part of the whole—VSS.

If we move to the HFA, we still see "exposure" of RH primary experience reflected in WM.

Here is an 8-year-old autistic boy's performance on a "Similarity" test (e.g., "What is the difference between the words 'tree' and 'bush'?") from Asperger's material.

> The bush, that is where the branches grow straight off the ground, completely jumbled up, so that it can happen that three or four cross over each other, so that one has a knot in one's hand. The tree, that is where there is first a stem and only then branches, and not so jumbled up, and rather thick branches. This happened to me once, that is where I cut into a bush, I wanted to make myself a sling, I cut off four branches and then I have an eight-part knot in my hand. This comes when two branches rub against each other, then there is a wound there, then they grow together (Asperger, 1991, pp. 53–54).

We can see here a visual image, very clear, exact, and factual, with all its physical features. There is a strict, photographic correspondence with real metrical parameters that betrays the "most objective" level—sensory-motor. At the same time, the image above all is subjectively felt—it is a picture of situations experienced, the image of this autistic boy's individual experience. This indicates RH situational thinking's significant role in the above two words' meanings.

Below are two more examples from Asperger of autistic children's performance on the question: *In which way glass and wood are different*?

> Glass is transparent. Wood, if you wanted to look through it, you would have to make a hole in it. If one wants to beat on a piece of wood then one has to beat a long time until it breaks, unless it's a dry twig. Then it would break easily. With the glass you need to hit only twice and then it's broken [eight-year-old boy with autism] (Asperger, 1991, p. 54).

Another boy, 7 years of age, answered the same question: "Glass breaks easily and wood doesn't. Glass is a mass, wood is sappy and damp. It has marrow in the middle. Wood burns to ash, glass stretches apart and then melts" (Asperger, 1991, p. 62).

Neither autistic boy explicitly used left hemispheric analytic processing nor operated with the categorical component of WM. Instead, these boys gave very rich observations of glass and wood's physical features. Their descriptions are experiential and very sensual, they convey visual, tactile, kinesthetic images (what they describe can be seen, its texture touched). These vivid and original descriptions are "expressions" of the right hemisphere at the situational (sensory-motor) level.

Thus, we can see that the cerebral organization of WM in these autistic boys is primarily right-hemispheric, and the functional level most pronounced is the situational (sensory-motor) level.

2.6.2 Is the Empirical Component of Word Meaning Impaired in Autism?

There are RH theories of autism, and they follow conventional logic: impaired development of the left brain with resultant right-brain compensation. In this case, however, one would first have to answer the question: *Is* the LH impaired in autism?

In this chapter, we considered word meaning connected with the major player of the *what* system, BA37. The neuropsychological syndromes specifically related to dysfunction of the left inferotemporal cortex include:

1. Visual object agnosia manifested by disorder of object naming.[16]
2. Ideational apraxia, loss of object meaning as a tool.
   Recall that it is the topological scheme (an object's "toolness") of the gnostic–praxic level that is indicated when we point to or name an object (see above in this chapter). Thus, the above two disorders are both related to the gnostic–praxic

[16] Patients with lesions in the right inferotemporal cortex misrecognize the objects due to fragmentation of the whole in visual perception (Kock, 1967).

level, the first includes deficit of empirical component of word meaning, i.e., disorder of naming in the visual modality.

3. Disorder of categorical recognition, failure to recognize the object as a member of a definitive category (Kock, 1967).
   This disorder affecting the symbolic level includes in itself deficit of the categorical component of word meaning.

Children with autism do not have visual object agnosia caused by LH dysfunction. They can recognize objects and name them. Moreover, a recent study showed enhanced object picture naming in autists compared with controls (Walenski, Mostofsky, Gidley-Larsom, & Ullman, 2008). For contrast, I will give an example of one of my patients, an 8-year-old boy, P., with a learning disability, who had a specific neuropsychological syndrome corresponding to a dysfunction of his left BA37.

P. spoke fluently, was friendly, and motivated to answer examiners' questions. His specific deficit was expressed in difficulty naming objects. For example, he was shown the picture of objects and asked to name them: lily of the valley—"tulip"; fly-agaric—"I don't know"—[Examiner] Is it eatable?—"No"—[Examiner] Why?—"Too much poison"; and watch—"second" (Glezerman, 1983, p. 165). Whenever prompted by the first syllable of the word, P. was able to name the object. His errors were not by chance, but words similar to the target word by meaning, thus the target word and its replacement (paraphasia) were from the same semantic field. This type of paraphasia and ability to name the object by sound prompting suggests a problem in word meaning, not in word sound. Similarly paraphasias were observed in P.'s spontaneous speech:

Patient P: "Let's remove the compass" (it was a stop-watch),
Examiner: "It is not compass, what is it?"
Patient P: "I forgot… second…"
Patient P: "Aren't the pictures in the purse? [referring to an envelope]"
Examiner: "It is not a purse."
Patient P: "Where you put letters"
Examiner: "What is the name of it?"
Patient P: "I forgot." (Glezerman, 1983, p. 165).

Autists have no such difficulties as did this patient of mine. Nor do autists lack understanding of objects' toolness, that is, they do not seem to have apraxia.[17]

Even in Kanner's material on LFA (some of them mute) an understanding of the functional meaning of objects remained intact. One autistic 5-year-old boy did not communicate with the examiner, but he would go after objects and use them correctly.

> He picked up a pencil and scribbled on paper that he found on the table. He opened a box, took out a toy phone, singing again and again: 'He wants a telephone', and went around the room with the mouthpiece and receiver in proper position. He got a hold of a pair of scissors and patiently and skillfully cut a sheet of paper into small bits, singing the phrase 'Cutting paper', many times (Kanner, 1943, p. 227).

[17] Patients with lesions in the left posterior brain who have apraxia do not know what to do with objects: how to strike a match, how to use a spoon, what a needle is for, etc.

However, in life, children with autism do not use objects in a conventional manner (according to the object's established proper function). Instead, they might be preoccupied with manipulation of certain objects, but this object use is not conventional but idiosyncratic, not functional but stereotypical.

2.6.3 Is the Categorical Component of Word Meaning Impaired in Autism?

What about the categorical component of word meaning? Because we are talking now about a higher level within the left hemisphere's BA37, it would be more appropriate to explore this question by studying HFA and gifted people with autism. Self-reports of HFA show that autists do not operate with abstract concepts, but instead achieve word comprehension indirectly, through visual associations. Let's look at one of Temple Grandin's accounts.

> My concept of dogs is inextricably linked to every dog I have ever known… if I think about Great Danes, that's what emerges: Dansk, the Great Dane owned by the headmaster at my high school, I visualize Helga who was Dansk's replacement, my aunt's dog in Arizona, an advertisement for Fitwell seat covers that featured that kind of dog…. There is no generic, generalized Great Dane (Grandin, 1995, p. 28).

What is very interesting is that she has intellectual insight into her peculiarities.

Take also the example of a 9-year-old autistic boy from Kanner's material, performing on a test involving word definition: "Tiger—'is a thing, animal, striped, like a cat, can scratch, eats people up, wild, lives in the jungle sometimes and in the forests, mostly in the jungle. Isn't it right?" (Kanner, 1943, p. 235). This autistic child could give an excellent answer, he used with ease categorical specifications of the word, but they are of little importance to him. What prevails in his mind, what his mind is preoccupied with, is the physical–sensual world. One wonders if this child used the words "thing" and "animal" not as a designation of a class (concept), but as a label, behind which is just another visual image of the singular animal. This possibility is supported by a recent study where subjects with autism showed significantly diminished differential fMRI activation to concrete versus abstract words compared to normal controls (Harris et al., 2006).

In her later publication, Grandin reveals how she was able to progress in her understanding of concepts. She continues to insist that words come secondarily to her. In order to understand both spoken and written language, she translates the words into visual pictures.

> All my thoughts are in photo-realistic pictures…. To form a concept from many specific photo-realistic pictures I have stored in my memory, I sort them into categories. Categorization of my specific visual memories was the beginning of concept formation. When I was a child, I categorized dogs from cats by sorting the animals by size. All the dogs in our neighbourhood were large until our neighbours got a Dachshund. I remember looking at the small dog and trying to figure out why she was not a cat. I had to find a visual feature that she shared with big dogs. I had to create a new category in my mind to differentiate. All dogs, no matter how big or small, have the same nose shape. My concept is sensory based, not word based.

Other ways of sensory-based categorization would be sound (barking or meowing) or smell (Grandin, 2010, pp. 141, 143). How Does Visual Thinking Work the Mind of a Person with Autism? by Temple Grandin, in: Autism and Talent, U. Frith and F. Happe, eds., 2010, Oxford University Press, New York. By Permission of Oxford University Press.

To fully appreciate what stands behind Grandin's experience, we need to clarify the term *category. Category* is understood as the categorical level of abstraction (see above in this chapter), where objects are related by common (categorical) signs. *Concept* is the verbal equivalent of category. Concept is an abstract, linear hierarchical sequence of categorical signs, which represents the categorical component of word meaning. It seems that in the literature *category* is sometimes understood as any grouping, which could be based on perceptual, functional, or categorical features. I believe (as was shown in beginning of this chapter) that the brain (LH) spontaneously "constructs" the world in parallel at concrete, functional, and categorical levels of abstraction. Categorical representations may be historically connected with modality-specific representations,[18] but cannot be reduced to the latter.

Grandin's extraordinary capacity to "see" and build a memory of innumerous visual combinations simultaneously ("my thinking is totally nonsequential"), with these RH representations being on the surface of her consciousness, allows her to closely approach something which is an equivalent to a concept at the lower brain levels. Still, the pattern she "extracts" is vicarious. Her compensation is likely achieved by the RH/LH interaction at the sensory-motor level, but what about abstract words that do not have corresponding visual representations in the brain? Interestingly, to understand abstract words, she uses RH visual situational thinking and even rudiments of RH visual situational-symbolic thinking. For example, for *peace* "an Indian peace pipe, or TV or newsreel footage of signing of a peace agreement" is called to mind; for honesty "an image of placing one's hand on the Bible in court…[a] news report describing a person returning a wallet with all the money in it provided a picture of honest behavior"; and for power and glory "a semicircular rainbow and an electrical tower" (Grandin, 1995, p. 33).

Now let's see what the literature says about the ability to categorize in autistic subjects.

Tager-Flusberg (1985) examined categories using pictures of common objects and reported children with autism do not have a specific cognitive deficit in categorization. She suggests children with autism have categorical representations but are unable to make efficient use of them.

More recent studies show sorting preferences in subjects with autism for concrete over abstract categories (Alderson-Day & McGonigle-Chalmers, 2011; Ropar & Peebles, 2007). Several studies found that in free recall autistic subjects did not group items to be recalled according to semantic-conceptual relations (Bowler, Gaigg, & Gardiner, 2008; Minshew & Goldstein, 1993). This is in accordance with Hermelin and O'Connor's (1967) earlier findings that autistic children fail to use

[18] For example, the supramodal temporal cortex, responsible for the phonological code of words, is historically built upon the auditory cortex, while the supramodal inferotemporal cortex responsible for word meaning is built upon the visual cortex.

semantic information to facilitate memory. Some studies have found that autists use semantic-conceptual associations if specifically cued to do so (explicit task) (Gaigg et al., 2008), and existing research shows some sensitivity to categorical recognition in autistic people, although such sensitivity is not as strong as that of the normal population. Toichi and Kamio (2001) found conceptual relationships for simple common words to be intact in autistic young adults, but suggest that semantic processing in subjects with autism might be qualitatively different from that in controls. The last statement is important, because all the above authors indicated that, instead of grouping recalled items using a preexisting semantic-conceptual network (the usual strategy to aid memory in normal people), autistic individuals grouped items idiosyncratically.

The neurophenomenological analysis conducted in this chapter "de-coded" what is called perplexing and idiosyncratic in autism and found it all to be a manifestation of RH associations at the sensory-motor (situational) level in the brain.

At the clinical-behavioral level, Toichi and Kamio (2001) come closest to this book's formulation.

> [A]lthough both groups showed similar performance in [the] task, the two groups might have employed different strategies. For example, individuals with autism may be more dependent on nonverbal strategies, such as visual imagery, which results in manipulating language differently from individuals without autism (p. 488).

All authors agree that concept formation, or the spontaneous generation of categories, is in some way abnormal in autism.

Thus, in autism, there is not a specific deficit (known neuropsychological syndrome) that can be attributed to a dysfunction of the left inferotemporal cortical area, BA37, and yet, the left BA37 is clearly not functioning properly.

2.7 Cerebral Organization of Metaphors and Autism

2.7.1 Metaphorical Language of Children with Autism Is of RH Origin

Kanner (1946) noted that the seemingly irrelevant and nonsensical utterances of autistic children are metaphorical expressions. However, their language becomes meaningful only if connection is established between the child's situational and emotional experience and his metaphorical utterance. Here is one of the examples Kanner gave.

> Paul G., while observed in our clinic at five years of age, was heard saying: 'Don't throw the dog off the balcony'. There was neither a dog nor a balcony around. The remark therefore sounded irrelevant. It was learned that three years previously he had thrown a toy dog down from the balcony of a London hotel at which the family was staying. His mother, tired of retrieving the toy, said to him with some irritation: 'Don't throw the dog off the balcony.' Since that day, Paul, whenever tempted to throw anything, used these words to admonish and check himself (Kanner, 1946, p. 242).

From the point of view of brain mechanisms, this observation can be explained as an exposed RH visual scene-situation, an indivisible whole including visual picture, action, and emotion. The utterance is part of the original visual scene-situation and becomes its emotional marker. When the child has that same feeling, the situations are identified, and the utterance reemerges. In other words, a singular situation experienced in the past and the currently experienced situation are identified by their common affect. Here is another example from Kanner's material.

> Jay S., not quite four year's old, referred to himself as 'Blum' whenever his veracity was questioned by his parents. The mystery of this 'irrelevance' was explained when Jay, who could read fluently, once pointed to the advertisement of a furniture firm in the newspapers, which said in large letters: 'Blum tells the truth'. Since Jay told the truth, he *was* Blum (Kanner, 1946, p. 243).

In metaphors, things from alien (remote) domains are usually joined together. In the above examples of autistic children's speech, metaphors are created by connecting alien domains through their instant identification. Kanner indicates that in case of Jay S., "analogy between himself as a teller of the truth and Blum does not differ essentially from the designation of a liar as Ananias, a lover as Romeo, or an attractive lad as Adonis" (p. 243). But in cultural symbols, according to Kanner, the listener is familiar with the analogy or "if the metaphorical reference to Ananias, Romeo or Adonis is not understood, dictionaries, encyclopedias or informed persons can supply the understanding" (p. 243). In contrast, to understand the metaphorical language of autistic children, one needs to know the source of the metaphor, which is an original and unique emotional–situational experience. For Kanner, the main distinction between autistic children's metaphorical expressions and cultural symbols is that the former are idiosyncratic (unique for each individual's experience). I would think that *any* metaphor has an idiosyncratic component. Cultural symbols are learned, but we understand them because we have an internal mechanism (right hemispheric) for symbol formation—identification. Not only in their creation but also in the process of their comprehension are metaphors perceived not just "logically" but also included into one's own emotional experience (and this is why and how we enjoy them), passing through the hearth of RH cognitive mechanism. This phenomenon is illustrated when LH dominance is weakened, and RH associations are revealed as a result. For example, if damage occurs to the left temporal region, responsible for word sound (phonological code of the word), LH word meaning of object reference and concept cannot be decoded, while RH associations, including aspects of figurative meaning from powerful emotionally loaded individual symbols, emerge on the surface. A patient of mine who suffered a lesion in the left temporal region (sensory aphasia) gave explanations of word meaning that illustrate the above point. When asked to define *resist*, he answered "Spartacus," and for enormous, "Gulliver" (Glezerman & Balkoski, 1999). We can see the patient's responses are far from direct explanations of word meaning, but instead they convey emotionally saturated visual images that are equivalents to these words' meaning. This example is also interesting because it shows that in the formation of individual symbols (RH equivalents of WM) cultural symbols as part of experience can be used.

RH content is idiosyncratic, because it is primary experience. In a way, each individual creates cultural symbols anew, although degree of idiosyncratic component varies significantly among individuals, as in the RH equivalent of WM.

As briefly mentioned in the beginning of this chapter, it is a well-known fact that children with autism are unable to understand metaphors. On the other hand, as discussed earlier, Kanner (1946) showed autistic children produce metaphorical expressions. In order to approach this contradiction we must first analyze how metaphors are processed in the normal human brain.

2.7.2 *Cerebral Organization of Metaphors in the Norm*

Neuropsychological studies have shown both cerebral hemispheres to contribute in characteristic ways to metaphoric competence. Winner and Gardner (1977) compared comprehension of metaphors in patients with unilateral, LH and RH lesions.[19]

All test subjects were presented with several sentences containing a simple metaphoric expression (e.g., "a heavy heart can really make a difference"). Each sentence was presented simultaneously with four pictures: one of the pictures represented the appropriate meaning of the metaphoric sentence (e.g., in the above case of heavy heart, a crying person was depicted); one was a literal representation of the sentence (a person carrying a large red heart and staggering under its weight); one depicted an object whose salient quality was described by the adjective (a five hundred pound weight); and one illustrated the noun (a red heart). Subjects were asked to point to the picture most fitting the sentence. After a pictorial choice had been made, the pictures were removed and subjects were asked for a verbal explanation of the metaphorical phrase contained within the sentence (e.g., "What does it mean to say a heavy heart?").

Table 2.1 shows that *double dissociation* was observed between patients with LH and RH lesions. Patients with the LH lesions could perform the nonverbal part of

Table 2.1 Comprehension of metaphors in patients with unilateral brain lesions (Winner & Gardner, 1977)

Heavy heart	Patients with left hemisphere lesion (intact right hemisphere)	Patients with right hemisphere lesion (intact left hemisphere)
Picture choice	A crying person	A person carrying a large red heart and staggering under its weight
Verbal explanation of metaphor	"It's heavy, the heart, a lot of weight"	"He's got many troubles"

[19] There were also two control groups in this study: patients with dementia (diffuse, bilateral brain damage) and normal subjects.

the test. When they were given a choice of four pictures to match with metaphoric sentence, they chose the appropriate picture. However, when they were presented with the same metaphorical phrase not accompanied by a picture, they gave a literal verbal explanation. Patients with RH lesions were able to give verbal explanation of the metaphorical phrase, but when they were given a choice of four pictures to match the phrase, they chose a literal pictorial interpretation of the metaphor. Unlike any other group in this study, patients with RH lesions did not find the literal depiction amusing or absurd.

Such experiments show that our ability to process metaphors is provided by complementary interaction of the two hemispheres, each having its specific contribution.

Metaphor itself, where usually remote terms are joined in a single figure of speech, is based on RH symbolic thinking. RH symbolic content is expressed through visual image, and, in RH thought, there is no content (meaning) without form (visual image). For example, in the metaphoric expression "heavy heart," a psychological condition is expressed through the physical image. However, the image (form) preserves its own value, so that RH symbol is always polysemantic. When the RH is damaged, polysemantics is lost. The patient perceives the object image as such—a red, heavy heart. On the other hand, the RH cognitive mechanism is limited. Within the RH symbolic system, objects (and situations) having different content are identified and are equipollent facets of an indivisible whole, with visual form and content making a single integrated representation. It is only through interaction with the LH that the symbol's components parts can be "dissected" and one can know that an object symbolizes another but is not that other. For example, in a poem line "A lonely sail gleams white in the blue mist of the ocean," we understand the image *sail* symbolizes a man through the metaphor *lonely sail*, and yet we also understand the sail is not a man. For RH thought, sail and man are interchangeable, identical in meaning. LH analysis of the words *sail* and *man* reveals that the categorical meanings of these words have only one categorical sign in common: objectness. Thus, these two perspectives are not confused, and we comprehend both literal (i.e., the image itself retains its meaning) and figurative meanings (Glezerman & Balkoski, 1999).

Figuratively speaking, the RH creates metaphors, but does not understand them. Isn't it characteristic for the metaphorical expressions of autistic children?

2.7.3 Autistic Children Create Metaphors but do not Understand Them

Let's return to case of Jay S., the 4-year-old autistic boy, who referred to himself as "Blum." Here, in a very condensed form ("Blum"), an emotionally saturated idea of truth, telling the truth, and who is telling the truth, are all expressed. Behind "Blum," there are situations, which are identified by affect becoming the symbolic system of meaning—an indissolvable whole. The example reflects not just polysemantic meaning, but also the undifferentiated wholeness of RH cognitive mechanism. I agree with L. Despert's conclusion about Jay S. (see discussion part of the Kanner article, p. 245).

> I wonder, however, whether the autistic child is not himself enmeshed in his own symbols, for while 4-year old Jay refers to himself as Blum, he does not say, and probably cannot say, "I am Blum" or "I am Blum because."

"Raw" RH associations of autistic children's metaphors are not really what we call metaphor. Although creative, their "raw" RH associations are not complete. Thus, metaphorical expressions of autistic children lack LH participation. This is why they are idiosyncratic, not directly communicable.

Is the RH needed at all for communication? I should say so, for without the RH, communication is severely impoverished, even prohibited. The LH organizes information into categories, the LH puts together linear sound sequences to make the phonological form of the word, and the LH combines words into sentences and phrases (grammar, syntax). On the other hand, knowledge about the world as primary experience is stored in the RH. How would the LH communicate, without the RH's experience? It was noted that when the RH is temporarily inactivated,[20] patients become very convivial (sociable) and verbose, and, although their speech is grammatically correct, it lacks substance and is full of unnecessary, repeated details—all of this is uncharacteristic for these patients and disappears after recovery (Balonov, Barkan, & Deglin, 1979).

The above examples of metaphor comprehension in patients with unilateral brain damage showed that patients with RH damage gave satisfactory verbal interpretation of the metaphorical phrase: heavy heart—"He's got many troubles," but chose a literal pictorial interpretation (a person carrying a large red heart, staggering under its weight). Direct connection between metaphorical phrase and its verbal explanation as a cliché is preserved (a product of LH verbal memory), but the RH polysemantic symbol, where meaning is expressed through an emotionally saturated visual image, is gone. Something similar can happen to cultural symbols. A visual image, singled out by the LH and connected with a LH monosemantic interpretation, can become a cliché LH sign (e.g., the dove as a conventional sign of peace). To remain *poly*semantic, cultural symbols need to be rooted in one's unique experience.

Now we are better able to resolve the proposed contradiction: severe disorder of spontaneous speech (8 out 23 children from Kanner's material remained mute) coexistant with a *creative attitude* toward language in autistic children. Asperger noted autistic children "are able to express their own original experience in a linguistically original form" (Asperger, 1991, pp. 70–71) and gave us several examples of such language use. One autistic boy (7-year-old) defined the difference between stairs and ladders as being "[t]he ladder goes up pointedly and the stairs go up

[20] Findings come from Balonov et al.'s (1979) examinations of depressed patients during recovery after unilateral ECT, when the involved hemisphere is inactivated for a short period of time.

snakedly" (Asperger, 1991, p. 71). Another 11-year-old autistic boy responded to a requested task: "I can't do this orally, only headily"; he described his sleep pattern as "long but thin" (p. 71); and once shared he did not like "the blinding sun, nor the dark, but best I like the mottled shadow" (p. 71). All these expressions are metaphorical, where physical, sensory, and visual–sensual characterizations refer to psychological aspects of experience.

To conclude, Kanner's observation that only digging out singular, original *situations* from an autistic child's past experience may give the key to the meaning of that child's metaphorical language is of extreme importance for understanding brain mechanisms in autism. It is not enough to say that in autism RH associations are on the foreground of consciousness. It is the situational level within the RH that gets exposed in autism.

Visual thinking of children with autism is visual–situational, not visual-symbolic. At the visual–situational (sensory-motor) level, situations are not united into symbolic systems, but remain "single." At the same time, the singular situation is a whole world in itself and all things inside the situation are identified based on common feeling.

2.8 Uniqueness of Right-Hemispheric Prevalence for Language in Autism

Current research, including functional neuroimaging (Boddaert & Zilbovicius, 2002; Garreau et al., 1994; Malisza et al., 2011; Muller et al., 1999; Ring et al., 1999), quantitative neuroimaging (De Fosse et al., 2004; Herbert et al., 2002, 2005; Rojas, Bawn, Benkers, Reite, & Rogers, 2002), and neurophysiological studies (Bruneau, Roux, Adrien, & Bartelemy, 1999; Dawson, 1988), gives overwhelming support for RH prevalence in autism. I will give a few examples concerning the area of interest in this chapter, BA37. Muller et al., 1999 fMRI study found the main difference in brain activation during sentence comprehension to be in the middle temporal area, roughly corresponding to BA37. Left middle temporal activation was four times less in autistic group than in controls, while right middle temporal activation was double in the autistic group compared to control. While sentence comprehension activates mostly two areas in the LH, the inferotemporal region, BA37 (comprehension of single words) and the inferior prefrontal region (comprehension of syntactic structure), in an autistic group, decreased activation in the LH and increased activation in the RH was mostly pronounced in one target area, which was BA37. Moreover, when autistic and control subjects were imaged while performing sentence generation task (more specific test for syntactic structure), activation in the left inferior prefrontal region was strong in both groups. Increased activation in the right occipital cortex (BA18,19) extending to BA37 was also observed in autistic subjects compared to the control group in fMRI studies during complex tasks of object recognition (Malisza et al., 2011; Ring et al., 1999).

Reversed lateralization in autism was not limited to visual modality related areas; it was reported as well in response to nonverbal and verbal auditory stimuli (Boddaert et al., 2000; Boddaert et al., 2001; Bruneau, Roux, Adrien, & Barthelemy, 1999; Garreau et al., 1994).

While these findings are important, they are not specific to autism. Reverse dominance of the cerebral hemispheres was found in other psychiatric disorders, such as schizophrenia, ADHD, dyslexia, SLI (specific language impairment), etc.

In this regard, the neurophenomenological analysis of word comprehension in autistic people conducted in this book turns out to be crucial for knowing what RH prevalence in autism means. While word meaning has complex cerebral organization, with separate representations of categorical and empirical components (LH), situational context, and visual image and visual-symbolic associations connected with a word (RH), for autistic people WM is reduced to its RH content. Moreover, WM for the autist is relegated to a deep historical level—the situational level. WM is situational-experiential, a word is not separable from an autist's unique experience. The RH situational (sensory-motor) level being the *only* one connected with WM is specific to autism, distinguishing autism from all learning disabilities. It also differs autism from schizophrenia where, I believe, RH *symbolic* level predominates. RH predominance in children is usually interpreted as compensation for early LH damage. The described above consistency with which children with autism use the RH situational level for word comprehension presents a problem for the interpretation that the RH prevalence in autism is secondary, a *compensation* for the primary LH deficit (in addition to there being no demonstrable specific neuropsychological LH deficit in autism).

Reorganization of brain functional systems as a reaction to the primary deficit can go only along "beaten tracks," established in evolution, of brain networks underlying human behavior. However, within this evolutionarily fixed pattern of cortical connectivity, each brain "chooses" its strongest link, the most well-developed region, to use for compensation, resulting in an individual-specific response. Considering the enormous individual variability of cortical regions in the brain (individual neuropsychological profile), many different responses are expected. Patients with LH lesions still may use intact zones in the left for compensatory measures, if their premorbid neuropsychological profile is such that their strengths lay in the LH, whereas others would have assets in RH that would rise up to make up for deficiencies in the LH. For example, in an fMRI study of children with early LH lesions, numerous compensatory functional reorganizations were observed with postlesion activation in response to language task showing lateralization to the RH as well as transfer to a perilesional area within the LH, in the left frontal area anterior to the lesion, and bilaterally (Liegeois et al., 2004).

If the RH is used for compensation, any level can possibly prevail: visual situational, visual situational-symbolic, or visual object-symbolic, depending on their prominence in the patient's premorbid neuropsychological profile.[21,22] Autism's specificity to *situational level* predominance is strong evidence against its being a compensatory phenomenon.

2.9 Conclusion

The paradox presented in the beginning of this chapter has been elucidated. The cerebral organization of word meaning presented here consists of LH empirical and categorical components while the RH contribution to WM are visual scene-situations and visual symbolic associations connected with the word. We found that autist's word comprehension is based mostly on the RH contribution to WM. Furthermore, it is the RH situational (sensory-motor) and not RH symbolic level that plays a leading role in word comprehension in autistic people. Knowing the RH origin of word meaning, one can explain idiosyncratic language in autism: RH associations are individual-specific. A lack (or deficiency) of LH participation in autists' cerebral organization of metaphors explains the dissociation between their creating metaphors, but not being able to understand them.[23]

---

[21] To illustrate reorganization of brain functional systems as a response to the primary, local deficit, I will use an example of *sensory aphasia* and contrast it with autism. Sensory aphasia is caused by a focal lesion in the left temporal region responsible for the phonological code of the word. Its primary deficit, therefore, is in word sound. When word sound becomes unstable, word meaning cannot be decoded as a result, and word comprehension is impaired. However, even though word sound has quickly slipped away, the instant it was heard an object image related to the word's meaning can still be evoked in the RH. The visual image, without support of word sound, is subject to RH rules and brings to the fore a particular group of associated holistic forms. Any one of a roundabout of images, similar in appearance but different in content, may come forth and push out a word sound. For instance, instead of the word "ditch" a patient may say "plate," instead of "wall" he might say "sheet," instead of "suitcase" he might say "well" (examples are taken from Bein, 1961).

Other patients with sensory aphasia may use different "layers" of the RH visual thinking to "compensate" for the primary deficit, such as visual-situational associations or even RH visual-symbolic thinking; in the latter, there is a peculiar narrowing of word meaning where the basic meaning of the word—concept and object reference—is lost, but figurative meaning remains intact. For example, a patient defined the meaning of "pipe" as "peace pipe," "dwarf" as "pygmean soul," "sharp" as "unpleasant, sharp tongue, everybody is afraid of it." Still another patient with sensory aphasia can still rely on his LH, and here the target word would be substituted under the rules of LH cognitive mechanism. For example, the concrete word "notebook" can be replaced by a word in a more abstract category: "stationary." Another such example is one patient's answer of "science" instead of "economics" (examples are taken from Bein, 1961).

[22] Applying the term premorbid neuropsychological profile to autism, a supposed neurodevelopmental disorder, means the particular individual's hypothetical brain as it would exist if the autistic disorder were removed.

[23] For the autist himself it is not metaphor but his primary experience.

What remains to be solved is why the left BA37 is not functioning properly in autism, when no specific dysfunction in this area can be found. Autism presents a unique developmental disorder in which language and emotional deficit are intrinsically tied in a specific pattern. It brings to mind the phylogenetic stage in the primate brain where communication, emotion, and cognition were one and the same whole. Is the deficit in autism coming from this deep root in the human brain's history? Is the left BA37 just a "performer" whose cognitive role in the expression of an active attitude toward the surrounding world cannot be realized in autism?

In this regard, the work of Ungerer and Sigman (1987) should be mentioned. These authors showed that while knowledge of perceptual and functional attributes of objects is a prerequisite for language development in normal and mentally retarded children, there was minimal correlation between the ability to sort out objects by color, form, and function and the development of language in young autistic children. In contrast, there was a significant correlation between verbal and gestural (pretend play) symbols in autistic children. The authors conclude that autistic children's disability is inherently cognitive and social and cannot easily be reduced to a singular dysfunction in either domain. We will explore this problem further in the following chapters.

Chapter 3
How Autistic Persons Perceive the World (Sensory-Motor Function Level in the Brain and Autism)

Well known are the paradoxical features in autism: fear of and fascination with moving objects; fascination with moving objects and desire for "sameness"; desire for "sameness" and hypersensitivity to change; hypersensitivity to change and fixation on parts of objects. To approach this puzzle, we must return to the RH situational (sensory-motor) level.

3.1 Right-Hemispheric Situational Level and Autist's Fear of and Fascination with Moving Objects

In the previous two chapters, we discussed that the unit for RH operations is the visual scene-situation (VSS). Also mentioned was how the RH processes information at the sensory-motor level: organizing and storing VSSs (moments of experience in the external world) in chronological order. Here, we will consider another aspect of the RH sensory-motor level—its contribution to consciousness. At the perceptual stage, with the RH constantly registering information, the content of RH consciousness is indeed a "flow of consciousness," that is, visual scene-situations representing a continuously changing impression of the outside world. Only a few stimuli from the external world are in the focus of attention. However, the RH "flow of consciousness" is an important background that is responsible for sustaining a so-called alert state, or state of attentiveness to external stimuli (Glezerman & Balkoski, 1999). On a broader scale, it is said that the RH gives a subjective view of the world, incorporating time and space, or a spatial–temporal background of one's world image (Cutting, 1990). With continuous flowing of situations, a constantly moving picture of the world, of "world movement," time as a RH subjective experience is perceived through a sense of objects' movement in external space. This is all determined by the holistic nature of the RH's cognitive mechanism. In normal individuals, RH experience does not reach conscious awareness. The LH analyzes the RH "flow of consciousness," distinguishing out content relevant to motivated,

T.B. Glezerman, *Autism and the Brain: Neurophenomenological Interpretation*,
DOI 10.1007/978-1-4614-4112-0_3,

goal-directed behavior and re-organizing this information according to its own cognitive mechanism. The LH determines the "focus of consciousness," which is the center of conscious awareness. "Focus of consciousness" relates to active, voluntary attention, an orientation toward objects important for future behavior. In some pathological conditions, RH experience overcomes its counterpart's interpretations and comes to the surface. This is what I believe is happening in autism.

A continuously changing impression of the outside world is revealed in an autistic individual's memory of his childhood:

> Confusion and terror...living in a frightening world presenting painful stimuli that could not be mastered.... Nothing seemed constant. Everything was unpredictable and strange. Animate beings were a particular problem. Dogs were eerie and terrifying. They were especially unpredictable. They could move quickly without provocation [example is from Cutting (1990), p. 384].

Is this child's experience a manifestation of an "exposed" RH "flow of consciousness"?

We do not have access to the subjective experience of low functioning autistic individuals (LFA), but behavioral observations of LFA might be a relevant equivalent to the above report of a high functioning autistic individual (HFA). Kanner in his original description of autism and all subsequent authors who observed LFA described intense fear of moving objects in autistic children. Here are a few examples of parents' reports from Kanner's material (Kanner, 1943):

> [WF, six y.o. boy:] He is afraid of mechanical things; he runs from them. He used to be afraid of my egg beater, is perfectly petrified of my cleaner. Elevators are simply a terrifying experience to him... (pp. 222–223).

> [HB, a five y.o. boy, was] tremendously frightened by running water, gas burners... (p. 231).

> [AL, nine y.o. boy:] He had many fears, almost always connected with mechanical noise—(meat grinders, vacuum cleaners, street cars, trains, etc.). Now he is afraid of the shrillness of a dog's barking. Usually he winds up with an obsessed interest on the things he was afraid of (p. 235).

> [EC, seven y.o. girl:] She was frightened by noises and anything moving toward her. She was so afraid of the vacuum cleaner that she would not even go near the closet where it was kept, and when it was used, ran out into garage, covering her ears with her hands (p. 240).

We can see the report of an HFA's subjective experience and the observations of LFA's behavior are complementary. Moreover, HFA's introspections may shed the light on the meaning of behavioral observations in LFA.

3.1.1 RH Disorders of Time and Autism

Studies of patients with the lesions in one cerebral hemisphere have shown that disorders of time perception arise mostly when lesions are located in the right hemisphere. They may be characterized as disorders in the "feeling" of time, disorders of perception of the "flow of time," while concepts of time sequences and temporal order expressed in language (before, after, first, second, etc.) remain intact because they are functions of the LH. Numerous quite extraordinary disorders of subjective experience of time in patients with damage to the RH are described in the literature. One patient reported momentarily feeling that "a woman I saw walking performed very slow movements, as though a film were running slowly, and I had a sense that time was moving down, or, more exactly, somehow the calculation of time was going backward." Another patient's experience was such that "[s]uddenly it feels that everything is stopping, everything is motionless as in a photograph" (Dobrochotova & Bragina, 1977, pp. 89–90). There was another patient with damage to the right parietal–occipital region who described his experience of time in the following way:

> Doctors and nurses were first of all moving with a measured step, conspicuously, as if on a film. Then the tempo of things became very erratic, sometimes coming at a furious pace, 'like moving pictures speeded up' as if the people involved were 'running a race'.... [Music] sounded very loud and very fast, as if 'several radios were all blaring away together'.... Sometimes, other people's speech seemed excessively fast and incomprehensible... [Hoff and Potzl (1934) cited from Cutting (1997), pp. 206–207].

A patient of Mullan and Penfield (1959) experienced the nurses in the hospital as moving so fast he could hardly follow them with his eyes; he also reported feeling that time was stretching out endlessly. Thus, we see various abnormal and unusual senses of time—time stops, time goes back, time stretches out, time is accelerating.

Note that the RH disorder in subjective experiences of time above involved *moment*, *speed* and *duration*, the very parameters that characterize *time* of the sensory-motor level's external spatial field in Bernstein's terms.

In the RH, time is not separable from movement and the object, i.e., moving object within the VSS, for the visual scene-situation might be just as well called the visual action-situation.

In the above examples, a distorted sense of time parallels distortions in sense of movement—acceleration and slowing, conjointly.

Distorted subjective sense of movement can range from the inability to detect movement or distinguish the fastness or slowness of motions, to movements seeming accelerated or slowed, to the inability to "catch" movement or progress in a general sense. The world then can be deprived of its most essential feature—changeability or transformation (Dobrochotova & Bragina, 1977).

It seems that individuals with autism suffer from *increased* "changeability": they cannot "stop the moment."

3.1.2 STG, Detection of Change and Autism

A particular region within the temporal lobe, the superior temporal gyrus (STG), has been identified as a major brain area involved in the automatic detection of any change in the physical features of stimuli. Gomot et al. (2002) performed an electrophysiological study to assess the detection of acoustic change in autistic and control subjects. They found STG activity in response to change started earlier in autistic individuals than in controls. Sensory cortical processes were normal and could not explain this unusual reaction to acoustic change in children with autism. Researchers concluded that subjects with autism have pathologically increased abilities in stimulus-change detection.

Studies have also shown STG involvement in the detection of moving objects. Movement as we just saw in the previous part of this chapter can be considered a particular case of change! In monkeys, single cell recording studies revealed the presence of cells in the temporal lobe, particularly the STG, sensitive to eye and mouth movement, to biological movement in general, and to rotational motion. In humans, the STG is activated when subjects view a face with moving eyes and mouth. According to Puce, Allison, Bentin, Gore, and McCarthy (1998), changes in direction of gaze activated the right STG more than the left STG.

Temple Grandin (1995) relayed one autistic person's report that "looking at other people eyes was difficult because the eyes did not stay still," and she speculated "some of the problems autistics have with making eye contact may be nothing more than intolerance for the movement of the other person's eyes" (p. 73). Indeed, in functional brain imaging study autists demonstrated significantly greater power of response in the STG than controls when viewing facial expressions from photographs of eyes (Baron-Cohen et al., 2000). These findings were not specific to faces. While listening to speech-like sounds, autistic children showed a significantly higher activation of the right STG and significantly lower activation in the left STG compared to controls (Boddaert & Zilbovicius, 2002; Boddaert, Zilbovicius, Belin et al., 2000). From the above studies, one may conclude that the STG is hyperfunctional in autism. In contrast, Blair, Frith, Smith, Abell, & Cipolotti (2002) believe that individuals with autism are dysfunctional in those regions of temporal cortex, particularly the STG, responsible for the detection of moving objects. These authors base this conclusion on their findings that autistic individuals had disordered recognition memory for objects consistently associated with movement, whether they were living (cats and horses) or nonliving (motorbikes and cars), while their recognition memory for nonmoving objects (buildings and leaves) was not only spared but superior compared to control. Blair and colleagues' argument was that objects consistently associated with movement come to be represented in the STG, while objects that consistently do not move would be represented elsewhere in the cortex. Their argument can be easily and strongly objected to on the basis of well-known brain functions. Object recognition is connected with the left inferior temporal region (BA37)—the *what* system. This same area stores information about objects in the form of *semantic* memory, while visual images of the objects are stored within the VSS, in the *right* inferior temporal region (the function of this region was analyzed in detail in Chap. 2). Objects are not stored (represented) in the STG, but rather the STG provides attention to moving objects. To explore this topic more, we must return to the problem of consciousness and attention.

The dichotomy of parallel/simultaneous RH processing versus linear/sequential LH processing corresponds to the lateralized attentional system in the human brain.

The traditional midline attentional system is more connected with the LH. The lower levels of this system control arousal: a continuum of sleep–wakefulness, with coma being the extreme pathological end of this continuum. The higher level of this system, the medial prefrontal cortex (MPC), is responsible for voluntary selective attention (i.e., selecting targets for goal-directed behavior)—the "focus of consciousness." The RH attentional system includes cortical areas within the right prefrontal and right superior parietal and temporal regions (Pardo et al., 1991). This anatomically distinct activating system in the RH mediates sustained attention to sensory stimuli independent of modality and laterality of sensory input.[1] It can operate independently of the midline anterior attention system (Pardo et al., 1991). Indeed, this system is a necessary condition for the RH "flow of consciousness"—sustained attention to the constantly changing sensory stimuli from the external world. STG then is a part of the RH attentional system registering movement as a particular manifestation of change, and its activity seems to be increased in autism.

3.1.3 RH Situational Level in Autistic Savant Drawing

To illustrate the hypothesis of RH sensory-motor level "exposure" in autism, the drawings of a uniquely talented autistic child, Nadia (Selfe, 1977)[2] will be examined from the point of view of Bernstein's brain function levels.[3]

Nadia was an autistic child with special talents. She demonstrated severe deficiencies in cognition and language and was at times nearly mute, at times with echolalic speech. She performed the drawings presented below between the ages of 3½ years and 6 years, 8 months of age.

First, let's look for a comparison at an example of drawings by a normal child aged 6 (Fig. 3.1).

For the normal child, the gnostic-praxic level is the leading one for drawing. This level provides a transition to an increasingly meaningful ordering of the external world. It is characterized by distinguishing objects for active use based on the functional (topological) scheme of the object. In Fig. 3.1, the horse and rider are functional schemes. The horse is drawn as an object with four legs, with a bridle suitable for riding. The rider's arms are drawn prominently to illustrate their usefulness in guiding the horse. This child does what the normal adult (who is not an artist) would

[1] Well known is attentional deficit (neglect) of the left side of space when the RH is damaged, but there is no neglect of the right side of space when the LH is damaged. This means the RH attentional system controls both sides of space while the LH controls the right side of space only.

[2] Nadia's drawings were originally examined and described by Selfe (1977).

[3] This neurophenomenological analysis of Nadia's drawings was previously presented in a poster, in shorter form, by Lucero and Glezerman 2004. The Right Hemisphere and Autistic Talent: a special state of consciousness. VIII Conference of Association of Scientific Study of Consciousness. Antwerp, Belgium, June 2004.

Fig. 3.1 Drawing by a normal child aged 6. From *Nadia. A case of extraordinary drawing ability in an autistic child*, 1977, by L. Selfe, London, UK: Academic Press; with the kind permission from Elsevier and Dr. Lorna Selfe

usually do, using the gnostic-praxic level as the leading one when drawing objects, that is he would draw schemes, not shapes.

In contrast, Nadia's horses (Figs. 3.2, 3.3, 3.4, and 3.5)[4] charge from the page—one can hear their hoofs beating, feel the hot breath emanating from flaring nostrils, and imagine the horses rushing by. Nadia is able to create an image that transcends its two-dimensional frame, so that the viewer experiences it in motion in a three-dimensional world.

In Nadia's drawings, the viewer can experience the heaviness of the rider, who is squeezing energetically the horse's flanks, fusing with the horse into a single whole

[4] The horse shown in Fig. 3.3 is drawn at approximately 5 years and 6 months, in Figs. 3.4 and 3.5—at approximately 5 years (Selfe, 1977).

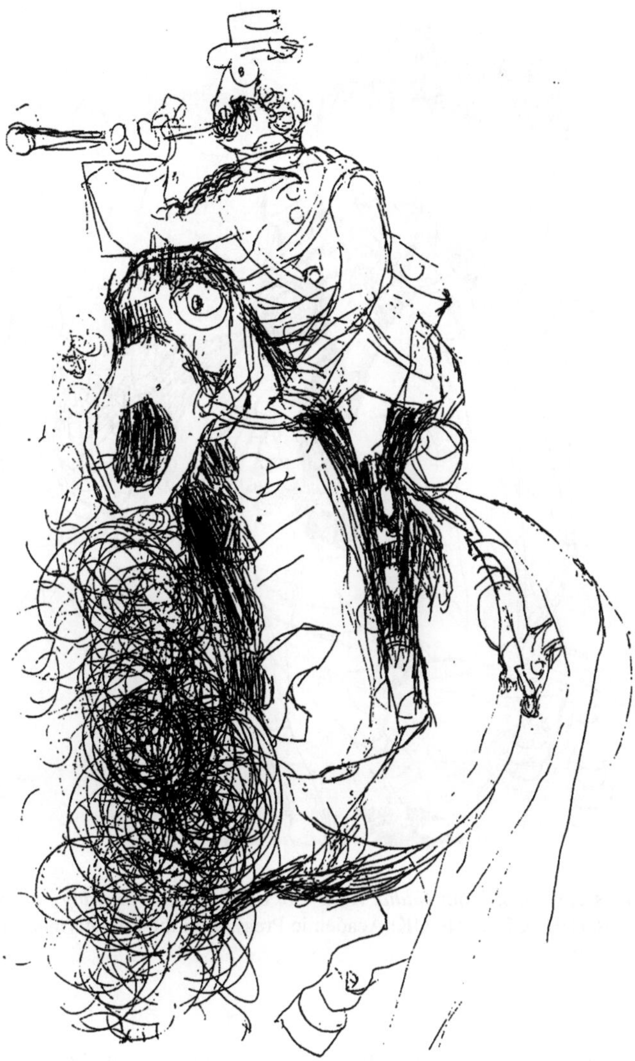

Fig. 3.2 Nadia's drawings. From *Nadia. A case of extraordinary drawing ability in an autistic child*, by L. Selfe (1977), London, UK: Academic Press; with the kind permission from Elsevier and Dr. Lorna Selfe

in its progressive movement (Figs. 3.3 and 3.5). She depicts with ease the complex muscular contractions that convey a tremendous sense of speed. This sense includes not only visual, but also a powerful kinesthetic experience, which we observe almost in a pure form in Fig. 3.6.[5] In this drawing, the viewer is awed by the horse at a gallop,

[5] This horse was drawn at approximately 4 years (Selfe, 1977).

Fig. 3.3 Nadia's drawings. From *Nadia. A case of extraordinary drawing ability in an autistic child*, by L. Selfe (1977), London, UK: Academic Press; with the kind permission from Elsevier and Dr. Lorna Selfe

which for an instant takes flight—the massive weight is in the air. This drawing "reconstructs" on paper the massive synergy of the thalamic level.

Nadia's depictions of progressive movement in the external world give the impression that this world is terrifying for her. It can be felt in the grinned muzzle of the horse, with puffs of breath bursting from his nostrils, in the rider's spurs piercing the horse's flanks, and in the bellicosely opened beak of the rooster depicting his cry (Figs. 3.2, 3.4, 3.5, 3.7, and 3.8).[6] We see here a manifestation of the fear of and fascination with movement characteristic of autism, already demonstrated in the examples of HFA reports and behavioral observations of LFA given earlier in this chapter.

[6] Roosters are drawn at approximately 6 years and 4 months (Selfe, 1977).

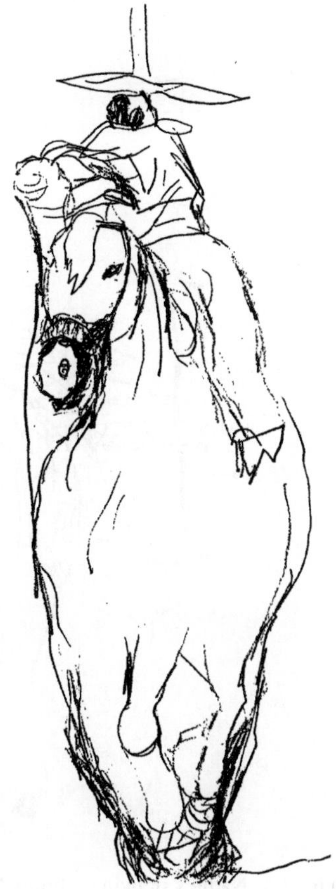

Fig. 3.4 Nadia's drawings. From *Nadia. A case of extraordinary drawing ability in an autistic child*, by L. Selfe (1977), London, UK: Academic Press; with the kind permission from Elsevier and Dr. Lorna Selfe

In Nadia's drawings we can see all characteristics of the RH "flow of consciousness": RH representations of action where movement is not separable from the visual scene. Figure 3.9[7] is remarkable in this regard. We see here the full human figure in action. The figure is offered in several views on a single page, legs kicking, arms thrusting, and torso twisting. The effect is cinematic. As they unfold over time, these multiple views create an impression of continuous motion. Just as the individual still images of a film come to life when they move through the projector, Nadia's figure comes to life in the viewer's imagination.

Indeed, this drawing does reflect RH "flow of consciousness": it simultaneously conveys the *moment of experience*, i.e., the visual scene-situation as a momentary segment of real time, and *continuous movement.*

The technical features of Nadia's drawings also indicate the predominant participation of the RH. In this regard, it is interesting to consider a study by Nikolaenko and Deglin (1984). In their study, subjects performed drawings while one of the

[7] Drawn at approximately 5 years (Selfe, 1977).

Fig. 3.5 Nadia's drawings. From *Nadia. A case of extraordinary drawing ability in an autistic child*, by L. Selfe (1977), London, UK: Academic Press; with the kind permission from Elsevier and Dr. Lorna Selfe

cerebral hemispheres was temporarily inhibited. These authors found that under conditions of relatively isolated RH functioning conveyance of the volumeness of objects was facilitated. It was also observed that the RH "sees" space from the perceptual perspective, the space limited to near the viewer. In addition, when the LH is inhibited (isolated functioning of the RH), depiction is strongly oriented relative to the vertical and horizontal axes. All these features can be observed in Nadia's drawings, especially in Figs. 3.2 and 3.4.

The movement, the "changeability," in Nadia's drawings is not exclusively expressed through the visual modality, for Nadia also provides a graphic representation of aural phenomena: the snorting of a horse, the crowing of a rooster (Figs. 3.2 and 3.10). In Fig. 3.2, a cloud of black lines emit from the horse's nostril—exaggerated in size and placed in the center of the drawing. These lines are the snort of the horse as he grasps for air, and the feeling of his hot, moist breath. The rooster's crowing is depicted similarly, with a tangle of lines emanating from its beak. Here Nadia's drawings evoke not just a visual experience, but an aural one as well.

Fig. 3.6 Nadia's drawings. From *Nadia. A case of extraordinary drawing ability in an autistic child*, by L. Selfe (1977), London, UK: Academic Press; with the kind permission from Elsevier and Dr. Lorna Selfe

In Nadia's earliest drawings (from ages 3–4), those of a train, she adds a face to the image of the train rushing downhill, its smoke billowing out behind it, and she surrounds the train with animal heads (Fig. 3.11). We wonder if Nadia is showing us that trains and animals are alike because they move. Perhaps the fact that the train is not alive and thus differs fundamentally from animals is less important for Nadia than the fact that they both share this capacity for motion. At the sensory-motor level, there are speed, moment, duration, and things (with mass and spatial parameters) that can move. These things are not objects yet. Objects as tools and members of categories come from higher levels. The idea of living things is a categorical sign that requires comparison of objects by their features (left hemispheric analysis at the categorical level). These higher levels are not in much use in autistic children.[8] Even more interesting, in this drawing of a 3-year-old child, we can see projection of temporal region STG functioning! That the STG is activated by moving things, whether it is an animal, car, train or eye gaze, was discussed above. Also noted was

[8] It is well documented that infants as young as 3 months are able to distinguish biological from nonbiological motion, fundamental for animate–inanimate category distinction. Nine-month-old infants can reliably distinguish between animals and vehicles [Mandler and McDonough (1993), Bertenthal (1993)—cited from Caramazza and Shelton (1998)].

Fig. 3.7 Nadia's drawings. From *Nadia. A case of extraordinary drawing ability in an autistic child*, by L. Selfe (1977), London, UK: Academic Press; with the kind permission from Elsevier and Dr. Lorna Selfe

that the STG, in particular the right STG, is hyperactivated in autistic people, according to recent brain imaging studies.

We may conclude that Nadia views the world through the "visual frame" of the sensory-motor level, free from higher levels' interpretations. Nadia's drawings convey information about the physical parameters of external space and its objects: distance and depth; mass, size, and three-dimensional shape; object localization and position; movements and interacting forces, all of which are characteristic for the sensory-motor level. Bernstein pointed out this level's greater degree of objectivity compared with higher levels and indicated the sensory-motor level's *metricity* and *geometricity* constitute the basis for precise and accurate movements. At the next level (gnostic-praxic) space evolves toward schematization, which results in gaining semantic organization and "order," but loss of the strict objective, photographic correspondence with real metrical parameters.

Again, it needs emphasizing that this is a child with a profound disorder of spontaneous activity and goal-directed behavior, with no sense of agentivity. She does her drawings without thinking, without any goal, without any voluntary intention to express something. She draws "without consciousness" or as one may say, in a "special consciousness."[9]

The sheer beauty is that her *RH sensory-motor level* is *projected* onto paper. Thus this production of an autistic child is invaluable for the scientific exploration of the autistic brain.

It is not a local area in the brain that is projected, but rather the whole historical stratum[10]—sensory-motor level with its specific spatial–temporal organization—that gets "exposed." The several modalities (visual, auditory, tactile, and kinesthetic) supplying afferentation for the sensory-motor level's exploration of the external spatial field are projected. Also, the powerful kinesthetic experience induced in the viewer by moving objects in Nadia's drawings suggests projection of what is deeper than the sensory-motor: the thalamic level.

Thus RH consciousness and its sensory-motor and thalamic levels are on the surface for Nadia, while in the norm these historical strata of consciousness are "covered" by the LH "focus" of consciousness and by the higher levels, gnostic-praxic and symbolic.

3.1.4 Preoccupation with Parts of Objects in Autistic Savant Drawing

When one looks at Nadia's drawings and their multiplicity of subjects, from horse and rider, to rooster to train, to football players, certain themes emerge. Nadia conveys with stunning accuracy those parts of the figure that are involved with *motion*. We can observe that Nadia is preoccupied with and fixated on parts of objects that are associated with movement. These "parts" are not signs of objects, not the result of left hemispheric feature analysis. They are by themselves experiential RH "wholes."

For example, her drawings of legs at rest (Figs. 3.12, 3.13, and 3.14) are paradoxically alive with movement. She offers the leg crossed on top with one definite

[9] Aren't her drawings a miracle! Nadia was not trained to draw, she had never seen a real horse. She was shown a children's picture book and asked to copy, but she did not do it immediately, but later, by memory. Her drawings have little resemblance with the book samples.

[10] Zeki and Lamb (1994) conducted a detailed analysis of so-called kinetic art (in which the perception of motion is strongly induced by a static figure), and the organization of the visual brain. They found that the artist's technique is very specific for optimal and selective activation of a small area within the visual cortex specialized for visual motion. They concluded that in creating kinetic art the artist "instinctively and physiologically" has unknowingly explored the organization of the visual brain. Here we see an "exposure" of the circumscribed area within the visual cortex in art. In contrast, in autism, the whole historical stratum of consciousness emerges on the surface.

Fig. 3.8 Nadia's drawings. From *Nadia. A case of extraordinary drawing ability in an autistic child*, by L. Selfe (1977), London, UK: Academic Press; with the kind permission from Elsevier and Dr. Lorna Selfe

outline but with several other lines that suggest multiple perspectives (Fig. 3.14). This is not a snapshot of a moment in time but rather a composite of the leg in several shifting positions. One can feel this pair of legs cross and uncross, fidget with nervous energy. The drawing projects a tension: this is a pair of legs now at rest, but will soon be in motion. Nadia's lines at the flexed knee joint demand it: they are dynamic and energetic—the expanding energy that will soon explode and extend the leg as if it were a piston in an engine.

Nadia's drawings (Figs. 3.7, 3.8, and 3.10) also show she is preoccupied with the rooster's beak where the action (creation of sound) comes from. Depiction of the

Fig. 3.9 Nadia's drawings. From *Nadia. A case of extraordinary drawing ability in an autistic child*, by L. Selfe (1977), London, UK: Academic Press; with the kind permission from Elsevier and Dr. Lorna Selfe

sound of crowing by the motion of the beak and a tangle of lines emitting from the beak gives a powerful impression of the rooster as animate, strutting through the yard and emitting his crow (Fig. 3.10).

Nadia's drawings provide insight into the preoccupation with parts of objects that is an important aspect of the autistic pattern. Nadia is able to grasp on paper what she was unable to grasp in her mind: a still moment in time. I believe Nadia's preoccupation with drawing "parts" is an attempt at spontaneous compensation for an "exposed" right hemispheric "world"—an attempt *to stop* and *to master the moment*. It is the left hemispheric cognitive mechanism that can "stop the moment." The RH provides constant, sustained attention to the continuously moving physical world. The RH allows us to catch minute details, subtle changes of sensory stimuli. The RH can fixate on the object, but one thing the RH cannot do is distinguish the object or its parts out of the situation. In accordance with RH processing, relations of *part/whole* are such that *part* is not separable from *whole* (VSS), and part is identified with the whole. Thus, *part* is representative of the unique, experiential VSS and, therefore, *part* is the same moment of experience as the *whole* (VSS).

Fig. 3.10 Nadia's drawings. From *Nadia. A case of extraordinary drawing ability in an autistic child*, by L. Selfe (1977), London, UK: Academic Press; with the kind permission from Elsevier and Dr. Lorna Selfe

Fig. 3.11 Nadia's drawings. From *Nadia. A case of extraordinary drawing ability in an autistic child*, by L. Selfe (1977), London, UK: Academic Press; with the kind permission from Elsevier and Dr. Lorna Selfe

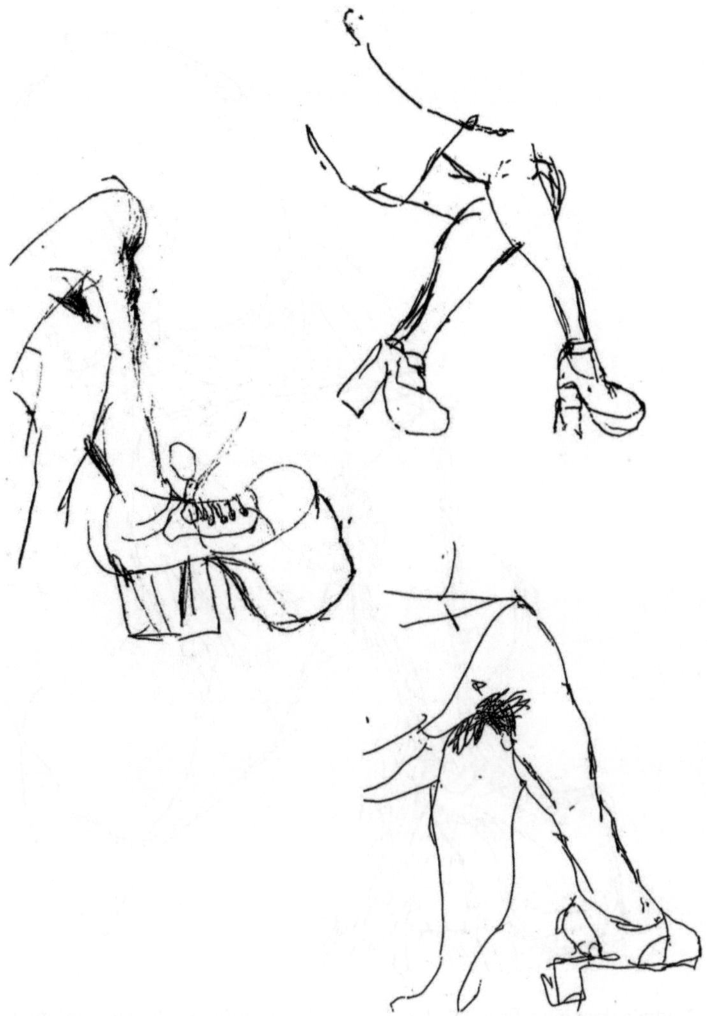

Fig. 3.12 Nadia's drawings. From *Nadia. A case of extraordinary drawing ability in an autistic child*, by L. Selfe (1977), London, UK: Academic Press; with the kind permission from Elsevier and Dr. Lorna Selfe

Recall that moments of environmental time characterize the time of the sensory-motor level, according to Bernstein. If in the RH these moments present the continuous flow of time not separable from movement, in the LH, there is an unfolding linear vector of temporal order in "synthetic" space. Time for the LH, at the sensory-motor level, is a sequence of discrete moments according to which progressive movement is organized in the external spatial field. Thus, the LH transforms RH continuous flow of time into discrete moments.

In Nadia's drawings of *parts* (take for an eloquent example her drawings of legs, Figs. 3.13 and 3.14) there are many separate lines as though they are repeating the

Fig. 3.13 Nadia's drawings. From *Nadia. A case of extraordinary drawing ability in an autistic child*, by L. Selfe (1977), London, UK: Academic Press; with the kind permission from Elsevier and Dr. Lorna Selfe

initial depiction, however, when looking closely at these drawings, we notice that each line is a new foreshortening of movement—its imprinted moment, rather than multiple repetitions of the same (perseveration). Namely these several sequential moments imprinted on paper give the impression that the *leg* is moving. We can also see successive images of discrete moments of movement depicted through several positions of the human figure in Fig. 3.9. By depicting separate moments of movement, Nadia's drawings demonstrate she used her LH to "stop the moment." It can also be seen that, preoccupied with parts of objects (legs, beak, etc.), Nadia draws them repetitively such that drawing parts becomes a stereotypical activity for her.

Fig. 3.14 Nadia's drawings. From *Nadia. A case of extraordinary drawing ability in an autistic child*, by L. Selfe (1977), London, UK: Academic Press; with the kind permission from Elsevier and Dr. Lorna Selfe

3.2 The Right-Hemispheric Situational Level and Autist's Hypersensitivity/Hyporeactivity to Sensory Stimuli

Markedly exaggerated subjective experience (increased sensitivity) and a hyporeactivity to sensory input are both consistently observed in autistic children. Many autistic children are thought to be deaf before they are diagnosed as autistic. Again, there is no access to the subjective experience of LFA, but we know about increased sensitivity to

Fig. 3.15 Nadia's drawings. From *Nadia. A case of extraordinary drawing ability in an autistic child*, by L. Selfe (1977), London, UK: Academic Press; with the kind permission from Elsevier and Dr. Lorna Selfe

environmental stimuli from reports of HFA. Grandin describes her hypersensitivity as visual and auditory overload, sensory mixing and jumbling. Oversensitivity is expressed in several modalities, mostly those that give information about the outside world (the external spatial field) such as visual, auditory, and tactile sense. Damasio and Maurer (1978) suggest these problems are due to attention disorder while sensory processing itself is not impaired. I would propose that sensory hypersensitivity in autistic individuals is due to a "hyperalertness" connected with exposed RH immediate experience—the constant stream of images in which content and time are indistinguishable. Very eloquent in this regard is the subjective experience reported by Therese Jolliffe, a high functioning autistic person: "Reality is a confusing interacting mass of events, people, places, sounds and sights. There seem to be no boundaries, order or meaning to anything" (Jolliffe, Landsdown, & Robinson, 1992, p. 16). Nadia's drawing (Fig. 3.15), showing us a projection of the chaotic RH world experienced by the autist, serves as an illustration to Jolliffe's report.

Jolliffe's description also gives a glimpse of what it feels like when the sensory-motor level predominates: terrifying dazzling brightness, clearness yet meaninglessness of the physical world. Remember that things as objects for use, the categorical classification of the world, logical relations between objects, and temporal analysis of sequence or temporal order are all connected with the left hemisphere and higher levels, gnostic-praxic and categorical. The very fact that HFA are able to give such insightful reports tells us that they do use LH higher levels—it is just that the sensory-motor level predominates. Reports of HFA also provide some insight into the nature of "the opposite side of the coin"—hyporeactivity to sensory stimuli. Here is Grandin's remembering of her response to sensory overloading.

> I could sit hours on the beach watching sand dribbling through my fingers. I'd study each individual grain of sand as it flowed between my fingers. Each grain was different, and I was like a scientist studying the grains under a microscope. As I scrutinized their shapes and contours, I went into a trance which cut me off from the sights and sounds around me (Grandin, 1995, p. 44).

Isn't this similar to Nadia's preoccupation with legs? Each grain is individual, unique, and not a member in a series, but each has its own unique shape and contour—we see here preoccupation with shape. Note the RH associations by holistic form, intense attention, and fixation on individual grains are attempts "to stop the moment."

Jolliffe, whose report was cited above, also explained her response to sensory mixing and jumbling. "A large part of my life is spent just trying to work out the pattern behind everything. Set routines, times, particular routes and rituals all help to get order into an unbearably chaotic life" (p. 16). Once again, this HFA has insight into her problems with sensory overloading and manages to employ LH "focus of consciousness" to compensate.

3.3 Right-Hemispheric Situational Level and Autist's Insistence on "Sameness"

Kanner (1943, 1951, 1958) distinguished desire for sameness as an essential feature of autism. In children with LFA, desire for sameness is expressed in a literal way: a certain way in which objects are placed in space, a specific sequence for the way things need to be done (an outing or bedtime ritual), etc. Here are some examples from Kanner's material (Kanner, 1943)

> [John, five-year-old:] Change of routine, of furniture arrangement, of a pattern, of the order in which everyday acts are carried out, can drive him to despair. When John's parents got ready to move to a new home, the child was frantic when he saw the moving men roll up the rug in his room. He was acutely upset until the moment when, in the new home, he saw his furniture arranged in the same manner as before. He looked pleased, all anxiety was suddenly gone, and he went around affectionately patting each piece. Once blocks, beads, sticks have been put together in a certain way, they are always regrouped in exactly the same way, even though there was no definite design (p. 245).

> [Donald, five year old:] If he spun a block, he must always start with the same face uppermost. When he threaded buttons, he arranged them in a certain sequence that had no pattern to it but happened to be the order used by the father when he first had shown them to Donald.
>
> ...
>
> Donald would not leave his bed after his nap until after he had said, "Boo, say 'Don, do you want to get down?'" and the mother had complied. But this was not all. The act was still not considered completed. Donald would continue, "Now say 'All right.'" Again the mother had to comply, or there was screaming until the performance was completed. All of this ritual was an indispensable part of the act of getting up after a nap. Every other activity had to be completed from beginning to end in the manner in which it had been started originally (pp. 219, 245).

The above illustrations of desire for sameness in LFA can be interpreted as an attempt to compensate by preservation of the spatial–temporal context of a situation as a moment of time.

3.4 Conclusion

Sensory hypersensitivity, idiosyncratic preoccupation with parts of objects, distress to variation that occurs in their surroundings and desire for sameness are among symptoms consistently observed in autistic individuals. Well known also is autists' fear of and fascination with moving objects. One study showed poor visual recognition memory in autistic individuals for the objects associated with movement (Blair, Frith, Smith, Abell, & Cipolotti, 2002). There is also evidence from the literature (Gomot et al., 2002) that autistic individuals have pathologically increased ability to detect change in incoming stimuli. These seemingly disparate manifestations, I have hypothesized here, have a common mechanism. In autistic individuals, RH content at perceptual (sensory-motor) level may be exposed. It implies that RH "flow of consciousness," visual scene-situations representing moments of environmental time and a constantly changing impression of the outside world, is not opposed by LH analytic processing or a LH "focus of consciousness." The desire for sameness and preoccupation with and fixation on parts of objects are attempts at a spontaneous compensation—"to stop the moment."

Chapter 4
How Autistic Persons Perceive Faces (Cerebral Organization of Face Recognition and Autism)

4.1 Research in Fusiform Face Area and Autism

It has been known for a century that damage to an area within the fusiform gyrus of the right occipito-temporal cortex results in a peculiar syndrome—*prosopagnosia.* The patient, otherwise normal, cannot recognize familiar faces. This area was labeled Fusiform Face Area (FFA). Modern brain imaging studies confirmed FFA's function: all people in the normal population activate the FFA during face recognition tests.

A hallmark study by Robert Schultz and colleagues showed that autistic persons did not activate their FFA when processing faces. Not surprisingly, Schultz's findings were sensational and met with the much enthusiasm in the autism research community. Lack of interest in the human face and limited amount of time engaged in face-to-face eye contact are considered important behavioral manifestations of autism (DSM-IV, American Psychiatric Association).

In Schultz et al. (2000) study, individuals with autism underwent fMRIs while performing a face discrimination test. They were presented pictures of same-sex pairs of neutral or nonexpressive faces. Subjects were supposed to indicate by pressing a button whether side-by-side images of faces were the "same" or "different." As a comparison, pairs of objects and patterns were also presented. There were no significant differences in accuracy of test performance between the autistic group and two control groups of normal individuals, but there was a significant difference in brain area activation. The evidence appeared simple and clear, hence its appeal: individuals with autism have dramatically reduced functional activity in the FFA during face processing.

When autistic individuals discriminated faces in this study, significantly greater activation was observed in the inferior temporal region (BA37, responsible for object recognition), together with reduced activation in the FFA, compared to controls. In the first series of experiments, activation was increased in the right inferior temporal region, but in the second series it was mostly increased in the left inferior temporal region.

T.B. Glezerman, *Autism and the Brain: Neurophenomenological Interpretation*, DOI 10.1007/978-1-4614-4112-0_4, © Springer Science+Business Media New York 2013

Attempting to explain the results of their experiments, the authors employed the popular psychological theory about autistic "cognitive style" (Frith, 1996).[1] According to this theory, abnormal cognitive function in autism results from a cognitive processing style that favors local processing or feature analysis over a global, holistic approach. Schultz and his colleagues' logic is as follows. Whereas face recognition in the norm is based on holistic processing implemented by the FFA, autists instead perform feature analysis when processing information and therefore activate a region more suitable for this—the inferior temporal region. This conclusion is built on a false premise that objects are mostly processed via feature analysis in the right hemisphere's inferior temporal region (the region activated in Schultz's first set of tests). The holistic versus analytic dichotomy belongs to the inter-hemispheric differences in cognitive mechanism, not to differences in brain regions. Faces are holistically processed in the right hemisphere (FFA), but objects are processed by both hemispheres: holistically in the right and analytically in the left.

The authors also suggest their findings could be a result of fundamental problems in autists' FFA, necessitating compensatory processing by neighboring regions. However, structural MRI scans of patients and controls in this study conducted by two experienced neuroradiologists, blind to participants' diagnosis, failed to find any abnormalities in the structure, symmetry, or gray matter intensity of autists' FFAs.

The year following Schultz's study, Karen Pierce and colleagues published their results from fMRIs of autistic subjects during a face perception task. Subjects with autism and controls were presented with neutral faces of males and females alternating with presentation of different shapes. The task was to press a button in response to female faces and in response to circles. In this study (Pierce, Muller, Ambrose, Allen, & Courchesne, 2001), subjects with autism were not significantly different from normal controls in terms of accuracy and response times on either the face or shape perception task. However, in striking contrast to virtually every existent fMRI study of face processing in normally developed humans, there was either abnormally weak or no activation in the autists' FFA in response to a human face. Considering the low FFA activation occurred in the context of autists' high accuracy and normal reaction times during the face perception task, the authors propose that autistic subjects "see" faces by utilizing different cortical regions. Functional foci of maximal activation or "hot spots" were identified for each autistic and normal subject. For every normal subject, a "hot spot" in response to faces was observed within the FFA. Each autistic subject displayed a "hot spot" in varied cortical areas, with the pattern of response not overlapping between autistic subjects: it was unique for each autistic individual. Some autists exhibited a maximal response to faces in the frontal cortex, while others did so in the temporal cortex, and yet others in the occipital cortex. This study confirmed Schultz et al.'s (2000) data that autistic people do not employ the FFA while perceiving faces; however, it did not confirm their findings of across the

[1] This theory is psychological; it is not directly related (congruent) to brain functional differentiation with its three dimensions.

board increased activity in the inferior temporal region in autists. Pierce et al. (2001) did not find any difference in the functional activation of the inferior temporal region between autistic and control groups.

Schultz et al. (2000) suggest that activation of the inferior temporal region instead of the FFA during face perception task is specific to autism. They also propose this change in neurofunctional organization (activation of inferior temporal region in response to faces) is caused by the "autistic cognitive style." However, in the Pierce et al. (2001) study, faces maximally activated a variety of individual-specific cortical areas in the autistic group. Thus the re-mapping of face processing does not appear to be a part of the autism brain pattern as such, but rather is a compensatory mechanism based on premorbid brain peculiarities of each autistic individual.

Hypoactivation of the FFA in autistic people during face perception was then demonstrated in several other studies, with a total sample size of 157 persons with autistic spectrum disorder (per Schultz, 2005). However, other studies found autistic subjects did not have significantly different activation in the FFA from that of normal controls in response to a familiar face (Pierce et al., 2004; Aylward et al., 2004; Hadjikhani et al., 2004).

For example, in Pierce, Haist, Sedaghat, and Courchesne (2004) study, the face perception test included personally meaningful faces, such as a mother or co-worker, as well as stranger's faces. Pictures of such faces were presented to autistic and normal control subjects, and the task was to press a button in response to all female faces. Significant FFA activity in response to familiar and unknown faces was found in both autistic and normal control groups. Individuals with autism also showed greater fusiform activity in response to familiar faces as opposed to stranger's faces, as well as the prototypical right hemisphere (and therefore *holistic* processing) dominance in response to both types of faces. Of note, nonautists' fusiform activity is also greater in response to familiar faces versus the face of a stranger.

Here we are now, back to the clinical picture alone as discriminatory. Autistic individuals do not look at faces and have poor eye contact, their faces show no emotion or abnormally modulated emotion, and their voices are monotonous or peculiar, not appropriate to situation. They appear uninterested in the social world. The DSM-IV states that children with autism may treat adults as interchangeable or may cling mechanically to a specific person. If autistic people are truly socially detached, there should be no difference in autistic brain activation in response to familiar faces versus that of a stranger. But Pierce's study showed autists indeed had increased FFA activation in response to a socially significant stimulus, i.e., a familiar face.

Let's return to the initial discovery of autists' lack of FFA involvement in face processing and the logical steps behind the authors' resultant conclusions (1) face processing is a primary feature of most human social interactions; (2) face processing is impaired in autism, i.e., a specific "face area" in the autist's brain is not activated (or significantly less activated) in response to faces; and (3) reduced activity of the "face area" in autism may be a neurofunctional marker of the disorder and may be the cause of autists' social deficits, the latter being a core deficit in autism according to the DSM (Schultz, 2005; Schultz et al., 2000). Schultz thus purports that FFA abnormalities are part of the causal mechanisms in autism. His resultant model of autism is rather complicated:

1. In early development, the trigger for autism is dysfunction of the amygdala, which in the norm assigns faces with emotional significance. Dysfunction of the amygdala leads a child to be uninterested in faces, to not pay attention to faces, to not look at faces.
2. Because the autistic child does not look at faces, he does not have experience with faces, and his perceptual skills connected with the FFA are undeveloped.
3. When the FFA is not used, it becomes dysfunctional. Dysfunction of the FFA closes the door to the autist's developing social knowledge and social skills. Here the author bases his argument on *conceptual* knowledge being strongly grounded in modality-specific *perceptual* systems. According to his model, seeing a face automatically and involuntarily activates FFA-based semantic knowledge about people and primes the observer for a social interaction. In those people with autism who have relatively impoverished social knowledge, there would be significantly less FFA activation.

This theory muddles ideas of causation and is quite mechanistic and reductionistic. Furthermore, it assigns the FFA with functions belonging to a whole network of face processing, not just this one area. FFA is part of the visual cortex and operates at the sensory-motor and gnostic–praxic levels. Conceptual knowledge belongs to the categorical, symbolic level. Cortical areas operating at the symbolic level are developed (in the history of the human brain) from and superimposed on the sensory modality-specific areas, but symbolic function and conceptual knowledge cannot be reduced to or fused with sensory experience. The parallel pathways to association cortex allow the simultaneous processing of information at the lower and higher levels, as illustrated by modern myeloarchitectonic studies.

And yet, the question remains: why is hypoactivation in the FFA during face perception observed in some autistic individuals? And why do some autists have FFA hypoactivation, and some do not?

Pierce et al. (2004) think a defect in the FFA is not likely a "marker" of autism. Instead, they conclude the autist's FFA is capable of responding to face stimuli, but whether or not it does so has to do with influences from other systems in the brain.

Furthermore, in *Brain Editorial*, Mary Phillips (2004) indicates several outstanding questions: Do face-processing deficits occur as a primary deficit or are they secondary to the social dysfunction in autistic individuals? Could there be no causal relationship between autists' facial processing impairments and their social dysfunction?

Trying to get around the conflicting data, authors attribute the discrepancies about FFA findings in autists to the various studies having used different MRI techniques, variances in the neuropsychological tasks presented, the wide age range of participants and degree of their social impairment, insufficiency of statistical power, etc. However, it seems no standard neuropsychological test is an appropriate tool for understanding autism. A different approach will be taken here. First, we will obtain a clear and coherent understanding about the function of various brain areas involved in face processing. From there we will explore what happens when these areas are damaged, and then we will look at how they function in autistic people.

4.2 The FFA, the Distributed Network of Face Processing and Autism

First of all, it must be emphasized that facial processing is implemented by a distributed network, meaning it is a concerted effort by several different brain areas located far from each other. Each area has its specific contribution, but all participating areas are activated simultaneously, with the end result being what one sees and feels about a face. The following statement by Zeki and Lamb (1994) can be well applied to the face processing network: "one should be enquiring not only into what the visual stimulus does to the cerebral cortex but also, and in particular, asking what the cerebral cortex does to the visual stimulus."

As already stated in this book, a particular cortical region's influence on a visual stimulus depends on its mode of information processing, based on the interplay of three dimensions (1) the cortical hierarchy to which the region belongs (historically newer areas are usually a higher function level); (2) the hemisphere (analytic versus holistic processing) to which it belongs; and (3) the modality to which it belongs or, if the area is supramodal, the modality on which it is superimposed (the "abstract" functional area within BA37 is built in evolution upon the visual modality). The FFA's specific contribution to face processing is determined by its being at the sensory-motor and gnostic–praxic level of the visual modality and by its use of RH holistic processing.

4.2.1 Damage to the FFA and Its Left Hemispheric Counterpart

The FFA is situated in the right inferior occipito-temporal junction. Cytoarchitectonically the FFA corresponds to the inferior component of BA18 and BA19 and, according to some authors (Damasio, Tranel, & Damasio, 1990), part of nearby BA37. BA18,19 are secondary fields of the sensory-motor and gnostic–praxic level. They are concerned with visual recognition of letters, numbers, color, and faces. It is interesting to note that visual object recognition "moved" in evolution from the visual cortex proper in the occipital lobe to the higher level BA37 in the temporal region. Letter and number agnosia occur with damage to the occipital region (BA18,19) of the left hemisphere, while a lesion in the symmetrical zone in the right hemisphere (FFA) results in disorder of face recognition—*prosopagnosia.* Color agnosia can result from a lesion in the occipital area of either hemisphere, but gets expressed differently depending on laterality of the lesion.

Prosopagnosia is the inability to recognize familiar faces in the absence of more generalized cognitive dysfunction. The face recognition impairment is relatively "pure" as it is confined to the visual modality and occurs in the setting of otherwise normal visual perception. The patients are unable to recognize familiar persons by their face alone, but might be able to identify them by their characteristic voice, gait, hairstyle, a distinguishing birthmark, etc.

Patients with prosopagnosia are able to recognize a face as a face, but they cannot recognize a face's individuality, the face of somebody they know well—spouse, child, doctor, co-worker, etc. This is a disorder of individualized recognition while a generalized approach to the face image is preserved. It is in contrast to letter agnosia, the syndrome of pure alexia that occurs with damage to the region symmetrical to the FFA in the left hemisphere. In letter agnosia, a patient recognizes individual features of a letter written in different styles, but fails to recognize it as a letter as such, a member of the alphabet system.

Most patients with prosopagnosia can recognize facial expressions and a face's gender and can make accurate estimates of a face's age (Tranel, Damasio, & Damasio, 1988), demonstrating that this deficit of face identity is highly specific.

4.2.2 Are Autistic People Prosopagnostic? Do Children with Developmental Prosopagnosia Become Autistic?

If autism's pattern included a primary dysfunction of the FFA and if said dysfunction caused autism's clinical picture, as has been suggested by Schultz (2005), the above two questions must be answered. There is no evidence that autists are unable to recognize faces of people whom they know well. One might ask whether there is a milder degree of deficiency (more subtle deficit) in face processing, which might be revealed in psychological testing? However, in the fMRI studies discussed earlier, autistic subjects (whether they do not activate the FFA at all, underactivate the FFA, or activate the FFA normally during face processing) were not significantly different from normal controls in terms of accuracy and response times on face perception tasks. The studies that investigated performance on familiar face recognition tests had contradictory results, some finding no deficits in autists (Teunisse & De Gelder, 1994), with another finding impaired recognition of faces (Boucher, Lewis, & Collis, 1998). In the Barton et al. (2004) study where authors used special tests more specific for prosopagnosia, one group of autistic subjects performed normally on all tests of face perception and recognition. The other group of autistic subjects was deficient in recognition, though they performed better than prosopagnostic patients. The "normal" and "deficient" groups did not differ in any parameters, including IQ and degree of social disability. The authors came to the inescapable conclusion that normal processing of facial identity is not incompatible with autism. The authors also concluded that social function does not correlate with the competency of face processing in autism.

Deficient face processing thus is not specific for autism nor does it belong to the pattern of autism.

There are individuals who have never possessed normal capacity for learning faces (developmental prosopagnosia), and cases of acquired, early onset childhood prosopagnosia have also been described in the literature (Barton, Cherkassova, Press, Intriligator, & O'Connor, 2003; Ellis & Young, 1988; Young & Ellis, 1989). Developmental prosopagnostics tend to be profoundly embarrassed by their deficiencies, they may become predictably socially isolated and withdrawn, but they do not develop the unique pattern of autism (Barton et al., 2003).

4.2.3 *Face as Right Hemispheric Gestalt: Face Is Holistic Form*

What is the individualized recognition that the RH provides? Faces share a basic configuration, a constant set of features (eyes, nose, and mouth) and the spatial relations of these features (eyes side by side above the nose, nose above mouth, etc.). This will be the topological scheme of a face, the "construction" of the LH gnostic–praxic level. Prosopagnostics recognize these characteristics of a face well.

Faces differ from each other only in subtle variations of this spatial structure; nevertheless, these variations make them distinctive qualities. Each face is a singular image.

Barton, Press, Keenan, and O'Connor (2002) showed faces that differed in a stepwise, quantitative fashion along the vertical mouth position (mouth displacement) and inter-ocular distance (eye displacement) to patients with FFA lesions. These patients were clinically prosopagnostic and severely impaired in discriminating changes in the spatial positions of facial features. Prior work had shown that the normal population is highly sensitive to such displacements. Later, Barton et al. (2003) examined, with the same method, three patients with prosopagnosia of childhood onset. They found that developmental prosopagnosia is similar to the adult-onset form in deficits of discriminating spatial arrangements of facial features.

In correspondence with the basic parameters of the FFA's functioning (its sensory-motor/gnostic–praxic level, visual modality, and RH holistic processing), faces are "treated" by the FFA as *holistic forms*. As discussed in Chap. 2, simultaneous recognition of holistic forms in different visual scene-situations (VSSs) can be considered the RH equivalent of the LH gnostic–praxic level. The RH holistic form (to be elaborated on more in the following chapters) is a continuous totality of unique qualities, the *whole* that is instantaneously perceived as relations of these qualities (*parts*). It can be called, therefore, pattern recognition. The power of *face* as holistic form assigns it a special place in visual experience. Benton and Tranel (1993) indicate: "Even the faces of persons whom one has met only on one occasion may be recognized instantly at a second encounter years later. There is no other category of nonverbal visual stimulus remotely like it in this respect" (p. 176).

4.2.4 *Face as RH Gestalt: What Is Familiar Face?*

Let's return to the notion of the FFA being of the sensory-motor and gnostic–praxic level. RH gestalt of face is not just a unique visual image. This image is included into a visual scene-situation (VSS), referring to a moment of experience in the external world. While holistic forms are simultaneously recognized in different situations (pattern recognition), they are also assigned with the emotion of their spatial–temporal context (VSS). Thus, when seeing a face one enters his own unique experience, a VSS, to which the visual image of the face belongs. What then is a familiar face? It is not so much the past experiences themselves with a particular face that give rise to a sense of familiarity as it is simply that there *has been* previous experience with that face. Thus, in recognizing a familiar face, two factors of RH cognitive mechanism are at work (1) holistic forms (faces) belonging to different situations are identified and (2) holistic forms (faces) are saturated with the emotion from the VSS to which they belong. Because emotions and affects are both determined by the visual picture and *determine* it in RH gestalt, we may ask whether assigning some meaning (emotion) to a face image will increase one's sensitivity to subtle visual variations of facial features.

In which way is a familiar face different from an unfamiliar face? I think it is not a qualitative but rather a quantitative difference. The moment one sees a new face, the visual image of the face is included into the VSS, a momentary experience in the external world, and, therefore, assigned with the emotion experienced at that moment. In familiar face recognition, there is more meaning (sense of knowing) assigned to the visual image of face because the underlying RH associative process (identification by resemblance of holistic forms and by common emotion) requires a greater number of neuron constellations than for discrimination of unknown faces.

Two facts are in correspondence with the difference between familiar and unfamiliar faces being a quantitative (not qualitative) one. Prosopagnosia usually affects familiar faces only (Young, Newcombe, de Haan, Small, & Hay, 1993) but it can also affect unfamiliar ones (Benton & Tranel, 1993). Furthermore, greater FFA activation occurs for processing familiar faces than unfamiliar, and the more familiar face the greater degree of FFA activation.

4.2.5 Disinhibited Right Hemispheric Processing of Faces

I will now discuss a unique case, described several years ago, of a patient with a circumscribed lesion in the left temporal–occipital area, symmetrical to the FFA in the right hemisphere (Vuilleumier, Mohr, Valenza, Wetzel, & Landis, 2003). This case demonstrates that a grotesquely exaggerated "sense of familiarity" can occur when, free from the "opposition" of its symmetrical counterpart in the left hemisphere, a disinhibited "face area" makes false identifications. This "natural experiment" teaches us a great deal about the FFA's function in the norm. The patient, J.R., was a 21-year-old woman, university student, admitted to the hospital for a cerebral vascular accident. An MRI of her brain showed a circumscribed area of infarction in the left hemisphere, close to the temporal–occipital junction. Detailed neuropsychological testing did not reveal any disturbance in language, visual or verbal learning, memory, object and face recognition, visual–spatial function, or any other cortical function.

> J.R. had no difficulty recognizing known people and learning new faces but spontaneously complained about continuously experiencing erroneous familiarity for unknown people's faces.[2] Her errors concerned other patients, visitors and hospital staff, and greatly

[2] Occasional feelings of false familiarity with unknown faces are not uncommon in normal people, though they are usually faint and are rapidly acknowledged as a trivial error (Young, Hay, & Ellis, 1985).

embarrassed her. Initially every person who entered her hospital room was familiar to her and she would smile at him or her, expecting the person to sit next to her bed and chat with her. It was quite frightening to her that many visitors did not notice her and continued towards her roommate, an elderly woman she knew she had never met.... During the first grand round on the ward (four days after admission), J.R. greeted one of the authors (T.L.) by using the familiar form of personal pronoun ('tu', an unusual response in French with an unknown physician), smiling at him as if he was somebody known to her, though she could not retrieve his name right away. Realizing her mistake, she excused herself and said: 'Sorry, you must be the professor, it got me again, I cannot trust my own perceptions. When you entered this door I thought I knew you well, well enough so that you would embrace me and call me by my first name. Apparently we do not know each other, but I still have that feeling of having met you many times'. She also explained: 'My eyes are caught by someone's face; I have the strong feeling of knowing him or her without being able to place him or her.... I have the feeling we have met in some place or talked together, but I cannot figure out where or when, or what we talked about. This happened to me occasionally before, as to everyone I suppose, but now it is present all the time and with everybody. This occurred with many nurses, I often asked them if we were together at school, or if we were living in the same neighbourhood. Also, I thought I had already met with some of the doctors, long before my illness, yet I don't know when'. By contrast, correct recognition was confident and associated with immediate retrieval of a specific identity. 'When I truly recognize someone, I have no doubt: I just know who the person is' (Vuilleumier et al., Hyperfamiliarity for unknown faces after left lateral temporo-occipital venous infarction: a double dissociation with prosopagnosia, *Brain*, 126: 2003, pp. 892–893. Printed by permission of Oxford University Press).

When J.R.'s disorder became less acute, the connection of false familiarity with the situational context was unmasked. A few weeks after the onset of her illness, J.R. experienced a feeling of familiarity toward unknown faces primarily in familiar places (on the university campus, in her neighborhood, in the hospital during follow-up appointments).

One may suppose that a lesion in the left hemisphere, symmetrical to the "face area" in the right hemisphere, leads to imbalance of a reciprocal inter-hemispheric relationship, with release ("unleashing") of right hemispheric process.[3] An increase in the RH's association of holistic forms results in identification of unknown faces with other faces that are known by personal experience. Unknown faces are thereby assigned with personal and affective meaning. A discrepancy in J.R.'s experience may shed more light on the FFA's function in the norm. Her judgment about people was not disordered, and she realized her errors quickly. She thus had an excellent "intellectual insight" as she complained spontaneously about false familiarity feelings, suffered great embarrassment by the inappropriate behavior caused by her errors, etc. Thus, at the same time she knew that feeling a face to be familiar was not "correct," she also could not help experiencing a strong, compelling feeling of knowing. This feeling of knowing was momentarily induced by the visual image of face—"my eyes caught by someone's; I have the strong feeling of knowing him or her..." We see then that knowledge about a person and a sense of knowing someone are dissociable, and therefore implemented by different parts of the brain. Logically organized, semantic

[3] If the left hemisphere has a compulsive desire to rationalize, the right hemisphere has a compulsive need to identify.

knowledge or "social knowledge" about people is stored in the left hemisphere and was readily provided to J.R. in the case of truly familiar faces: "When I truly recognize someone, I have no doubt: I just know who the person is."

4.2.6 *Social Knowledge and Sense of Knowing*

Schultz (2005) proposed the "social knowledge" hypothesis of FFA's function, suggesting that "seeing a face automatically and involuntarily activates FFA-based semantic knowledge about people and in some sense primes the observer for a social interaction" (p. 142). He believes FFA abnormalities can cause autistic deficits in social behavior, disabling autists with fundamental impairments in deriving and retaining social semantic knowledge from face-to-face experiences. "Activity within the FFA represents both perceptual and social conceptual processes and in this way may represent a core mechanism for the pathobiology of autism" (p. 144).

We can now see how this hypothesis is reductionistic (conceptual processes are reduced to and fused with perceptual processes) and incompatible with what is known about the FFA's functions.

FFA function, even exaggerated as in the earlier example, includes a "sense of knowing" but does not contain any knowledge about social interactions or social values. A sense of knowing has nothing to do with social knowledge. In contrast, a sense of knowing refers to one's own internal and idiosyncratic experience, namely VSS saturated with affect. The feeling of familiarity has not so much to do with the face itself, as it is a glimpse of the immediate RH experience to which that face belonged; it is a feeling "of having experienced before," a feeling of belonging, and, eventually, a feeling of self.

By what I have already discussed in previous chapters about pattern recognition and RH holistic processing, the FFA function would not be disordered in autism. In fact, the opposite seems to be true: it may be a strength, not a weakness, in autism. As we saw in the above example, the sense of familiarity may range from subjective certainty of knowing a person to a strong feeling of personal closeness and intimacy. One can come to a truly exciting hypothesis then that people with autism are capable of experiencing closeness and intimacy.

4.2.7 *The Left Amygdala Role in Face Processing Network*

The amygdala, an almond-shaped structure located at the depth of the medial temporal lobe, is of paramount importance in autism and will be discussed in Chap. 7. I will only mention here its contribution to facial processing. Several studies have shown the left amygdala may be critically involved in identifying facial emotional expression, mostly from the eye area (Baron-Cohen et al., 2000). Note this "division of labor." The right hemisphere is responsible for facial identity, whereas the left hemisphere is responsible for recognition of a face's emotional expression. The right hemisphere assigns a face with the unique, idiosyncratic, experiential quality, but the left hemisphere's perception of emotional expression from another's face serves qualitatively different and complementary purposes. Emotional expressions are conventional. Human facial expressions for fear, surprise, anger, disgust, and happiness are universally recognized (Ekman, 1982). They play a direct communicative role, which is what drove their evolution (Darwin, 1872).

Emotional expressions are made by the contraction of facial muscles, with a certain pattern of muscles involved for each emotion. Thus, it is still the visual modality (visual image) through which we recognize emotion in another person's face (the amygdala is not a visual area). Muscle contraction patterns for particular emotional expressions form subtle alterations in the spatial relations of facial features superimposed on a unique face. While the FFA will recognize it is Mr. N *who* is smiling, the left amygdala identifies his *smile* as a sign of happiness. The amygdala's role here is analogous to the extracting of conventional phonemes from the "river" of sounds in someone's speech and the combining of these abstract phonemes into words to comprehend the meaning conveyed by the speaker.

Prosopagnostics, therefore, will recognize the *smile* from Mr. N's face, but they will not "catch" the uniqueness[4] of Mr. N's smile.

In summary, one of the left amygdala's functions is to extract emotional, socially relevant information from a face. In general, social knowledge, social reasoning, and the means of communication (language, motions, and actions) are connected with the left hemisphere. The respective roles of the right occipital–temporal region and the left amygdala in the face processing network fit this paradigm.

4.2.8 The STS Area Role in Face Processing Network

Facial muscle contractions result in movement of facial parts, especially the eyes and mouth. The movement of facial parts changes a face's spatial configuration and the spatial relations of its features. Indeed, facial expressions are made by movements, especially by the moving gaze and mouth. As discussed before (see Chap. 3), the superior temporal sulcus (STS),[5] a region on the lateral surface of the temporal lobe, is involved in the detection of change in general as well as the detection of

[4] To say a prosopagnostic person can understand emotional expression from a face is not fully correct. He is able to understand the conventional part of emotion. The uniqueness of Mr. N's smile is made up of more than just fixed-in-evolution, "conventional" facial movements. Such movements belong to the sensory-motor and gnostic–praxic level. The thalamic level also contributes background movements for emotion, reflecting one's internal state and idiosyncratic RH experience. These thalamic emotional expressions will be discussed in Chaps. 7 and 8.

[5] I used the term superior temporal gyrus (STG) in Chap. 3, not superior temporal sulcus (STS). In the literature, both terms are used interchangeably, referring to the similar functions. It seems that STS is more specific to moving eyes and mouth.

moving objects. Studies of primates found neurons within the STS sensitive to gaze and head direction, and some cells in the primate STS also responded to moving views of the head and body (Hasselmo, Rolls, Baylis, & Nalwa, 1989; Perrett, Hietanen, Oram, & Benson, 1992; Yamane, Kaji, & Kawano, 1988). Interestingly, while both the FFA and the STS are activated by faces in humans, the activation of the FFA to whole faces is greater than it is for isolated eyes (Allison et al., 1994). The STS's activation is larger to isolated eyes than to whole faces (Bentin, Allison, Puce, Perez, & McCarthy, 1996). FMRI studies of normal people viewing faces whose eyes repeatedly changed direction of gaze or whose mouth opened and closed demonstrated a discrete area in the STS to be activated preferentially by moving eyes and mouth, and this area was suggested to be responsible for the perception of eye and mouth movement (Puce, Allison, Bentin, Gore, & Gregory McCarthy, 1998). Moreover, in humans, perception of eye movement and that of mouth movement are dissociable functions. For example, some patients with temporal lobe lesions are deficient in determining gaze direction, whereas others can no longer lip-read (Campbell, Landis, & Regard, 1986). There is also evidence that the right STS is more involved in eye movement, while the left STS is more involved in mouth movement (Pelphrey, Singerman, Allison, & McCarthy, 2003; Puce et al.).

Summarizing the findings about the FFA and STS's roles in the facial processing network, Allison, Puce, and McCarthy (2000) suggest the FFA's contribution to face perception is the image of permanent facial features, while the STS provides moment-to-moment changes in facial configuration.

4.2.9 The STS and Autists' Lip Reading Versus Eye Tracking

Klin, Jones, Schultz, Volkmar, and Cohen (2002) studied autists' viewing patterns of social scenes using eye-tracking technology. Precise measurements of a subject's visual focus were superimposed over the dynamic images of viewed film clips. Autistic and control subjects watched short clips from *Who's Afraid of Virginia Wolf*. Each film clip was chosen for its rather intense interpersonal interactions, with screen shots dominated by close-up faces. Control viewers' visual focus shifted from eye to eye, following a heated dialogue between characters. In contrast, the focus of the autistic viewer shifted from mouth to mouth. Furthermore, during dialogues the subjects with autism focused almost solely on the speaker's mouth without monitoring the listener's reactions.

Authors of this study emphasized that, in real social situations, many crucial social cues occur very rapidly. If the autist does not focus on eyes, he will miss salient features such as the emotional reactions of speakers, which may lead to a general failure in assessing the meaning of the entire situation and preclude adaptive social reactions. Such a disadvantage starting early in life, they propose, may explain the development of social disability in autists. The authors go on to argue that the described behavior—not looking at eyes—is a phenotypic trait of autism, i.e., a primary, core manifestation that is inherited, or indeed the genetic vulnerability to autism.[6]

The findings of this study, notwithstanding the reductionist conclusions it gave rise to, are interesting. What impressed me most is that autists consistently looked at the mouth. People with autism do not use language for communication, but they look at the mouth (most likely lip reading) to extract the social context from speech. That autists did not look at eyes could have been predicted by anyone with clinical experience, but still the question remains: "Why does the autist not focus on eyes?" The introspection of high functioning autists can provide valuable information for this inquiry. Recall Temple Grandin's remark about autistic people not looking at eyes (see Chap. 3).

> [S]ome problems autistics have with making eye contact may be nothing more than intolerance for the movement of the other person's eyes. One autistic person reported that looking at other people eyes was difficult because the eyes did not stay still (Grandin, 1995, p. 73).

We may consider then that the cause of not looking at eyes is the same as the cause for young autistic children appearing deaf: overwhelming hypersensitivity to physical movement (change) in any modality. I already explained this sensitivity as stemming from the autists' exposed RH "flow of consciousness," the RH sensory-motor level (see Chap. 3).

The above authors also examined the autists and normal comparisons' responses to physical cues. Viewers with autism responded to the physical cue more quickly than the normal comparison. The autist's visual focus, in fact, moved faster than the camera itself, showing a clear understanding of, and quick response to, the physical cue. With tension resulting from physical and social cues occurring together, the normal comparison viewer was "distracted" from physical cues and attended to social cues—the salient display of emotion. However, the viewer with autism was constantly and "goal-directedly" attentive to physical cues, and his focus followed the leading edge of the camera, and not the social–emotional turmoil of the characters (Klin et al., 2002). This finding is also in line with the special state of consciousness in autism, which I have argued is the physical, constantly moving world of RH "flow of consciousness."

I believe that autistic people's problem with a "moving gaze" is a factor in the pervasive lack of eye contact in autism, but it cannot fully explain it.

Klin et al. (2002) results were confirmed by some studies (Joseph & Tanaka, 2003; Spezio, Adolphs, Hurley, & Piven, 2007), but several other studies did not support their conclusion that autists are less fixated on the eye area and more on mouth than controls (Bar-Haim, Shulman, Lamy, & Reuveni, 2006; Pelphrey et al., 2002; van der Geest, Kemner, Camfferman, Verbated, & van Engeland, 2002). This discrepancy in the eye–mouth fixation in autism in different studies can be explained

[6] Using brain imaging split from the tradition of various schools of psychiatry and neuropsychology leads to such a conclusion where perceptual visual deficit or genes of "not looking at eyes" are the etiological determinants of a mental disorder.

by the type of stimuli presented. Two factors in Klin et al. (2002) study might increase STS sensitivity to the presented stimuli (1) emotional intensity of characters (intense emotional expression requires more significant facial movements); and (2) a "moving" face image (emotion was expressed by real moment-to-moment changes in gaze direction and eye configuration). In contrast, if we take, for example, the Bar-Haim et al. (2006) study that showed no difference in attention to eyes and mouth between HFA and control group, subjects in that study were presented with static and emotionally neutral images.

Although lack of eye contact is a significant symptom in autism, its value for understanding autism is limited. A recent study illustrated this statement. Spezio et al. (2007) examined response accuracy, face gaze, and information from the face the viewer relied on to make judgments about face.[7] This study showed that although HFAs had significantly decreased use of information from the eye region, they performed a task (judgment of basic emotion from the face) at the same level of accuracy, and with the same reaction time, as did controls. Furthermore, while HFA participants showed a significantly less use of both eyes, they were no different from the controls in time they spent fixating on the left eye.[8] The authors conclude that differences in use of facial information are not fully explained by differences in face gaze by HFA participants, thus, "the participants with autism process faces abnormally, above and beyond their fixation to them" (Spezio et al., p. 930).

Dalton et al. (2005) examined the relation between gaze fixation and brain activation during face processing in individuals with autism. They found that activation in FFA was strongly and positively correlated with the amount of time spent fixating on the eye region for the autistic group but not the control group. In this study, participants with autism who showed average eye fixation durations similar to those shown by control nevertheless showed decreased activation of FFA (see Dalton et al., 2005, Fig. 7). Spezio et al. (2007) conclude that theirs and Dalton and colleagues' findings taken together allow for the suggestion that brains of people with autism treat facial information differently even when direct eye gaze is the same.

4.3 Conclusion

To summarize, first recall the network of face processing in brain:

1. The right FG (FFA) is responsible for face identity.
2. The left amygdala is responsible for perception of face emotional expression.
3. The right STS is responsible for perception of moving gaze.

---

[7] To determine those areas of the face that were in fact relied upon by the subjects in this study, the authors used the "Bubbles" technique, analyzing then the areas of the face that were revealed as a function of participants' accuracy.

[8] This finding is very interesting by itself. Normal fixation on the left eye, i.e., the left visual field, indicates better use of facial information by the right hemisphere than by the left in autistic participants.

Having analyzed the autistic brain's face processing network, we may suspect one player being a "hot spot" for insight into autism: STS.[9] If indeed the RH sensory-motor level is on the foreground of consciousness for autists, the physicality of a moving gaze is exaggerated, more actual and dominant in an autist's consciousness than the emotional meaning assigned to facial expressions by the amygdala. On the other hand, we so-called normal people perceive emotional expression from the eyes on a conscious level, but are not usually aware of said emotion's physical equivalent, namely the moving gaze. Not looking at other people's eyes simply (literally) because "they did not stay still" may explain autists' inconstant mistakes when performing face perception tests as well as the variable results of FFA activation while viewing faces, with autists potentially capable of adequate performance or "normal" FFA activation.

Is the emotional meaning of eye movement, assigned by amygdala, then disordered in autism or is it just forced to the background of consciousness? This will be discussed later in the chapter devoted to the amygdala. In this chapter, we have come to the inescapable conclusion that autists' problem with *face* goes beyond the limits of the face processing network.

[9] In this chapter, I did not analyze the amygdala's functioning in the autistic brain's face processing network.

Chapter 5
How Autistic Persons Act in the World (Cerebral Organization of Movements and Autism)

While restricted repetitive and stereotyped patterns of behavior, interests and activities are united into one domain in behavioral classification (DSM-IV, the American Psychiatric Association), these phenomena represent a heterogeneous group of behaviors that stem from different hierarchical levels and areas in the autist's brain. Both stereotypical behavior and an autist's preoccupation with parts of objects (so-called "restricted interests") contribute to the working hypothesis of this book, i.e., prevalence and exposure of the right-hemispheric sensory-motor and thalamic levels. Thus both phenomena will be discussed in this chapter and their significance for the autistic clinical-brain pattern explored.

5.1 Hierarchical Organization of Movements in the Brain

5.1.1 Bernstein's Levels

Bernstein's theoretical model of movement construction in brain will be briefly revisited before comparing the stereotypical movements and actions of autists with those of the normal population and of patients with brain lesions, for autists differ from both. Recall that Bernstein discovered movement organization in the brain to be a vertical hierarchy with necessary participation of several specialized, autonomous function levels. When moving and acting consciously and voluntarily, we are fully aware of what we are doing in correspondence with our goals and purposes. What we are not and cannot be aware of, no matter how hard we concentrate, is how much work must be done in the subcortical levels for realization of our desired actions. Whether it is a seemingly mundane walk up a flight of stairs or the leap of a gymnast onto a narrow balance beam, Bernstein understood that movements require split-second integration of contributions from several specialized and hierarchical levels in the brain. The level that determines the end goal for movements and actions is usually the cortical level. It is called the leading level and it is what

T.B. Glezerman, *Autism and the Brain: Neurophenomenological Interpretation*, DOI 10.1007/978-1-4614-4112-0_5, © Springer Science+Business Media New York 2013

one is conscious of. All other lower levels participating in movement creation are called background levels, and one is not aware of the complex work that takes place in these levels.

This background complexity lies in the periphery's constantly relaying information about the space where movement is performed, and the body's position in it, such that movements are really continuous near-instant corrections of where the moving part needs to be in a certain space in order to execute the desired action.

5.1.2 *Different Space of Bernstein's Levels Determined by the Intended Action's Meaning*

Bernstein's different levels are characterized by the different "spaces" they master and senses by which that "space" is explored. For example, for a circular movement of the hand, the leading level (with its particular dominant sensory input) would differ depending upon the movement's task. A person could make a circle in the air while doing a gymnastic exercise or as a part of choreographic movement. In this case, movement is performed under the control of proprioceptive sense from extremities because the "space coordinates" by which it is performed are that of one's own body. On the other hand, the visual sense would control a person's movement when she is tracing a circle engraved on paper or drawing a circle by imitation. In this task, the spatial coordinates to be worked within are in the external spatial field, which are known by vision. This latter "space" belongs to Bernstein's sensory-motor level, attuned to orientation with the outside world. Continuing up the strata of movement levels, a person may also make circular movements with her hand while sewing. These circles are made under sensory control (visual, tactile, and proprioceptive) that enables one to manipulate the parts of the needle relevant to its function. The "space" one must be able to make a mental construction of for this task is that of the object's topological scheme (gnostic-praxic level). Finally, a person may draw the circle on a chalkboard while proving a geometric theorem for her students. In this case, the drawing of the circle itself is not the main goal (task). The leading factor here is the abstract idea of the relation between the drawn circle and other elements of the mathematical illustration. If the mathematics teacher distorts the correct geometric form of the circle here, it would not evoke correctional impulses in her motor system. However, the correctional impulses would definitely emerge in her motor system if she were an art teacher with different goals of instruction.

5.1.3 *Walking as an Example of Bernstein Level's Integration*

It must be remembered that there is always integration of these autonomous levels in movements. Walking is a good example of this. Walking implies several tasks—"where to," "along which surface," "against which obstacle." The leading level and the corresponding sensory control will be those that are most appropriate for this movement's task. For walking, the leading level must be one whose sensory control is mostly the visual sense, providing information (sensory feedback at each and every moment of movement) about the external spatial field where movement is taking place. However, walking itself requires simultaneous contraction of multiple muscle groups. Thus, "a synergetic concert of walking" must be provided by the lower, background level the "space" of which is the spatial coordinates of one's own body with sensory input coming primarily from the extremities' muscles.[1]

Thus, a movement's orchestrations are never an either/or phenomenon in regard to levels' involvement but instead a dominant, yet also dependent and integrative, one.

In the norm all levels participate in an integrated manner to realize movement. In pathology, however, there are situations where the cortical region is damaged, and the leading level is not functioning. In such cases the level just below the leading one is exposed, which means it becomes conscious and is now the leading level.

5.1.4 Object Actions and Their Topological Schemes

Object actions are "movements" at a leading gnostic-praxic level. As any other movement, object action is based on feedback from senses. In the case of object action, it is mostly the visual, kinesthetic, and tactile senses that provide information about the "space" where the action is to be performed, and here the "space" is the object itself, or more exactly, the topological scheme of an object (i.e., the idea of the object as a tool). The topological scheme of an object is a "frozen" mould of action to be performed with the object, and it is stored in the posterior associative cortex. For example, of all stimuli ever received from the environment concerning a cup (how it looks visually, how it feels, its particular weight in the hand, etc.) only "useful," functional features of a cup (solid sides, bottom, open upper part, and handle) coalesce into the synthetic image of its topological scheme. The cup now has a *meaning*—it is an object for drinking. The "space" of the cup is explored by several modalities, so that there is a visual image of its topological scheme and a kinesthetic (muscle sense) image of the positions of the manipulating hand—how that hand must organize itself to pick up the cup by the handle in order to drink from it. The static scheme of the object action is translated into a dynamic scheme of action in the left frontal (premotor) cortical region. In the left premotor region the image of inner programming, a "plan" for unfolding the spatial scheme in time, gives a sequential, temporal order to the positions of the manipulating hand during object action. Thus the complex coordinated movements required to pick up one's morning cup of coffee from the table and drink from it are choreographed by the left frontal lobe.

---

[1] This is the thalamic level at which nothing is known about the outside world; this level's focus is singularly that of one's own body.

5.2 The Making of Automatisms and Pathological Stereotypy

In rational behavior, one has a goal and the motivation to achieve it. One then uses internally stored programs to execute the necessary actions. Complex object actions involve a successive chain of movements—"kinetic melodies"—in which discrete movements are united by the meaning of the task (semantic sequence) that cannot be reduced to just moving things in space and overcoming external forces (the latter is characteristic of movements where the sensory-motor level is leading). Examples of semantic sequences would be the succession of movements for starting and driving a car or the sequence of movements for articulations that produce a word. One builds these sequences from separate units of movements, which get unfolded as a linear structure. Selection of units for the semantic sequence will be made depending on the task at hand.

During individual development, in the process of acquiring skills, technical components of certain object actions are transferred to lower levels, in particular to the "next-in-command" sensory-motor level (level of possession of the external spatial field). This process is called automatization. Automatized components (automatisms) are movement patterns stored at the lower level, but they do not *originate* at that lower level. The motives for their realization come from a higher level (Bernstein, 1947, 1990). Automatisms of object action (gnostic-praxic level) are stored in the subcortical center of the sensory-motor level, the striatum. The striatum can be viewed as the repository for patterned motor sequences.[2]

These sequences were originally formed at the gnostic-praxic level from separate movement units, but once they are posited in the striatum, they are stored there (and re-activated when called into play) as complete wholes or ready-made patterns. Once automatisms are formed at this lower level, motor patterns are removed from consciousness (unloading the field of consciousness). Acquiring new motor skills engages prefrontal regions, linked with attention to action, and basal ganglia and cerebellum. As automatic task performance takes over from effortful, attentive acquisition, activity in prefrontal regions falls away (Berns, Cohen, & Mintun, 1997; Passingham, 1997; Zeman, 2001). Note that conscious attention to action corresponds to linear unfolding of separate movement units.

Again, automatisms can be complex in composition, but they are not initiated at the lower level where they are stored. Higher-level choices must be made about the performance of automatic pre-set packages of behavior. For example, the leading level for driving a car is the gnostic-praxic level, but the spatial skills of driving are connected with automatisms stored in the striatum. The more skillful the driver, the more automatisms will be involved in his driving. These set patterns are manipulable only as whole entities, and the intervention of consciousness or attention might actually disrupt their smooth implementation (Graybiel, 1998).

[2] There are various terms for these motor patterns in the literature, which I will use, such as *ready-made patterns, cliché, behavioral macro* or *"chunk," automatic and pre-set packages of behavior.*

Thus, there are two key features of automatisms (1) once formed, they may be difficult to break apart and (2) they are removed from conscious awareness, enabling the human brain to be very efficient and economical.

Our actions can be more or less "conscious" depending on how much automatic performance contributes to it. When speaking, a person expresses his thoughts by building sentences. Generation of a new sentence—combining words (units) into a linear structure by using syntactical rules—is the job of the left prefrontal region. In many conversations, there are often only few newly formed sentences without any use of lower-level patterns of speech. Actually a significant part of conversation is "filled" by habitual speech combinations or ready-made utterances. "How do you do," "good job" or "stuck in traffic" are examples of simple speech clichés, but there are also more complex combinations that are still nonoriginal automatisms used in communication. Even though these clichés are not being consciously generated, in normal people they are appropriately combined with newly generated sentences, under control of the higher level. The speech of different individuals varies significantly as to proportion of newly generated (from the cortex) versus ready-made utterances retrieved from the striatum. There are people who are especially "fluent" in manipulating and combining speech clichés thus they would have a larger percentage of automatisms in their communications.

5.2.1 Pathological Stereotypy

In certain pathological situations, the striatum is hyperactivated and takes on a life of its own, free from the control of higher levels. This results in automatisms being present out of context of goal-directed behavior. When this release happens, automatisms not only arise out of context, but they are often repeated. Thus, the automatism becomes stereotypical, i.e., repetitive nonfunctional behavior. Ridley (1994) defines stereotypy as the excessive production of one type of motor act, or mental state, which necessarily results in repetition.

There are experimental and clinical data that support the role of the striatum in stereotypical behavior. The neurotransmitter dopamine is involved in the activation of the striatum. Infusion of dopamine into the caudate nucleus of a rat's striatum produces stereotyped oral behaviors such as grooming and gnawing (Fog & Pakkenberg, 1971).

It has also been described that when humans repeatedly and chronically ingest large doses of amphetamine (which increases the release of dopamine in the brain), a particular psychotic episode can result that is characterized by a form of stereotypy referred to as "punding." In this condition, the person engages in long, complex and repetitive behaviors such as sorting collections of small items or dismantling pieces of mechanical equipment. He or she also becomes "cut off" from other events, especially interactions with other people. While the person can be distracted for a short time, he or she becomes distressed if prevented from pursuing this activity (Rylander, 1972). Stereotypic effects of large doses of amphetamine were also described in rodents, monkeys, and marmosets (Randrup & Munkvad, 1967; Ridley, Baker, Owen,

Cross, & Crow, 1982; Scraggs & Ridley, 1978). Furthermore, dopamine-blocking drugs infused into the striatum can prevent amphetamine-induced stereotypical behavior (Costall, Naylor, & Olley, 1972).[3] On the other hand, frontal lesions in rats exacerbate the effects of amphetamine treatment (Iversen, 1971) and frontal lesions and stimulation of the dopamine system produce very similar behavioral and cognitive effects in monkeys (Ridley and Baker, 1982). Ridley (1994) suggests that striatal dopamine regulation by prefrontal cortex plays a role in the frontal cortex's inhibitory influence on striatal activity.

The general rule is that stereotypical behavior emerges when, for one reason or another, goal-directed behavior is not possible. The most typical case of stereotypical behavior is the result of frontal lobe damage, where dysfunction of the higher-level brain center results in disinhibition and pathological hyperactivation of the lower, next-in-command brain center, the striatum.

In summary, the exquisite skills people acquire from experience in object action are due to automatization, i.e., formation of complex, patterned motor sequences stored in the striatum. The striatum, however, can be pathologically hyperactivated and these technical components of action are then actualized autonomously, without correction and feedback from the higher gnostic-praxic level. A classical example from neurological literature is a demented patient who knitted a stocking with 12 heels. Movements in this case are well coordinated, but irrelevant. Stereotypical behavior thus represents the technical components of object action re-activated on their own, bereft of meaning or goal, so that there is no direction from the gnostic-praxic level to keep the striatum on task.

5.2.2 *The Drive Behind Stereotypical Behaviors*

One wonders what might be the cause for the repetitiveness of pathological stereotypy? Patients usually have an irresistible desire to perform these motor sequences again and again. Ridley (1994) indicates that although there is no obvious external incentive involved in stereotypy, it is important to consider the internal motivational state of an animal or person exhibiting stereotyped behavior. Drugs like amphetamines, which induce stereotyped behavior, are also drugs of abuse and engender repeated self-administration in rats, pointing to their influence on motivation. It has

[3] I had a patient in whom a medication that decreases dopamine in the brain was tapered off too quickly. This patient had quite significant frontal lobe dysfunction, and thus weakened control of subcortical levels, at baseline. He developed an acute stereotypical reaction, which I called a "Sisyphean toil." He would spend hours rapidly performing and then repeating motor sequences. For example, he was constantly arranging and re-arranging his slippers into a line at the foot of his bed. He also spent hours taking change out of his purse, arranging it on his night table, putting it back in his purse, and then taking it out again and so on, until he was exhausted. He gave the impression of being under some imperative to continue these movements, ceaselessly. He was also unable to stop, even when aware of wanting to stop. He was able to respond to questions, but would quickly be distracted and look away. This stereotypical behavior was decreased and disappeared when he was given a powerful medication that decreased dopamine in the brain.

been suggested that stereotypical behavior is reinforcing "in itself" and therefore "of itself." Ridley stated that stereotypy might be driven by excessive internal motivational states even if these are not related to the environment in the usual adaptive "purposeful" way. Kinsbourne (1980) also considers stereotypical behavior as being related to bodily needs and largely endogenously rather environmentally triggered. In his opinion, it has a de-arousing function, serving to protect the organism from *information overload.* Whatever the reward may be, the striatum's "pseudoactivity" creates an internal motivational state that in turn drives one to repeat the behavior. This motivation behind stereotypy will be further explored when speaking about these behaviors in autism.

5.3 Autist's Nonrhythmical Stereotypical Movements: Role of the Striatum

Children with autism do not have apraxia (a disorder where comprehension of the object as a tool is lost or performance of the motor sequence of object action is impaired). Autists can perform object actions, which means Bernstein's gnostic-praxic level is intact, if they "choose" to use it. Children with autism know the functional meaning of objects, they know how to use objects, they just don't. Instead, they employ the sensory-motor's sublevel (the striatum) and engage in nonfunctional repetitive motor sequences. Examples of such stereotypical behavior include filling and emptying sequences with liquids and solids, or collecting and dispersing things about a room. Autists' stereotypies also include verbal expressive behaviors such as repetitive ordering, repetitive telling and asking, and any other repetitive use of words, sentences or parts of dialogues that have no obvious communicative purpose (in speech cliché not only content, but also intonation is preserved in automatisms from the striatum).

Nonrhythmical stereotypical behavior is not specific to autism. Besides neurological cases of frontal lobe damage, where the striatum is autonomously "acting out" because of its release from the control of the higher levels, hyperactivation of striatum and its release of patterned motor sequences is a nonspecific reaction to illness (to distress) in a variety of psychiatric disorders: obsessive–compulsive disorder, schizophrenia, mental retardation, etc. In these cases, seemingly meaningless repetitive behavior represents internally motivated replacement activity that serves the purpose of "discharging" information overload, with a concomitant dearousing, anxiolytic effect.

Repetitive behaviors may also be observed in healthy human or animals caused by external factors. For example, so-called confinement-stereotypy is described in animals kept in small enclosures with pacing up and down in one part of the enclosure. People confined in prison or waiting for appointments may also exhibit the to-and-fro pacing characteristic of confinement-stereotypy. This behavior is a response to an abnormal environment, and it disappears when the environment has changed (Ridley, 1994).

There are also cultural analogies of stereotypy that aids analysis of its broader meaning for the human psyche. Examples are the endless repetitions of a one-line verse, various alliterations and assonances in the texts of primitive cultures, comic word reduplication seen in all languages (Humpty-Dumpty, Handy-Spandy), repetitive words in songs, and repetitive movements in dance.[4] Bernstein even speculated that the attractiveness of repetition, stemming from the striatum, may be the origin of the verse rhyme.

Cultural stereotypies have a playful, stress reducing character. In contrast to pathological stereotypies, they are voluntary and under the control of the higher levels, and they help us to function normally. Pathological stereotypies, on the other hand, can exacerbate the dysfunction caused by illness.

To conclude, although nonrhythmical stereotypical behavior is not specific for autism, 100 % of subjects with autism exhibit at least one type of stereotypy (Bodfish, Symons, Parker, & Lewis, 2000). This behavior is diagnostic for autism, reflecting one aspect of autism's brain pathogenesis.

5.4 Autist's Preoccupation with Parts of Objects and Its Underlying Brain Network

Children with autism are not interested in the shared/conventional meaning of objects and do not use objects in a conventional manner or according to the object's established proper function. At the same time, children with autism might be fascinated with certain objects and their parts and preoccupied with manipulation of certain objects, however, this object is not conventional, not functional, but idiosyncratic. Parts of objects are explored as physical, sensory things (referring to the sensory-motor level), and these parts are by no means functional features of the object.

Persistent preoccupation with parts of objects is included into the domain of repetitive behavior in DSM-IV. It is usually objects or parts of objects involved in motion. Typical examples include fascination with light switches, water taps, drain pipes, electricity pylons, burglar alarms, vacuum cleaners, washing machines, trains, planes, clocks, etc. (Baron-Cohen & Wheelright, 1999).[5]

In Chap. 3, by looking at Nadia's (autistic savant-artist) preoccupation with drawing legs (Figs. 3.12, 3.13, and 3.14), an attempt was made to show that the "parts" autists are preoccupied with are not details, i.e., results of the LH analysis (or feature analysis as most authors consider it). They are not the result of breaking down the whole into component parts according to any of the LH principles: concrete, functional, or categorical. The parts are "wholes" by themselves, experiential and idiosyncratic—moments of experience. We deal here with RH processing where "whole" (VSS) and "parts" are *simultaneously* recognized and identified. As proposed earlier, in autism, the leading functional level is the sensory-motor level, mostly in the RH. Preoccupation with and fixation on parts was considered to be a

[4] Interestingly, Bernstein noted that in evolution of the animal world, songs and dance only emerge when the striatal level is well developed, namely, in birds.

[5] The same objects were enumerated in Chap. 3 as inducing an intense fear in a child with autism.

manifestation of the autistic brain's spontaneous compensation for the "exposed" and constantly moving RH world—*to stop the moment* (and to master it). Fixation, i.e., intense attention to "part," requires the RH attentional system (see Chap. 3), that is responsible for sustained attention to sensory stimuli (the RH prefrontal and posterior parietal/superior temporal cortex). "To stop the moment," to distinguish discrete moments of experience out of the RH continuous flow of time, requires participation of the LH (mostly the prefrontal cortex).[6] To "master" the moment means for the autistic brain to repeat: *to stop* (the moment) *and repeat.* The *parts* become automatisms, involving the striatum in their retrieval—repetitive manipulation with *parts*.

Preoccupation with objects and parts of objects is expressed in different forms: repetitive touching, repetitive moving (manipulating) objects or parts, collecting, and drawing in LFA; drawing in autistic savants; and classifying in HFA. No matter what form preoccupation with parts manifests as, it creates a special state similar to some kind of meditation.[7]

In particular, the type of meditation comes to mind where subjects focus and sustain their attention on a particular object. This form of meditation is designed to lead one into a subjective experience of absorption with the object of focus. Interestingly, brain fMRI studies during meditation showed increased activation in the prefrontal (PC) and medial prefrontal (MPC) regions concerned with attention and motivation, more so in the RH. At the same time, decreased activation was observed in the posterior cortex (Newberg & Iversen, 2003). Curiously, among neurochemical changes during meditation, these authors found increased dopamine levels in the striatum.

While preoccupation with parts in autism relates to a spontaneous brain response, intense focus on the object in meditation is voluntary and motivated. It is interesting in this regard that even if the act is conscious and voluntary, focus on the object without a motive for goal-directed behavior (not concentrating on functional features of the object) will lead to the predominant activation of the RH. Moreover, decreased activation of the posterior cortex where VSS content is stored suggests that subjects' concentration is focused on the temporal aspect of experience, stripped of content. This is what may be happening in autism: concentrating on the moment.

---

[6] The following example of the experience of a patient with damage to the RH might be an illustration of fixation on the object, analogous to what is observed in autists:

> If he looked out of the window at such moments [his disorder was intermittent], pedestrians in the outermost field of vision seemed to be moving at a speed of about 30 mph and cars would 'shoot by'. At the same time, if he fixated on any of the pedestrians, they appeared to be moving along 'as in a funeral procession' [Hoff and Potzl (1938) cited from Cutting (1997), p. 208].

Is this the explanation for the autist's peripheral gaze? Is it the RH's way of looking?

[7] We may recollect here Temple Grandin's report about her childhood when she could sit hours on the beach "studying" each individual grain of sand as it flowed between her fingers. "As I scrutinized their shapes and contours, I went into a trance which cut me off from the sights and sounds around me" (p. 44; see also Chap. 3).

Thus, the network underlying preoccupation with *parts* (this term is used generically for all forms, from simple to complex) might be outlined as: *(RH BA37*[8]*)–RH PC–LH PC–Striatum.*

I believe that preoccupation with parts and its brain pathogenesis as hypothesized here is specific and unique for autism.

5.5 Autist's "Sameness Behavior" and Its Underlying Brain Network

The desire to preserve sameness in autism was discussed in Chap. 3. Here we will consider "sameness" as part of autistic repetitive stereotypical behavior. In DSM-IV, "sameness" is replaced by "inflexible adherence to nonfunctional routines/rituals," and in the current literature sameness and rigid ritualistic behavior are often used interchangeably. This removes the poignant specificity of the term Kanner coined. What is interesting and has eluded the attention of modern research is that Kanner from the very beginning considered "sameness" in the context of *wholes-parts* perception in autistic children (Kanner, 1951). Therefore, he indeed explored autistic children's internal state, the specific subjective experience of their world, underlying the behavioral manifestations of the preservation of sameness:

> The autistic child desires to live in a static world, a world in which no change is tolerated…. The totality of an experience that comes to the child from the outside must be reiterated, often with all its constituent details, in complete photographic and phonographic identity. No one part of this totality may be altered in terms of shape, sequence or space. The slightest change of arrangement, sometimes so minute that it is hardly perceived by others, may evoke a violent outburst of rage (Kanner, 1951, p. 23).

Kanner's description of *what* is desired to be preserved exactly corresponds to the RH VSS, a moment of experience in time, in this book's formulation.

The RH VSS as a moment of experience is stored as perceived, including its parts with all their fine details and arrangements, but the parts are not to be distinguished out of the situation (RH holistic processing). The examples below demonstrate that desire for the preservation of sameness is desire to preserve the *whole* or to-complete-to-the-whole (RH VSS):

> Among the toys laid out for John F. were two dolls, one of which had a cap, while the other was bareheaded. Generally John paid little or no attention to dolls. When he noticed that the cap of one of the dolls was missing, he immediately asked for 'the hat,' picked up the doll and ran up and down with it, shouting for the hat. He was not reassured until the cap was produced. He made sure that it fitted, then put the doll down and lost all interest in it…. Susan T. noticed some cracks in the office ceiling and walls. She kept asking anxiously and repeatedly who had cracked the ceiling and could not be calmed by any answer given to her. She was obviously unhappy and every time she was in the office, she kept exclaiming: 'Who cracked the ceiling?' 'How did it crack itself?' Anthony F. became aware of the same cracks and asked almost literally the same questions as Susan. He touched some of the

[8] RH BA37 is put in parenthesis here to emphasize RH PC's larger contribution to this network.

cracks within his reach and said, very seriously: 'I don't know whether it's right or not—the wall' (Kanner, 1951, p. 24).

We see here that "sameness behavior" is similar in its genesis to preoccupation with and fixation on the parts of objects, i.e., to "stop" the moment of experience in the external world. Again, the autist is preoccupied with the object that is not functional and with *parts* that are idiosyncratic, *wholes* by themselves, and symbolic representations of the VSS, i.e., moment of experience. In the case of "sameness," the accent is not on fixation on *parts* but rather on preservation of the content of the *whole*, the VSS—its completeness and spatial–temporal organization. The above examples illustrated the desire for "completeness." Examples of autistic children's desire for preservation of spatial arrangement and retention of sequences (temporal organization) were given in Chap. 3. Here is another example from Kanner's material. "When Elaine C. was sent to fetch a specific object, she always brought it if it was in the place where she knew it usually to be; if it was not there, she would not bring it even if it was very near and plainly visible" (Kanner, 1951, p. 25).

Meaning of the *part* is indistinguishable from the *whole*, the VSS within which the *part* was originally perceived.

In Chap. 3, an example of Donald, a 5-year-old autistic boy, from Kanner's original 1943 work was given. Here is an example of the same Donald taken from Kanner's later work when Donald was 9 years old:

…Donald had been going to school…. On one afternoon, the session had been dispensed with; no one in the family knew about this. Donald went to school as usual. Though no other child was in the classroom, he sat down in his seat, took out his books, did some writing, and left when the bell rang. He evidently could not accept an 'irregular' free afternoon contrary to established routine. The part had to be made to fit with the accustomed whole… (Kanner, 1951, p. 26).

We can see here that, with multiple repetitions, the unique, idiosyncratic, and experiential VSS becomes an automatism.

Accent on the content of the VSS, predominantly its completeness and spatial–temporal organization, requires participation of the right posterior cortex, mostly BA39,40.[9] The making of "automatisms" suggests involvement of the right striatum. Thus, the likely network involved in "sameness behavior" is *right TPO (BA39,40)—right DLPC—right Striatum.*

It seems that sameness behavior and its brain network is specific for autism.

5.5.1 Repetitive Behavior in Autism and Compulsions in Obsessive–Compulsive Disorder

It is interesting to compare repetitive behavior in autism with compulsions in obsessive–compulsive disorder (OCD). While typical compulsions of OCD patients are hand washing, cleaning and counting, these behaviors are not common in autists

[9] BA39 and BA40 will be discussed in detail in the next chapter.

(McDougle et al., 1995). "Compulsions" in autism are represented by repetitive phenomena belonging to "sameness behavior": creating and maintaining patterns, insisting furniture and other objects remain in the same place, rituals in going to bed, etc. (Prior & Macmillan, 1973).

Compulsions in OCD are responses to obsessions, intrusive thoughts caused by pathologically exaggerated harm concern. Compulsions, thus, represent "an emergency response" to the imagined danger by retrieving object action automatisms stored in the left striatum. "Sameness behavior" seems to be a response to the constantly changing RH world by the retrieval and preservation of the VSS stored in the right striatum.

Nonrhythmical stereotypical behavior in autism, as described above, may include retrieval of object action automatisms stored in the left striatum (e.g., filling and emptying sequences with liquids and solids), however, this behavior is not a response to obsessions and cannot be called a compulsion. In general, although repetitive behavior in autism and compulsions in OCD have a common link in their underlying networks (striatum), they do not share other parts of their networks, and they are qualitatively different phenomenologically.

5.6 Autist's Rhythmical Stereotypical Movements: Role of the Thalamo-pallidum System

The thalamic level seems to be a source for the rhythmical stereotypies of the autist. As discussed in earlier chapters, the thalamic level's main contribution to complex movements is to use its mastery of one's own body to unite into a common rhythm the multilinked "pendulums" of the extremities with their diverse and complicated spectrum of oscillation frequencies. Constant flow of kinesthetic sense to the thalamus allows a finely choreographed chorus of muscle contractions that forms a background for all complex movements initiated at higher levels. This level knows nothing about the external spatial field. The space of one's own body is the only space the thalamic level knows, thus it is aptly referred to as the level of the possession of one's own body (Bernstein, 1947, 1990).

If one could extract thalamic level movements in a pure form, nonintegrated in higher-level movements, they would be rhythmical, rocking, repetitious movements with a complicated, but exactly reproduced, rhythmical pattern unique to each individual. This rhythmical pattern is why, while all gymnasts can do backflips, no one gymnast's flip can ever be identical to another's. In addition, this level also assures that the rhythmical pattern of each individual asserts itself exactly and invariably each time. A gymnast's coach instructor will be able to recognize the unique signature each gymnast brings to her backflips.

When the striatum is damaged, the "next in command" effector center of the thalamic level, the pallidum, becomes disinhibited and pathologically hyperactivated (the neurological condition most typical of this is Huntington's chorea). The result is hyperkineses—involuntary, nonfunctional repetitive movements that are fragments of thalamic-level synergetic movements. Bernstein described them as devoid of any meaning, grotesque "backgrounds without figures."

Autists' rocking and jumping bodily and limb movements as well as their mannerisms (hand flapping, tiptoe, etc.) fit Bernstein's conceptualization of the thalamic level. Here are a few illustrations of such movements in children with autism:

> [Ten-year-old Hans] He skipped around humming and waiving and flailing his arms in the air.... These rhythmical movements, particularly the skipping and arm waiving were repeated several times day after day so that his counselor quite spontaneously spoke of a 'movement phase'.... Rotating played an important part in his rhythmical movements. He liked to do head-over-heels, rolled sideways back and forth on the lawn...he lay awake...at night talking to himself and making circular motions with his hand (Bosch, 1970, pp. 7, 8, 13).

> [Nine-year-old Dieter] He waived his hands up and down, and this up-and-down movement was transmitted to his entire body increasing in intensity until he was moving from a tip-toe position to a crouch, sweating hard all the time (Bosch, 1970, p. 23).

> [Seven-year-old Fritz] [I]n his first year...he began to make unusual rotating and waggling movements with his hands, often directly in front of his face. When he learned to walk he tended to skip rhythmically, turning his hands with rapid pronatory and supinatory movements or flapping them in a sort of waving gesture (Bosch, 1970, p. 31).

These movements are stereotypical, with "chased reiteration" of rhythmical patterns. They are introverted, without any connection to the environment. The autist's complex postures and hand mannerisms are like fragments of synergies, as though the thalamic level is acting independently, disconnected from higher levels and "making" meaningless "backgrounds." These postures in autism are analogous to movement disorders in neurological patients with pathological hyperfunction of the thalamic level as a result of damage to the striatum. The difference between the pathological bodily movements in autistic subjects and that of brain-damaged patients is that, in autism, the pathological movements are dynamic and may disappear; they are not completely involuntary, but instead are self-motivated, self-induced, as a reaction to some distress. The autist's state while performing stereotypical movements of the thalamic level is often one of ecstatic excitement:

> He ran around in circles emitting phrases in an ecstatic-like fashion. He took a small blanket and kept shaking it, delightedly shouting, 'Ee! Ee!' he could continue in this manner for a long time and showed great irritation when he was interfered with. All these and many other things were not just repetitious but recurred day after day with almost photographic sameness (Kanner, 1943, p. 228).

There is an environmentally induced stereotypy described in the literature that might represent thalamic-level body movements: rocking the torso, banging the head on the wall, and other types of self-stimulating and self-damaging behavior. It is called deprivation-stereotypy and is observed in animals reared alone. It has been suggested (Ridley, 1994) that deprivation-stereotypy is not a direct response to the immediate environment but represents a degree of detachment from that environment, possibly because an appropriate interaction with the environment has not developed during a "critical period" in infancy. This theory of deprivation stereotypy can also be poignantly applied to autism, with the difference being that the "deprivation" is not due to the autist's environment, but to his internal inability to relate to it in a conventional manner.

Deprivation-stereotypy is much less reversible than confinement-stereotypy (which, as discussed above, represents exposure of the sensory-motor level). The latter may dissipate rapidly if the environment returns to normal. Applying these observations to autism, it may be suggested that the relative involvement of the sensory-motor versus thalamic level in stereotypical behavior is an important indicator of the degree of severity and prognosis in autism. The more a deeper, thalamic level is involved, the more severe the disorder and poorer the prognosis.

As is true for nonrhythmical stereotypical behavior, rhythmical stereotypical behavior is diagnostic, but not specific, for autism, reflecting a particular aspect of autism's brain pathogenesis.

5.7 Conclusion

Thus, four types of stereotypical behavior were distinguished based on their distinct underlying brain networks (1) nonrhythmical stereotypical movements; (2) preoccupation with objects (or their parts); (3) "sameness behavior"; (4) and rhythmical stereotypical body movements and mannerisms.

Described above, different types of stereotypical behavior and their underlying networks in autism reflect what we have already observed in Chap. 2 (autist's way of word comprehension) and Chap. 3 (autist's perception of the world): exposure of the sensory-motor and thalamic levels in autism, with the higher levels not damaged but nevertheless not functioning properly.

According to Bernstein, if in construction of movement a higher level is not functioning, the lower level next to it becomes the leading and, therefore, conscious one. However, when subcortical levels become "conscious," the degree of conscious awareness and voluntary control are not the same as cortical level consciousness. Degree of awareness and voluntary control increases with the leading level, along the vertical axis of brain structures. In Bernstein's model of hierarchical levels in the brain, each level is characterized by its unique spatial–temporal organization. In a broader sense, we can assume, each level represents a different spatial–temporal organization of experience (qualitatively different content of consciousness). If a lower level is "exposed," it is its *space* that one becomes aware of. When the thalamic level is "exposed," there would be a "switch" in consciousness toward an increased awareness of one's own body-space with a subjective experience of euphoric excitement, cut off from the outside world. Preoccupation with parts of objects is an eloquent example of increased awareness of the external space of the sensory-motor level, intense attention on the object as a physical thing, on the slipping-away moment of experience. From HFA reports, we learn that this intense concentration leads to some kind of meditative state (see above report of Temple Grandin in this chapter). A similar state could be observed in Nadia, the autistic savant-artist:

> [A]fter surveying intently what she had drawn she often smiled, babbled and shook her hands and knees in glee. However, it was at this point that she was most distractible and she could fall into a staring reverie perhaps lasting for several minutes before she continued drawing... (Selfe, 1977, pp. 8–9).[10]

The physical world of the sensory-motor level has quite a different "consciousness" than the "meaningful" world of the higher levels with their abstract *space* and *time.*

---

[10] Here we can think about an increased awareness of both the internal space of the thalamic level and the external space of the sensory-motor level.

Chapter 6
How Autistic Persons Perceive Space and Spatial Relations (*Where* System in the Brain and Autism)

In this chapter, we will examine the conflicts and contradictions in the autist's abilities to comprehend the spatial relations between objects and speculate what brain patterns may underlie those peculiarities. We will also explore the meaning this may have in the context of the whole autistic clinical-brain pattern.

To do all this, we will travel along the dorsal visual pathway (DVP), the brain network responsible for the *where* system. As opposed to the *what* system connected with visual recognition of objects *per se*? the *where* system is concerned with objects' location and their spatial relations. Both systems receive visual information, and their pathways begin in the visual occipital cortical region, but then they diverge: the *what* system pathway continues to the temporal-occipital region (BA37) and then to the orbitofrontal cortex (OFC), while the *where* system pathway heads for the parietal-occipital area, BA39 and BA40, and then on to the dorsolateral prefrontal cortex (DLPC) (Mishkin, Ungerleider, & Macko, 1983).

Situated between temporal, parietal, and occipital fields,[1] BA39, 40 represent the phylogenetically youngest part of the parietal cortex. From the point of view of cytoarchitectural hierarchy, BA39 and 40 are considered intermediate, secondary-tertiary fields, which functionally corresponds to the sensory-motor/gnostic-praxic levels and transition to the symbolic level.

6.1 The Right Parietal-Occipital Region, Topographical Memory, and Autism

The spatial function of BA39,40 is modified in each hemisphere. The left hemisphere's *analytic* interpretations create a different phenomenon than that of the right hemisphere's holistic processing. Furthermore, within the right hemisphere, the

[1] This is why this region is often referred to in literature as temporal-parietal-occipital region (TPO)—a term which will also be used in this book.

T.B. Glezerman, *Autism and the Brain: Neurophenomenological Interpretation*,
DOI 10.1007/978-1-4614-4112-0_6,

visual-scene-situation (VSS) is construed differently in its *what* system ("object vision") of BA37 versus its *where* system ("spatial vision") of BA39,40 (Glezerman & Balkoski, 1999). In the right inferior temporal region (BA37), VSSs are represented as a single whole, including objects saturated with affect. This situation-with-object is a unit for operations in symbolic thinking. In the right-hemisphere's *where* system, the visual situation is interpreted as a purely spatial *gestalt*. BA39,40 construct the space of the visual scene—its dimensions and contours—and the objects within it are "spatial forms" in their unique locations and positions in which they were perceived. Thus, in the right parietal region, it is "singular spatial" situations that are represented (Glezerman & Balkoski). Information stored in BA39 and 40 is "setting the scene in a matrix of spatial coordinates in the same way as we would perceive the scene if it were actually happening" (Cutting, 1990, p. 38). Whereas in the LH there is one continuous external spatial field as an arena and coordinate system for moving and acting, the RH creates singular spatial situations. There are as many spatial situations stored as are perceived, and each situation is unique in its configuration of objects within it (Glezerman & Balkoski).

Damage to the TPO (BA39,40) in the right hemisphere results in a phenomenon called "topographical amnesia." Patients with such lesions experience difficulties in visual recognition of landmarks with their unique orienting value, and they lose a sense of locality. They cannot accurately describe or visualize familiar routes, and they are unable to find their way in a previously familiar environment (their neighborhood, their street, or their room in the hospital).

As a contrast to topographical amnesia, the phenomenon of extremely expressed right-hemispheric ability comes to mind—the astonishing topographical memory of aborigines, with their extraordinary sense of location and direction, as described by anthropologists. These peoples possessed exhaustive knowledge of the territory they occupied and each unique landmark therein. "After hearing a shower they know where, in which rock, some water remains; they will know the location of that particular crevice where water will remain longer" (Levy-Bruhl, 1930, p. 74). Having once passed through a place, the aborigines remembered it with exquisite detail. They recognized *places* as we recognize familiar *faces*—as *gestalts*.

Strangely, descriptions of autistic peoples' memory for locations and routes are reminiscent of this unique topographical memory of aboriginal people. What follows are some examples of autists who demonstrate impressive topographical memories:

> There was not a location, not an orientation in her world that she did not have memorized…. By age two, Elly[2] was suspected of being developmentally delayed… Speechless, uncomprehending, unable to care for her physical needs, Elly was retarded in every functional way…. But even as Elly fell further and further behind her peers, signs of a remarkable memory began to appear…. Elly was two year's old and had been walking for only three months when one day she disappeared. After a frantic search she was finally found,

[2] "Elly" is an autistic woman described several times in various writings by different authors since the time she was a child. In some writings she has been called "Elly," in others, "Jessica." Her case will be discussed several times throughout this work, and, for consistency, she will always be referred to as "Elly."

> crawling on the painted cross-stripes and arrows of a parking lot several blocks away. Elly had never walked to this place before and had traveled the route only once, pushed in a stroller by her mother. She had been fascinated then by the bold markings on the ground—enough, it seemed, to undertake an expedition of her own. One trip and the route had been firmly fixed in her memory—well enough for successful navigation on her own wobbly legs. And so it was in general. One trip, one exposure to something new, and the information seemed stamped in the mind of this child who seemed oblivious to much of the world around her…she appeared to have the tiniest physical details of her environment imprinted indelibly on her mind (Bogyo & Ellis, 1988, pp. 265–266).

There is also the 18-year-old autist, J.D., who had an IQ of 58 but possessed a remarkable ability to remember routes he had traveled:

> When his parents took him someplace in the car and told him where they were going, J.D. would be very upset and agitated if the driver took a route different from the one taken before. And the reverse was true as well: if family started up its trip on a familiar route, J.D. would say the name of the place the route led to and would have difficulty if the driver altered that route (Waterhouse, 1988, p. 329).

J.D. was given an experimental test of gestalt memory for object location in which the experimenter places "junk objects" in unique positions on a large black plywood board. The subject sees each of ten or more objects for only 2 s. After this quick presentation, the subject is given the objects one by one and asked to place each junk object exactly where it had been originally placed by the experimenter. J.D. performed very well on this test, better than any normal adult control (Waterhouse, 1988).

In conclusion, topographical memory, an ability specifically connected with the right parietal-occipital region, is not only intact in autistic individuals, but also often outstanding.

6.2 The Thalamus, Intrapersonal Space, and Autism

Here we will discuss the representation of intrapersonal space in the brain. In order to do so, our travels must take us down to the thalamus, the main collector of sensory information in the brain. According to Bernstein, kinesthetic sense is the afferentation for the thalamic–pallidum level. While the majority of the kinesthetic pathway's fibers passing through the VP[3] terminate in the sensory-motor cortex, some fibers terminate within the thalamus, likely in the integrative LP[4] nucleus. Connected with

---

[3] VP (ventral posterior nuclei group) is the principle relay station for somatic sensations (Kandel & Schwartz, 1985).

[4] The LP (lateral posterior nucleus) is the associative thalamic nucleus, receiving afferents from the VP (the somatosensory relay thalamic nucleus) and containing extensive reciprocal connections with the inferior parietal cortex. Although the LP's function remains largely unknown, it has been suggested that it is concerned with complex somesthetic association mechanisms related to various parts of the body (Malcolm, Carpenter, & Sutin, 1983).

the parietal-occipital region (TPO), the LP is the major thalamic contributor to the thalamo-parietal body scheme system (Kandel & Schwartz, 1985). It was suggested (Glezerman & Balkoski, 1999) that the LP is concerned with integration of kinesthetic sensations according to body spatial coordinates, making, in Bernstein's terms, the afferentation of one's own body.

As noted before, the constant inflow of kinesthetic information about the body's ever-changing positions is crucial for the organization of movements. For our purposes in this chapter, it is important to note also that from this constant flow of kinesthetic information, an *invariable* entity is built: the space of one's own body.

Primarily kinesthetic sensations from the body define its boundaries (just as the visual modality defines the external spatial field). This continuous image of one's own body space belongs to the thalamic level, it is the *space* of this level (in Bernstein's terms).[5]

It was suggested elsewhere that in the formation of one's sense of self, the thalamic level contributes the subjective experience of the boundaries and continuity of one's body, as defined by kinesthetic sense. This kinesthetic body image was termed *I-space* (Glezerman, 1986; Glezerman & Balkoski, 1999). Translated into the psychical metaphors of higher function levels, the thalamic level provides a fundamental experience of self wholeness. The thalamic level thus exists as a background and is intimately incorporated into the higher levels such that the experience of kinesthetic-body-space is not normally part of conscious self-awareness. We learn about this *I-space* when it is weakened, like in the patient described in Oliver Sacks' "Disembodied Lady" (1987). This patient suffered from acute sensory polyneuritis resulting in the disruption of kinesthetic inflow to her brain. The patient felt her body was dead, not real and not hers, stating: "It's like something's been scooped out of me, right in the center" (p. 51). It was the patient who found this word—*disembodied*— to describe her experience. Schizophrenia seems also to be an instance of disordered integrity of *I-space* where distorted kinesthetic feeling surfaces in consciousness (Glezerman & Balkoski, 1999). Take for example a patient with schizophrenia who conveyed his experience: "My body feels too short for me; my whole body feels like a woolen suit which was left in the rain and became wet, then dried out and became short. When you move it stretches and binds" (Angyal, 1936, p. 1036). Nothing like this is happening in autism. In fact, not only is the thalamic level *not* disordered, but it is strong enough in autism to be used for compensation. Interestingly, subjects with autism unknowingly and instinctively explore *I-space* by seeking pressure for their bodies, thereby increasing kinesthetic and tactile flow to the thalamus. According to Temple Grandin, parents often report that their autistic children love to crawl under mattresses, wrap up in blankets, or wedge themselves in tight places. She herself was "a pressure seeker":

---

[5] Unique for each of us is the sensual kinesthetic image of our body, stored at the thalamic level. Only at this level, kinesthetic (muscle) sense is directed toward one's own body (bodily feelings). Recall that kinesthetic sense from pathways that pass through the thalamus, onto the somatosensory cortex, has a different meaning than that of the thalamic level. Kinesthetic sense in the sensory cortex is used for evaluation of external stimuli, their weight, consistency, and texture. This fits with somatosensory cortex belonging to the sensory-motor level.

> [W]hen I was six, I would wrap myself up in blankets and get under sofa cushions, because the pressure was relaxing. I used to daydream for hours in elementary school about constructing a device that would apply pressure to my body. I visualized a box with an inflatable liner that I could lie in. It would be like being totally encased in inflatable splints (Grandin, 1995, p. 63).[6]

Eventually, Temple Grandin invented a "squeeze machine." The idea for this machine came from squeeze chutes where cattles are placed for vaccination. She once entered such a machine herself: "I asked my aunt Ann to press the squeeze sides against me and to close the head restraint bars around my neck.... This was the first time I ever felt really comfortable in my own skin" (p. 63).

Grandin copied the design of this cattle chute and built her own human squeeze machine. "Entering the machine on hands and knees, I applied pressure to both sides of my body...using the squeeze machine on a daily basis calms my anxiety and helps me unwind" (pp. 63–64).

In communication with other people with autism, Grandin learned that many crave similar pressure and have devised methods to apply it to their bodies. For example, one person with HFA designed a special suit "to feel better." "He made a pressure suit consisting of a wet suit with an inflatable life jacket under it. He can adjust the pressure by blowing air into the valve on the jacket" (Grandin, 1995, p. 65). It is truly amazing that this autistic man's device looks like a literal translation of the above schizophrenic's metaphor for describing his intensely distressing experience ("My body feels too short for me; my whole body feels like a woolen suit which left in the rain and became wet, then dried out and became short. When you move it stretches and binds."). It is an eloquent illustration of the cardinal differences between autism and schizophrenia at this deep and fundamental level of self. What makes an autist feel better ("the first time I ever felt really comfortable in my own skin") is the source of distress in schizophrenia. Being "in touch" with kinesthetic bodily feelings accentuates the autist's experience of *I-space*. As suggested before, the fundamental deficit in schizophrenia is on *this* thalamic level—the integrity of physical self is compromised (Glezerman & Balkoski, 1999) and manifested clinically by bodily sensations of distinct kinesthetic quality (pressure, stretching, pushing, shrinking) coming to the surface.[7]

Interestingly, rhythmical stereotypical movements of LFA and instinctive pressure-seeking of HFA have the same target, the deepest physical foundation of self: body space. These two groups of autists differ in their level of consciousness. In LFA, the thalamic level becomes the leading one and, therefore, "conscious." However, the introverted thalamic level by itself knows only about body space, so while doing

---

[6] From THINKING IN PICTURES by Temple Grandin, copyright © 1995, 2006 by Temple Grandin. Used by permission of Doubleday, a division of Random House, Inc.

[7] Patients with schizophrenia and autistic individuals are similar at a superficial, behavioral level: they do not communicate with the external world. Some psychiatrists considered autism as childhood schizophrenia. Indeed, in children it is sometimes difficult to differentiate autism from schizophrenia. Even the term "autism" was taken from Bleuler's autistic thinking as a core symptom of schizophrenia. However, both Kanner and Asperger emphasized the specificity of their described syndrome and that it had nothing to do with autistic thinking.

rhythmical movements, the LFAs' degree of awareness of the outside world and their voluntary control is low. HFAs "get in touch" with the wholeness of physical self, while retaining awareness and voluntary control.

Suffice it to say, *I-space*, the physical foundation of self, is intact in autism.

6.3 Returning to the Right Parietal-Occipital Region: Thalamo-Parietal System and Integration of *I-Space* and *Non-I-Space*

The thalamic kinesthetic body image is projected to the parietal-occipital region (BA39,40). This makes a specific network, the thalamo-parietal system (*RH LP*[8]*–RH TPO*). A key function of the right parietal-occipital region is integrating *I-space* with multiple visual-spatial situations, the latter have been termed *non-I-spaces* (Glezerman, 1986; Glezerman & Balkoski, 1999). The concept of this integration is essential for understanding the nature of RH gestalt connected with BA39,40. *Non-I-spaces* are subjectively felt (part of self) because they have been integrated with *I-space*. RH spatial situations are singular, and, therefore, they are "separate spaces" that do not connect and do not cross one another. There are as many *non-I-spaces* as there are singular spatial situations represented in the right hemisphere, but there is only one *I-space*, so that identification of *I-space* and *non-I-space* results in multiple unconnected, subjectively felt spaces, multiple "selves" that are not divisible from the outside world.[9] Thus, if we were able to separate this part of the RH self, we would experience multiple selves, indivisible from others or from physical surroundings. It is from this layer in the brain that people get a "sense of belonging"[10]—mystical experience.[11] Usually we are not aware of this layer of our experience, it is incorporated into higher levels of self. Our conscious experience is of one separate, individual self. There is, however, a unique natural model where RH gestalt of identified *I- space* and *non-I-space* appears to be exposed.

Early twentieth-century French anthropologist Levy-Bruhl (1930) described so-called "collective representations" of indigenous societies. This term, collective representations, refers to an attitude toward perceived objects where feelings, objects,

[8] The lateral posterior nucleus of the thalamus.

[9] We will see in the following chapters that it is the LH that gives us a sense of having *one* separate self.

[10] "Sense of belonging" as used here does not equate to the familiar connotation of this phrase: of being a member of a group—a separate but related part (for example, as one feels he belongs to a particular social group). Instead, I am using "sense of belonging" to refer to the right hemispheric experience of identification with the group and all its parts, animate and inanimate, including the surrounding space. It is an experience of an indivisible continuous whole, infused with meaning.

[11] A mystical experience is in fact a special state of consciousness connected with exposure of the particular part of the brain; it occurs rarely and in exceptional situations.

and subject are inseparably fused. For the Australian aborigines, as Levy-Bruhl describes it, space is not something homogenous and uniform, indifferent to what fills it, nor is it devoid of its own features and in all parts identical to itself. Each social group is mystically connected with the territory it occupies and moves upon. Between land and social group there are relations of mutual participation, equal to some kind of mystical property, which cannot be altered, taken away, or conquered. Each place's characteristic landscape (its rocks, trees, springs, sand dunes, etc.) is mystically connected with visible and invisible creatures, with personal "spirits."

Levy-Bruhl suggests there is a co-participation—a kind of symbiosis experienced by tribe members based on identification. Each individual is himself or herself and at the same time is his or her reincarnated ancestor. Totem (the clan to which an individual belongs) is also identified both with the individual and with plants or animals whose name the totem carries. Furthermore, the totem is identified with a part of space whose borders are clearly cut and which is populated with spirits of totemic ancestors.

In Chap. 2, I have already given an example of a nineteenth-century Native American tribe who claimed that although they were human beings they were at the same time red parrots (Levy-Bruhl, 1930). I explained it by the mechanism for the formation of the RH symbolic system: *identification* by common emotion.

Levy-Bruhl explained such identifications by his co-participation law, which assumes that things are assigned some kind of mystical power in the consciousness of the aborigine, and it is this that things have in common despite differences in their physical features. Levy-Bruhl further emphasized a continuity of things that were united by these mystical forces circulating in objects and creatures. The source of this magical continuity was one of felt vitality: all animate and inanimate objects are imbued with continuous vitality resembling that vital power which they felt in themselves (Levy-Bruhl, 1930). Thus, the aborigine can be someone who lives now and at the same time be his reincarnated ancestor, his own totem, and the animal whose name the totem carries because all these things are connected by the continuity of mystical power that circulates among them. In rituals such as animal dances of the totem (bison dance, snake dance, deer dance) this continuity, this indivisibility of the collective representation, is reproduced in action. These rituals are the intercourse in which the living individual, the ancestor transformed in him, and the animal that is the totem of the individual are merged together.

From the point of view of brain mechanisms, we see here the "signature" of the RH. The RH gestalt of identified *I-space* and *non-I-space* is on the foreground of consciousness and experienced in a more or less pure form in these cultures (Glezerman & Balkoski, 1999). Moreover, RH representations are not only externalized and exposed but are embodied in societal rules and played out and reinforced in rituals and ceremonies. Through these societies we can "touch," "see," and describe RH representations, specifically gestalts connected with the right thalamo-parietal system: *RH LP–RH TPO* (BA39, 40).

In this spatial gestalt, we encounter a new organizing principle for RH identification: *non-I-space* is subjectively felt because it is identified with *I-space*, which has its own rhythmical pattern, unique for one's physical body.

Thus, in the aborigine rituals, the RH parietal "version" of self—integrated *I-space* and *non-I-space*—seems to be externalized. It is the projected subjective experience of "thalamic" rhythm, the rhythm of *I-space*, that is perceived as a mystical power and unites individual—ancestor—totem into an indivisible and continuous whole. This "mystical power" is not a religious belief that aborigines have, it was not developed by thinking or working through, but it is their direct, immediate experience of meaning, of symbiosis, of the continuity of life.

Having proposed that collective representations of aborigines are outward manifestations of the RH cognitive mechanism, I will attempt to illustrate below the analogy between characteristics of the RH "spatial" gestalt and Levy-Bruhl's collective representations.

1. RH spatial gestalt: *non-I-space* includes visual images as spatial forms, exactly preserving their size and dimensions as well as their unique location in space as they were perceived. The spatial situation (*non-I-space*) is also imparted with the individual rhythmical pattern of one's body as a result of the integration of *I-space* and *non-I-space*.
   Collective representations: part of the land on which some social group (totem) lived *is* the totem itself. However, this identification of space and objects should be understood as their indivisibility and continuity, but not as their loss of uniqueness: individual, ancestor, and totem constitute continuous whole without losing their triplicity (Levy-Bruhl, 1930).
2. RH spatial gestalt: meaning for right hemispheric gestalt does not reside in objects united by their common objective features and signs (as in logical thinking). In the RH spatial gestalt, the internal rhythm of the kinesthetic space of one's body "cements" things into a continuous whole.
   Collective representations: totem is a dynamic totality of mystical interactions where objects, people, animals/plants are only conductors (bearers) of these mystical forces (Levy-Bruhl, 1930).
3. RH spatial gestalt: because objects are united in the RH spatial gestalt not by their physical, objective features but as "conductors" of individual-specific rhythmical pattern, their physical appearance must preserve absolute consistency as unique qualities, including their position and location in space. This brain mechanism has an interesting application in the aborigines.
   Collective representations:

   > A particular position of parts of the drawing may have one meaning if it is used in connection with one totem, and have an absolutely different meaning in connection with another totem.... Even man-made objects were assigned with mystical properties.... Observers indicated that aborigines demonstrated extraordinary dexterity in making various objects, but they always reproduced them as the same up to the smallest details in form or ornament. From generation to generation, they reproduce things with a striking consistency and never improve them (Levy-Bruhl, 1930, pp. 23, 82).

This last example teaches us that one must be careful using words like "detail," "part," or "part of the whole," when examining RH representations. RH gestalt is a totality of unique qualities where the whole is directly and immediately perceived as the relations of these qualities. Thus, Levy-Bruhl aptly observed that the aborigines perceive individual in collective and collective in individual (Levy-Bruhl, 1930).

The formation of the RH symbolic system connected with the right inferior temporal region (IT) was described in Chap. 2. It is based on *identification* by common emotion and resemblance of holistic form (network: *RH BA37–RH DLPC*). In this chapter, a new organizing principle for the RH identification is introduced—the rhythm of one's kinesthetic body space (network: *RH thalamus (LP)–RH TPO–RH DLPC*).

6.4 Spatial Representations and "Qualitative" Quantity

Temporal and quantitative concepts originated in history from spatial representations. Levy-Bruhl (1930) indicated that the modern concept of number is not possible in thought without numerals in language. However, he emphasized that man was able to count long before numerals in language and the concept of number were developed. As described by Levy-Bruhl, in the collective representations of aborigine cultures (and in purely RH thinking!) objects are not thought of as separate from each other *or* from the space where they are located. A certain number of objects are felt as a spatial totality where each object is not a quantitative unit in a series of others, but rather a unique quality of space, exactly preserved in memory. Nevertheless, such a collective representation implies an exact sum of objects as a "quality" of the representation that differs it from others. Take the following observation for instance:

> When they return from the hunt for wild horses, nobody asked them: 'how many horses did you bring?', instead they are asked: 'how much space does the herd of horses you brought occupy?' When they are going hunting, while already in the saddle, they look around, and if even one of the numerous dogs is missing, they are calling her by name. Observers were very surprised how these people who are not able to count, immediately recognized that among a significant pack of dogs only one is missing (Levy-Bruhl, 1930, p. 121).

Even when the first numerals emerged in language (numbers one to ten), the "umbilical cord" with the right hemisphere was not torn asunder. Included into collective representations, numerals are assigned with the mystical properties, which belong solely to a given number, secondary to its mystical co-participations. Each number has its own individual physiognomy, some kind of a mystical atmosphere, some kind of a "magnetic field." Each number is felt in its own special way, different from other numbers. From this point of view, numbers did not constitute a homogenous sequence and, thus, were not fit for even the most simple of arithmetic operations.

6.5 The Left Parietal-Occipital Region, Grammatical Constructions in Language and Autism

Now we will leave BA39 and 40 in the right hemisphere for a while to explore their counterparts in the left hemisphere. In the left hemisphere "RH gestalt"—the visual-spatial image—is analyzed, and its signs of spatial relations are distinguished. For example, the LH dissects out signs of directedness, spatial coordinates, quantity, and also the spatial-temporal signs of approaching–receding. When this function is disordered because of damage to BA39 and 40 in the left hemisphere, patients may orient well in the environment yet experience difficulties in performing tasks requiring schematic concepts of space: drawing maps, spatial planning, and any other schemes reflecting spatial relations among objects. Typical mistakes include reversing left–right directions. For example, in performing the "clock test," in which a patient is asked to draw a clock designating a certain time, patients with left parietal-occipital (BA39,40) lesions (in contrast to patients with right parietal-occipital lesions) preserve the whole spatial image of the clock but may reverse the directions of the clock hands 180 degrees (mirror reversal).

The left hemisphere's *what* system derives from its processing the functional and categorical signs of objects. These signs characterize objects as such and reveal the intrinsic, logical relationships between objects. On the other hand, meaning for the left hemisphere's *where* system is brought about by "geometrical" analysis of space. This analysis distinguishes spatial-temporal signs and serves to characterize the outer relations between objects—their contiguity in space and time. External spatial relations between objects are linguistically expressed in so-called spatial words (adverbs, prepositions). For example, "Draw a cross *under* a circle" or "Draw a cross *to the right* of a circle." Spatial words allow for abstraction of the location and position of an object in space from the object itself. It did not happen immediately in the history of language, for in some languages the RH spatial gestalt of objects' unique positions in space was reflected in content words' meaning. For example, in the Yagham language of Terra del Fuego, examined in nineteenth century, the pronouns *he* and *she* had many different words that expressed the location of the man or woman: whether he or she was located inside the wigwam or to the right or left, and so on (Levy-Bruhl, 1930). Only with further development of language were spatial relations expressed by special *grammatical constructions*, separate from the meaning of content words.

Let us now consider how the autist understands the language of spatial relationships. We will do so by looking at another of Temple Grandin's personal accounts:

> Spatial words such as 'over' and 'under' had no meaning to me until I had a visual image to fix them in my memory. Even now, when I hear the word 'under' by itself, I automatically picture myself getting under the cafeteria tables at school during an air-raid drill, a common occurrence on the East Coast during the early fifties.... I can remember the teacher telling us to be quiet and walking single-file into the cafeteria, where six or eight children huddled under each table.... I can remember the teacher scolding me after I hit Alfred for putting dirt on my shoe. All of these memories play like videotapes in the VCR in my imagination. If I allow my mind to keep associating, it will wander a million miles away from the word 'under' to submarines under Antarctic and the Beatles song 'Yellow Submarine'. If I let my mind pause on the picture of the yellow submarine, I then hear the song. As I start humming the song and get to the part about people coming on the board, my association switches to the gangway of a ship I saw in Australia (Grandin, 1995, p. 30).

Is the left parietal-occipital region, responsible for grammatical words "over" and "under," disordered in Grandin? It is not possible to tell here because RH visual-situational associations overwhelmingly dominate her consciousness.[12]

[12] Note, however, that Grandin has insight into and partial voluntary control over this associative flow. This is the influence of the left frontal lobe, which will be discussed later in its relation to autism.

The so-called non-root morphemes used to convey spatial relations and ultimately personal relations are also of interest in relation to autism, and some of Grandin's experiences, in particular. Examples of non-root morphemes include inflectional endings (e.g., the regular forms of plural), suffixes, and prefixes. When a non-root morpheme is added to a root morpheme (a content word), the word acquires an additional semantic mark. This grammatical mark gives reference to specific meaning within a morphological paradigmatic series just as a book's code indicates its proper place on a shelf in a library (Katznelson, 1972). Based on the signs of spatial relations, morphological paradigms are stored in the left inferior parietal region and organized by the logical principle of opposition or similarity. Here is a example of a paradigmatic series of prefixes and prepositions, organized by simple oppositions (1) quantitative, that is, *in*crease–*de*crease, negation–affirmation (e.g., *under*value—*over*value; desirable—*un*desirable); (2) quantitative–spatial (e.g., build *under*—build *over*); (3) spatial (e.g., *under* ground—*above* ground); (4) direction of movement in space, approaching–withdrawing (e.g., walk *in*—walk *out*, move *to*—move *away*).[13]

The inflectional endings in the declension paradigm reflect spatial relations, as illustrated by the linguist Roman Jakobson (1971). Indeed, semantic marks of the different cases represent particular signs of spatial relations. For example, according to Jakobson, the genitive case expresses the relation between *part* and *whole* of an object as well as represents a sign of volumeness, the extent to which the object taking part in the message is less than the full volume of the object: "part of the house" or "hand of the brother." Spatial relations expressed in grammatical constructions of the genitive case "grow" from literal contiguity in space ("part of the house") into the more abstract ideas of belonging and possession ("equipment of the worker"), and then into the even more abstract idea of relatedness per se ("father's brother").

Patients with damage to the left parietal-occipital region speak fluently and effortlessly. They understand the meaning of isolated words, both concrete and abstract, yet they are unable to comprehend a two-word combination like "father's brother" whose meaning goes beyond that of the component words. Such word combinations imply relations between the two words that are expressed by the grammatical construction (e.g., the possessive form of the noun). The following is an example of the performance of a patient who suffered from damage to the left parietal-occipital region as a result of a stroke (Glezerman, 1986). The patient was asked whether the expressions "brother's father" and "father's brother" are the same or different:

> [I]t may be different people...I have a brother, Ivan...the brother to the father...one... brothers can be different...completely different...father has a brother, neighbor.... Examiner: 'what kind of relative is father's brother?' Patient: 'grandfather. Father's

---

[13] Another group of non-root morphemes is suffixes. They, in contrast to other non-root morphemes, do not just mark but are closely interwoven into lexical meaning. The suffix's grammatical meaning reflects spatial relationship of the parts within a single object. Noun's suffixes may express the idea of form, e.g., suffix *-ium* marks the sign of containment, receptacle—planetar*ium*, sanitar*ium*. Suffixes may also reflect representation of a great number of single objects "molded" into one form, giving to this quantity unity or collectiveness, e.g., human*kind*, brother*hood*.

> brother...it is my brother, and it is his father. One needs to think it out, just a moment, I will say. Well, then they have two brothers. I have father, and he additionally has brother'. Examiner: 'what kind of relative is brother's father?' Patient: 'It is their grandfather, and they have two brothers' (Glezerman, 1986, p. 173).

We can see that the patient is completely lost and absolutely unable to grasp the relationship expressed by the possessive form. At the time of examination, this patient had an average IQ, and his erudition, the ability to abstract verbal information, and vocabulary were above average. Thus, his verbal deficit was highly selective and specific and could not be explained by a general intellectual deficiency.

Do people with autism perform as did the above patient with damage to the left parietal-occipital region? They do not, and we can return to Grandin's story for proof:

> Personal relationships made absolutely no sense to me until I developed visual symbols of doors and windows. It was then that I started to understand concepts such as learning the give-and-take of a relationship.... At that time I still struggled in the social arena, largely because I didn't have a concrete visual corollary for the abstraction known as 'getting along with people.' An image finally presented itself to me while I was washing the bay window... I had no idea my job would take on symbolic significance when it started. The bay window consisted of three glass sliding doors enclosed by storm windows. To wash the inside of the bay window, I had to crawl through the sliding door. The door jammed while I was washing the inside panes, and I was imprisoned between the two windows. In order to get out without shattering the door, I had to ease it back very carefully. It struck me that relationships operate the same way. They also shatter easily and have to be approached carefully. I then made a further association about how the careful opening of doors was related to establishment relationships in the first place. While I was trapped between the windows, it was almost impossible to communicate through the glass. Being autistic is like being trapped like this. The windows symbolized my feelings of disconnection from other people and helped me cope with the isolation. Throughout my life, door and window symbols have enabled me to make progress and connections that are unheard of for some people with autism (Grandin, 1995, pp. 34, 36–37) (see footnote 6).

What is she doing? Or, more exactly, what is her brain "doing"?

Unable to relate to people emotionally, Grandin developed "ideas" about personal relationships, and she did it exactly as the idea of relationship developed in the history of language—through spatial relations.

Using the RH VSS, she is able to access the LH spatial-temporal signs: approach–withdraw, add–subtract, relevant to the idea of relationship ("give-and-take," "move slowly," "open carefully"). It helped her, in turn, to make some "bridge" to more abstract meaning of grammatical forms, *relation* as such, including personal relationships. This is how the meaning of grammatical constructions that expressed relationship developed in the history of language: from literal contiguity of objects in space to contiguity of objects in semantic space.

I was awestruck when I read Grandin's account of her bay window experience. It is supposed that the doors of the phylogenetic roads have been forever closed in modern man. They are not closed for the autist's brain! It seems that Grandin's brain spontaneously re-organized by re-wiring and reviving hence "overgrown" historical pathways.

Grandin's starting point for understanding personal relationships was the visual-spatial situation. An example of the historical stage when grammatical constructions were not developed yet in the LH and relationships were expressed directly through the visual-spatial situation came to mind while I marveled at the implications of Grandin's bay window experience: the derivation of the Native American word *utotama. Utotama* meant "older brother" but its component words *u-mani* and *ota* stood for the personal pronoun "his" and "moving along a surface, performed with effort, slowly, like crawling," respectively. Thus, the word for older brother literally meant "his to move with effort." Indeed, the older brother in this Native American culture was the guardian of the child, his teacher and tutor in the clan (Blonsky, 1935, p. 96). The idea of continuous responsibility and care by the older brother is represented by this visual-situational rendering of the word.

To express a relationship linguistically, one chooses the corresponding grammatical construction out of a morphological paradigmatic series (*LH TPO*) and then puts it into "action," to build a semantic structure (*LH DLPC*); the brain network, therefore, is *LH TPO–LH DLPC*. We do not think explicitly about nor are we consciously aware of the spatial relations underlying language's grammatical code. What is more, we have lost contact with the visual scene-situation that was this code's predecessor.[14]

Grandin, however, reproduces the whole history of the grammatical construction of relationships. For her intellectual understanding of relationship, the brain network is *RH TPO–LH TPO–LH DLPC*. Finally, why is the above-mentioned patient with local damage to the left parietal-occipital region not able to compensate as Grandin does? The patient had a specific language deficit and lost the "key" to the grammatical language code. The patient's right-parietal area is intact, but it does not come to his aid as did Grandin's. The history of language is not reproduced in this patient's brain. This contrast seems to hint that the autistic brain is special in this way. Also important to note is the deficit in the patient with focal brain damage is selective, involving only the *linguistic* aspect of relationships. For this patient, emotional relatedness and the concept of relationship (based on the left amygdala—frontal lobe system) are intact.

In conclusion, Grandin's left parietal-occipital region is not disordered and functions properly, if her right-parietal region is initiating.

One outstanding question remains: How could an appeal to the spatial origin of grammatical constructions compensate for the deficit in emotional relatedness? Is there some connection between these two in the history of language? We will return to this question in Chap. 9.

[14] It is interesting, however, that when we hear the above example of word meaning "moving along a surface, performed with effort, slowly, like crawling" which implies "his older brother," we do not find it irrelevant. It resonates in us with some kind of emotional pleasure, like that of a good metaphor.

6.6 More About the Left Parietal-Occipital Region and Autism: Numbers and Calculation

A.R. Luria (1966/1980) showed that the spatial-analytic function of the left TPO is the basis for the concept of a number and for calculating operations. Coordination of quantity and order in modern numeration is possible because our brains have an "inner spatial schema" for it. In the numeration system, decimal-rank principle is expressed by spatial position: the meaning of each figure in a number is determined by its position in the number's linear structure. Luria compared the meaning contained in numbers' structure to the grammatical meaning of two-word combinations. The separate numerals in a number have their own meaning, as do the separate words in "father's brother."[15] However, the general meaning of a number is determined both by the numerals and their relationship to each other, just as "father's brother" derives meaning from the individual words as well as their grammatical relationship. In numbers, however, spatial relations are more literal, with meaning derived directly from the relative positions of numerals, from left to right, in the linear structure of a number. Damage to the left parietal-occipital region, especially to BA39, results in so-called primary, "parietal" dyscalculia. Primary dyscalculia is diagnosed not just by inability to calculate, but also by the selective character of the disorder (on the background of normal intelligence) and by the specific character of mistakes. The type of mistakes made in primary dyscalculia reveal the spatial origin of the deficit and show it is a disorder of one's inner spatial coordinate system. For example, subjects with primary dyscalculia may not be able to understand the decimal-rank structure of the number, especially in more difficult tasks where some ranks are expressed by zero. For example, N., a 9-year-old boy with an IQ of 100, writes 10001 and reads it as "thousand and one" when asked to write 1,001. Typical mistakes in calculations include mirror reversals in direction. When asked to do the calculation 25 – 16, N. performs as following: 20 – 10 = 10 and 6 – 5 = 1 and answers "11" (Glezerman, 1983, p. 171). We see here mirror reversals (180°) in direction. Subjects with primary dyscalculia may evaluate numbers by their individual numerals' value, ignoring their positional rank. As a result, 1,897 may be evaluated as larger than 3,001 (Luria). In primary dyscalculia, difficulties in calculation and in grasping the decimal-rank of the number can overlap with difficulties with quantitative and comparative concepts. For example, X., a 9-year-old girl with an IQ of 93 performed the task "The brother is 10 years old, he is 2 years older than his sister. How old is the sister?

X.: "10 + 2 = 12, sister is 12 year old, sister is older."

Examiner: "If the brother is older than his sister, so, the sister is?"

X.: "Younger [perplexed], older, younger… 10 – 2 = 11… 8, sister is 8 years younger than brother" (Glezerman, 1983, p. 60).

[15] Recall the example from the prior section of the patient with a lesion in the left parietal-occipital area.

Mistakes in calculations made by patients with lesions in the left parietal-occipital region are not chance mistakes, they are systematic, predictable, and specific. Also, these patients are highly motivated to learn and can achieve the correct answer by trial and error. This clinical picture has nothing in common with what we see in autism. Both Kanner and Asperger's material, as well as subsequent observations, have shown that autistic children are often interested in numbers and calculations, and able to perform complex calculations without being taught. Take, for example, a 6-year-old boy, Fritz, who was one of Asperger's cases:

> The mother reported that at the very beginning of schooling he set himself the problem—what is bigger 1/16 or 1/18—and then solved it with ease. When somebody asked for fun, just to test the limits of his ability, 'What is 2/3 of 120?', he instantly gave the right answer, '80'. Similarly, he surprised everybody with his grasp of the concept of negative numbers, which he had apparently gained wholly by himself; it came out with his remark that 3 minus 5 equals '2 under zero'. At the end of the first school year, he was also fluent in solving problems of the type, 'If 2 workers do a job in a certain amount of time, how much time do 6 workers need?' (Asperger, 1991, p. 45).

However, the same child who astoundingly solved complex math problems could not learn simple calculation methods at school. Thus, Fritz clearly illustrates Asperger's ideas of autistic originality: autists create everything out of their own thought and their own experience. The other side of autistic originality, as seen in Fritz, is that they are unable to learn in conventional ways. Asperger gave an example of another autistic boy who, when just starting school, could pose and solve the problem of how many seconds are in 2 h. However, when asked to work out $5+6$, he said, "I don't like little sums, I'd much rather do a thousand times a thousand." He then proceeded to perform "spontaneous" calculations for a while, until it was insisted that he solve the given problem. He then presented his original, but awkward, method: "Look, that's how I work it out. 6 and 6 equals 12 and 5 and 6 is 1 less, therefore 11" (Asperger, 1991, p. 75). This autistic boy was not using a LH "inner" spatial coordinate system (a linear, from left to right scale where numerals represent sequential series), nor did he work within decimal rank, adding to tens: $6+4$ and then adding the leftover units: $10+1$. But the question remains, how *does* he do it? First, consider another example to further illustrate autists' mathematical ways. Here is the explanation of Harro, an 8-year-old autistic boy, as to how he solved the problem $27+12=39$:

> "2 times 12 equals 24, 3 times 12 equals 36, I remember the 3 [he means 27 is 3 more than 2 times 12], and carry on". The use of conventional methods that are taught at school, for example, starting with tens and then units when subtracting, did not occur to Harro (Asperger, 1991, p. 55–56).

Such mathematical processes in autists seem cumbersome, however, they perform them very quickly, much faster than a non-autist would be able to do it the same way. Analyzing these and other cases of autistic children, one can see that, although their "method" seems awkward, by no means is it a chance one. Autists have one thing in common: they do not calculate at all, but instead *play* with numbers and with each number's individual *qualities*. For example, Fritz plays with the qualities of the number six: 2 times 6 equals 12, and 6 is 1 more than 5. To solve the problem

"27 and 12," Harro ponders the qualities of the number twelve: 2 times 12 equals 24 and 3 times 12 equals 36.

When reading these cases, the introspections of extraordinary calculators such as Aitkin came to mind. Aitkin (a phenomenal calculator and professor of mathematics) reports that almost every number appears to him in the multitude of its attributes. For example, Aitken once heard the year 1961 mentioned and immediately apprehended it as "37 times 53, *and* 44 squared plus 5 squared, *and* 40 squared plus 19 squared" (Smith, 1983, p. 56).

Taking all this into account, it seems autistic individuals do not learn in a conventional way precisely because they do not utilize a left hemispheric inner spatial schema or spatial coordinate system (linear, from left-to-right, quantitative-positional) as their framework.

Instead, autists must use some kind of combinatorial ability. Their combinations are of the patterns of a number's "behavior" in relation to other numbers. The totality of these various patterns creates a number's own unique repertoire.

6.7 Numbers for Autistic Individuals Are "Qualitative" Quantities: They Are of Right-Hemispheric Origin

A keen interest in numbers is accompanied by autistic children's peculiar perception of numbers. For the autist, numbers are seen as individual entities, having their own indwelling worth: "Frederick, 6-year-old autistic boy, can count up to the hundreds and can read numbers, but he is not interested in numbers as they apply to objects" (Kanner, 1943, p. 223). Numbers are also charged with intense emotion. Park and Youderian (1974) write about Elly,[16] an autistic girl and an extraordinary calculator:

> Elly's interest in numbers as individuals burst out at the age of 12. Suddenly, 30 or 40 integers, all the small ones and few surprisingly large, emerged with special names, invented by her, and special characteristics. Some she liked, some she hated. Cars whose license plates bore the number 75 were kicked ritually with passion. Numbers changed without warning, amid the tears of real pain, from 'good,' as she put it, to 'hate'.... At 13 she knew the prime factors of all the numbers from 1 to 1,000 and beyond, and would write them on request—write them because she would not pronounce them. The numbers, along with certain other words, were 'too good'; they had passed into a state in which they were so charged with joy and excitement that their names could not be pronounced. Even 4 years later, after months of work on verbal 'desensitization', there are many such 'magic words' which she tries to avoid saying and cannot pronounce without a smile (Park & Youderian, 1974, p. 314).[17]

The individuality and experiential nature of numbers as shown in this example and the others noted before suggest that numbers are RH gestalts for autists.

[16] Elly's outstanding topographical memory was discussed in reference to the right parietal-occipital region at the beginning of this chapter.

[17] Numbers for Elly are not only "emotionally charged individuals," they are assigned with almost mystical power. The intensity of her experience of numbers—awe and excitement, nearly ecstatic joy—suggests involvement of the thalamic level: *RH LP–RH TPO*.

However, autists also can use numbers and know well their names in language, so they may be using their left hemisphere, too. Although numbers as quantitative concepts originated from RH spatial representations in history, once the concept was formed, the LH developed its own functional system for numbers and calculation, working autonomously within the frame of its own cognitive mechanism. Furthermore, once the concept of number is formed in the LH (in history and in individual development), numerals as part of speech develop: the words "twenty-five" or "one hundred forty-three" in visual (written) and auditory form and "25" or "143" in visual image form, for example. Numbers' visual images are stored in both hemispheres, but they have different meaning in the LH versus the RH.

This difference in "interhemispheric meaning" warrants additional explanation. We must remember that the cerebral hemispheres differ not by what material they process, but by their "method," i.e., how they process information. Consider the example of word sound. In the LH, a word's sound is a part of the language code system. The code consists of a linear sequence of separate units (signs), which are phonemes. This sound code is connected with the semantic code of the word, also a LH representation. The RH counterpart of phonological code is a continuous, whole, individual rhythmical-melodic image, inseparable from the situation in which it was first heard, or from a situation that was especially emotionally charged that included the word. In the right hemisphere, a word sound is thus identified with a situational context and its concomitant emotion. Word sound can also be included into any RH gestalt whose roots go into the history of the word's meaning. In the history of language, some speech sounds originated from imitation of natural sounds: the rush of wind and waves, the rustle of leaves, etc. Such "primary sound complexes" had their own inherent meaning, but when the phonological code was formed, these complexes were re-constructed and codified by the left hemispheric cognitive mechanism. "Fragments" of these complexes, once meaningful unto themselves, might become a part of modern words. Being a part of the ancient sound complex and carrying "traces" of its meaning, these fragments could then be combined with other fragments. Incompleteness of such sound images, with their implicit, ancient meaning, creates unlimited possibility for RH sound-symbolic associations. Sound symbolism plays a significant role in art, especially in poetry. Thus, word sound in the RH is a typical RH gestalt that can be fodder for the chasm of RH associations.

The same, I suggest, is true for numbers. As already discussed, numerals and quantitative concepts derived from right hemispheric spatial representations. Furthermore, the first numerals to emerge in language were different from the modern numeral, each number of the former was emotionally charged and had its own individual physiognomy—clearly RH qualities.

It seems as though in autism the RH equivalent of a number is dominant, while the LH "serves" the needs and purposes of the RH.

In Elly, a series of situations are reduced into condensed form, where markers of identified situations are on the surface, while the individual situations themselves are "behind the scenes." For example, "the moon is the number 7, and so is the sun and so, apparently, is a cloudless sky" for Elly (p. 316); this prime is a bearer of such intense emotion, it is unutterable. This is still RH gestalt, an indivisible whole, charged with affect, but in condensed symbolic form.

Music induced powerful visual images in Elly:

> ...[N]o clouds at all, the sky is the radiant image of a pleasure so intense that to bring it down to a really bearable level would, she showed us, require 4 closed doors between her and the phonograph. It is she explains 'too good' but she can bear it for a while. The music changes; the rapture abates by one degree: one cloud, three doors... (Park & Youderian, 1974, p. 315).[18]

It should be noted that Elly is influenced by VSSs (*what* system, BA37 of the RH) along with visual-spatial gestalt (*where* system, BA39,40 of the RH). Interesting to note here is that Elly also has an extraordinary drawing ability, and, as opposed to Nadia, exact geometrical forms and color played a significant role in her talent.[19]

Here is a striking example of how numbers (not objects) are direct markers of emotion in RH gestalt for children with autism:

> Anthony, 5 year old autistic boy, was preoccupied with the seemingly arbitrary number '55'. The puzzle of his frequently expressed fondness for '55' was solved when, on one occasion, he spoke of his two grandmothers. It was known that one of them had shown little interest in him, while the other had reared him with much patience and attention. Anthony said: 'one is 64 and one is 55. I like best 55' (Kanner, 1946, p. 243).

6.8 Exceptional Calculating Ability, Autism, and the Right Hemisphere

Outstanding mental calculating ability can occur in people with very low, with average or with very high IQ. Among phenomenal calculators with very *low* IQ are the famous autistic twins whom Oliver Sacks called "calculators who cannot calculate" (Sacks, 1987, p. 197) who lacked even the most rudimentary powers of arithmetic. There is also a case described by Hermelin and O'Connor (1990) of a young autistic man who was a phenomenal calculator despite lacking all language abilities, not even able to comprehend or communicate through signs. Those with very high IQs and extraordinary calculatory abilities are also great *mathematicians*, among them are Gauss and Aitken. We have already explored professor Aitken's introspection on his mathematical abilities. Another phenomenal calculator, Wim Klein, gave a noteworthy self-reflection as well.

> 'Numbers are friends to me, more or less. It doesn't mean the same for, it does it–3,844? For you it's just a three and an eight and a four and a four. But I say, "Hi! 62 squared."' (Sacks, 1987, p. 208).

Whatever the variability of LH participation (verbalizing or otherwise expressing the results of calculations in a conventional way), it seems that the common denominator

---

[18] Interestingly, the visual-spatial symbols of doors are important for Elly and also for Temple Grandin as we will see in Chap. 8.

[19] BA40 of the parietal lobe and BA19 of the occipital are areas mostly specialized for shape and color, respectively.

for the phenomenal calculators of any IQ is involvement of RH immediate and direct experience.

This reality is captured in Sacks' ponderings about the twins:

> They are not calculators, and their numeracy is 'iconic'. They summon up, they dwell among, strange scenes of numbers; they wander freely in great landscapes of numbers; they create, dramaturgically, a whole world made of numbers (Sacks, 1987, p. 206).[20]

Preoccupation with numbers and constant practice go hand in hand with special talent for calculations. However, child prodigies can carry out complex calculations at speed equal to those of adult calculators. Thus, before familiarity or practice could have had much opportunity to produce the skill, it is apparently fully developed (Hermelin & O'Connor, 1990). The speed itself is so high that it speaks of immediacy, of perceiving numbers and their relations in a flash and simultaneously, rather than conducting LH sequential, logical calculations:

> A box of matches on their table fell, and discharged its contents on the floor: '111', they both cried simultaneously; and then, a murmur, John said '37'. Michael repeated this, John said it a third time and stopped. I counted the matches—it took some time—and there were 111. 'How could you count the matches so quickly?' I asked. 'We didn't count,' they said. 'We *saw* the 111.'… They saw their number, as a whole, in a flash. 'And why did you murmur '37,' and repeat it three times?' I asked the twins. They said in unison, '37, 37, 37, 111' …. 'How did you work it out?' I said rather hotly. They indicated, as best they could, in poor, insufficient terms—but perhaps there are no words to correspond to such things—that they did not 'work it out', but just 'saw' it in a flash. John made a gesture with two outstretched fingers and his thumb, which seemed to suggest that they spontaneously *trisected* the number, or that it 'came apart' of its own accord, into these three equal parts, by a sort of spontaneous, numerical 'fission'. They seemed surprised at my surprise—as if I were somehow blind; and John's gesture conveyed an extraordinary sense of immediate, *felt* reality. Is it possible, I said to myself, that they can somehow 'see' the properties, not in a conceptual, abstract way, but as *qualities*, felt, sensuous, in some immediate, concrete way? And not simply isolated qualities—like '111-ness'—but qualities of relationship? (Sacks, 1987, pp. 199–200) (see footnote 20).

Is not the twins' reaction to the number 111 evidence that they are experiencing a RH gestalt, analogous to the gestalt behind the aborigines' collective representations described above? Recall these collective representations were a continuous totality of unique qualities where the whole is directly and immediately perceived as the relations of these qualities. Furthermore, does not the twins "seeing" one hundred and eleven matches strike a chord similar to the aborigine's "counting" horses by how much space a herd of horses occupies, where space is not divisible from content—each individual horse in its unique position in space? The twins instantaneously "see" the whole space—totality (the 111-ness), and, simultaneously, they "see" that totality as a relation of unique qualities within it (37×3).

From the point of view of abstract thinking and mathematical knowledge, the twins were factorizing, dividing a compound number into three equal parts, receiving primes—37:

> [T]hey had then gone on to 'factor' the number 111—without having any method, without even 'knowing' (in the ordinary way) what factors meant. Had I not already observed that they were incapable of the simplest calculations, and didn't 'understand' (or seem to understand) what multiplication or division was? Yet now, spontaneously, they had divided a compound number into three equal parts (Sacks, 1987, p. 200) (see footnote 20).

The twins evidence no LH concept of number,[21] so what meaning do these numbers hold for them? One may suppose, for the twins numbers are not units in a series, nor are they a homogenous sequence (LH representations). Instead, numbers are more like RH spatial gestalt. In such gestalt, an *image* is not divisible from the *space* it occupies. Based on these observations, we can make the following speculations. Ideas of quantity in the RH are repetitious (and as such should be very attractive for the autist) elements of space. Recall the gesture one of the twins made:

> John made a gesture with two outstretched fingers and his thumb, which seems to suggest that they spontaneously *trisected* the number, or that it 'came apart' of its own accord, into these three equal parts, by a sort of spontaneous, numerical 'fission' (Sacks, 1987, p. 199).

So what did the twins do? They trisected space, but was it the space of a number, the number as space, or some kind of spatial image? Strikingly, they immediately saw the internal properties of one hundred and eleven—its primes.

Nearly all autists who are interested in numbers are particularly preoccupied with primes.

Primes are numbers that cannot be divided by any whole number except themselves and one. Again, in the usual mathematical sense, the twins did factorizing, the process by which a given number is divided into its composites until primes are reached and it can be divided no further.[1]

6.8.1 *Autistic Calculators: How Do They Do It?*

As supposed before, autists do not utilize the LH concept of number as an abstract notion of quantity, separate from the objects counted. Thus, for the autist, it is easier to add "4 apples to 8 apples" than to solve the problem "8+4=?" The twins were even more concrete: "They can subtract apples but not dollars" (Horwitz, Kestenbaum, Person, & Jarvik, 1965, p. 1078). "They cannot do simple addition or subtraction with any accuracy, and cannot even comprehend what multiplication or division means" (Sacks, 1987, p. 197). At the same time, the twins play easily with primes:

> John would say a number—a six-figure number. Michael would catch the number, nod, smile and seem to savour it. Then he, in turn, would say another six-figure number, and now it was John who received and appreciated it richly. They looked, at first, like two connoisseurs

---

[21] However, they do name numbers, presumably a function of the left hemisphere.

> wine-tasting, sharing rare tastes, rare appreciations....[John] brought out a nine-figure number; and...his twin, Michael, responded with a similar one....John, after a prodigious internal contemplation, brought out a twelve-figure number. I had no way of checking this...because my own book—which, as far as I know, was unique of its kind—did not go beyond ten-figure primes. But Michael was up to it, though it took him five minutes—and an hour later the twins were swapping twenty-figure primes, at least I assume this was so, for I had no way of checking it (Sacks, 1987, pp. 201, 203) (see footnote 20).

One can see the twins indeed operate with numbers in isolation—they do not need "apples or dollars"—and so in fact they are not concrete at all. However, they are not abstract either, in the usual meaning of LH logical processes. The twins must be utilizing some kind of "RH abstraction."

Here we are entering uncharted territory, something unimaginable—such psychic processes have not been described by modern psychology. The savant "stands as a landmark of our own ignorance and the phenomenon of Idiot-Savant exists as a challenge to our capabilities"—says Arthur Holstein (from Horwitz et al., 1965, p. 1078), a discussant on the case of the twins. Holstein's statement stands true to this today. I also agree with Darold Treffert's (1988) assertion that no model of brain function will be complete until it can account for this rare but spectacular condition.

Although we cannot fully understand savants' capabilities, using the framework of RH cognitive mechanism we can outline how savants may "see" the numbers they are so adept at playing with. Recall that the *where* system, with its major players BA39,40,[22] is the "geometrical" part of the brain; its spatial vision requires efficient analysis of shape properties and spatial relations (Ullman, 1995). The right hemisphere is "preoccupied" with holistic forms, and it "obsessively" identifies objects with similar holistic forms (shape). Indeed, simultaneously recognizing holistic forms (pattern recognition) from different visual scene-situations may be the RH's way of perceiving the spatial organization of the physical world. For the autist, different quantities (repetitious elements of space) may be directly "seen" as different spatial configurations, i.e., holistic "shapes." This might explain why autists like the prime numbers so much, for perhaps primes represent unique qualities, "shapes," with all other numbers being "impure" as they are different combinations of primes.

Many autistic children express a great interest in and are extraordinarily deft at mentally manipulating shapes and geometrical figures. Elly, the autistic girl referred to in earlier chapters, is a case in point:

> At age 5, when Elly was using only a handful of words, she incorporated into her tiny vocabulary the terms 'square,' 'triangle,' and 'hexagon'....Elly knew 'pentagon,' 'hexagon,' 'heptagon,' and 'octagon' well before she knew 'sister' or 'cousin' or could produce a simple sentence (Bogyo & Ellis, 1988, p. 269).

Consider also Donald, a 6-year-old autistic boy who "when asked to subtract 4 from 10, answered: "I'll draw a hexagon" (example from Kanner, 1943, p. 222). There is also the phenomenal calculator, described by Hermelin and O'Connor (1990),

---

[22] BA39 appears to be specialized in spatial relations between objects, while BA40 is concerned predominantly with *shapes* (Glezerman, 1986).

who evidenced unexpectedly high skills with shapes. This man was given a two-part test in which non-representational shapes[23] were presented, among which "the odd one out" had to be selected; the same had to be done with representational objects. His responses to the first part of the test, the shape problems, were correct 73% of the time whereas his responses in the second part of the test involving pictures of objects were right only 23% of the time. This man's non-verbal IQ was 67 (his verbal IQ was impossible to determine, because he was not able to speak or even to communicate through signs). However, on the non-verbal Raven's Matrices test, his performance was equivalent to an IQ of 128. Raven's Matrices consists of a series of visual patterns from which a piece is missing. For each item, the subject is given from six to eight pictured inserts of which one contains the pattern's missing piece. They must choose among several options the piece that completes the whole. Simple problems in this test consist of homogeneous patterns from which a piece has been "cut out"; more difficult tasks contain simultaneous variations in several different dimensions, such that shape, size, and orientation must be attended to. This test is designed to measure non-verbal analogical reasoning. However, the Raven test's tasks can also be solved as a typical RH spatial gestalt: a totality of unique qualities where the whole is directly and immediately perceived as the relations of these qualities. This is, I think, exactly how the autists "see" them. Autists' exceptional speed in performing simply does not allow for sequential logical operations to "extract" the ruling principles. They just instantaneously "see" the whole and its "parts." Elly's performance on the Raven test is illustrative:

> She needed no instruction and quickly began to turn the pages, pausing only to glance briefly at the patterns and to point immediately at the missing piece....We ordered the Advanced Progressive Matrices, a set constructed specifically to assess the skills of gifted individuals. Again we watched with amazement as Elly turned the pages more quickly than we could consider the choices (Bogyo & Ellis, 1988, p. 271).

Frequently, autistic individuals are not able to explain their "strategy."

> [Elly was] able to name the relevant dimensions and features, to articulate the rules governing their progressive alterations.... It was clear that to Elly these rules and regularities were obvious, self-evident in the designs themselves. She seemed puzzled that the solutions needed any explanations at all—as if we had asked her what shaped peg would best fit in a round hole (Bogyo & Ellis, 1988, p. 271).[24]

Finally, here is one more curious observation about "shapes" and autism. Hurlburt, Happe, and Frith (1994) designed experiments to examine autistic individuals' reports about their inner experience. In their study, three Asperger syndrome adults reported their inner experiences to be predominantly visual. One of them attracted my attention, in particular, with his peculiar responses. Nelson, a 24-year-old man with average IQ, described an image as being "the shape of my thoughts."

[23] Non-representational is defined as that for which a name does not exist.

[24] There is variability among autists in degree of LH involvement, but the core "player," I believe, is the RH, and this constant belongs to the pattern of autism per se.

"Nelson was looking at the brick wall of a house near his home…[h]e was visually focused on three or four of the bricks…[h]is thoughts had 'taken the shape of' the bricks, following the outline of the individual bricks" (p. 390). Is not this "outline" a holistic form, what I have argued to be the unit of RH operations?

> Nelson described the bricks as 'looking out at him' to capture the fact the bricks had taken over the focus of his attention for the moment, seemingly of their own accord; that is, he was not aware of directing his own attention towards the bricks[25]…. Nelson was feeling a 'little bit sentimental' about the bricks; he had seen them many times before, had grown up with them. He could not describe how this sentimental feeling presented itself… (Hurlburt et al., 1994, p. 390).

Doesn't Nelson describe here RH identification of holistic form?

6.8.2 Autistic Calculators: What Are They Doing?

In discussing Sacks' observations of the savant twins above, I did not address a very important aspect that is ignored by current researchers.[26] What I want to draw attention to now is the *experiential* character of the twins' activity, which Sacks keenly picks up on:

> [T]hey were seated in a corner together, with a mysterious, secret smile on their faces, a smile I had never seen before, enjoying the strange pleasure and peace they now seemed to have. I crept up quietly, so as not to disturb them. They seemed to be locked in a singular, purely numerical, converse. John would say a number—a six-figure number. Michael would catch the number, nod, smile and seem to savour it. Then he, in turn, would say another six-figure number, and now it was John who received, and appreciated it richly. They looked, at first, like two connoisseurs wine-tasting, sharing rare tastes, rare appreciations. I sat still, unseen by them, mesmerized, bewildered. What were they doing? What on earth was going on? I could make nothing of it. It was perhaps a sort of game, but it had a gravity and an intensity, a sort of serene and meditative and almost holy intensity, which I had never seen in any ordinary game before, and which I certainly had never seen before in the usually agitated and distracted twins. I contended myself with noting down the numbers they uttered—the numbers that manifestly gave them such delight, and which they 'contemplated', savoured, shared in communion (Sacks, 1987, pp. 201–202) (see footnote 20).

Sacks' question ("What were they doing?") can be rephrased: "What were the twins *experiencing*?" The answer to this latter question, I believe, is that they were experiencing a sense of ***belonging***. They were in that special state of consciousness where the RH gestalt of identified *I-space* and *non-I-space* is exposed and on the foreground of consciousness. Theirs was a profound mystical experience "of belonging." They were identified with each other and with numbers, numbers for them being "representatives" of spatial situations and "bearers" of "mystical" power. It is the subjective experience of "thalamic" rhythm, the rhythm of their *I-space* that was

[25] Is this meditation? Recall Temple Grandin's experience while looking at grains of sand.

[26] Kanner and Asperger, however, noted it.

exposed and projected outside, uniting the twins and numbers into a continuous whole and creating a mystical sense of belonging.

A comparison to the twins' experience with numbers inevitably came to my mind: the sacred objects from the collective representations of aborigine cultures! From the point of view of their brain mechanisms, these sacred objects are externalized holistic forms. According to Levy-Bruhl, sacred objects are pieces of stone or wood of the oblong form decorated with some sort of ornament. Each totem had its own sacred objects. They were perceived as bearers of mystical power, for they simultaneously represented the spirit of each individual, the spirit of the ancestors, the substratum of totemic existence and the vessels of vital power. As a result, these objects were treated with reverence and awe, and individuals of the totemic group experienced a special connection with these objects, with some magic force passing from the objects to them and from them back to the objects. Here we thus see, externalized and fixed in cultural rules, the RH gestalt, a continuous and indivisible whole.

The twins exchanged prime numbers as aborigines exchanged the sacred objects. Sacks catches this reality in his insightful observations.

> They do not approach numbers lightly, as most calculators do. They are not interested in, have no capacity for, cannot comprehend, calculations. They are, rather, serene contemplators of number—and approach numbers with a sense of reverence and awe. Numbers for them are holy, fraught with significance (Sacks, 1987, p. 208) (see footnote 20).

6.9 Autistic Calendar Calculators and the Right Hemisphere: "Soul in Numbers" and "Memory Without Consciousness"

After traveling along the path of the *where* system, looking at its inter-hemispheric differences, excursing into its history, analyzing how it works in normal people, how it struggles in brain-damaged patients, and how it unusually expresses itself in autism, we can now turn to the so-called calendar calculator autistic savants. I will argue that this mysterious and obscure ability is also specifically connected to the right inferior-parietal region (BA39,40).

Calendar calculating is the ability to give the correct day of the week for a given date. The twins discussed at length above were also phenomenal calendar calculators:

> The twins say 'Give us a date—any time in the last or next forty thousand years.' You give them a date, and, almost instantly, they tell you what day of the week it will be. 'Another date!' they cry, and the performance is repeated. They will also tell you the date of Easter during the same period of 80,000 years (Sacks, 1987, p. 197) (see footnote 20).

Not surprisingly, the twins' extraordinary abilities in calendar calculation have been examined and documented by several other interested investigators:

> Their calendar calculations go far beyond the range of any hitherto reported....G. can identify instantaneously the 15th of February 2002 as a Friday, August 28th, 1591 as a Wednesday.... [T]hey can calculate dates for which no formal calendar exists such as the years in which a certain day in a certain month falls on a Sunday....[F]or example, when asked in what years April 21st will fall on a Sunday, each will answer correctly 1968, 1957, 1963, 1946, etc. (Horowitz et al., 1965, p. 1075).

As did Sacks, these authors emphasize that the twins cannot add, subtract, multiply, or divide single-digit numbers. "[T]he product of 3×6 might be given as 8. Although they cannot add up to 30, when given your birth date, they can accurately tell you it is 30 weeks until your next birthday or 13 weeks since you last had a birthday" (p. 1075).

Dave, a 14-year-old autistic boy with an IQ of 50, is another calendar calculator who was described by Michael Howe and Julia Smith (1988). When given the task "What is 1,981 minus 1,963?" Dave replied "9,000" and then "3." It seems Dave's answers are random or, rather, echolalic, but it is clear he could not use an inner structure to do a simple calculation.[27] However, the situation radically changed when calculation tasks were presented to him in the context of a calendar. He then solved the following problems with speed and ease: "If I was born in /say/1908 how old would I be in /say/1973?" or "If I was born in 1841 how old would I be in 2304?" (p. 378). He must have had *some* kind of internal structure to perform these tasks. One must then ask what this structure could be.

Musical, artistic, and mechanical savants' phenomenal abilities can all be compared with the corresponding talent in the normal population. Calendar calculation "talent" on the other hand is idiosyncratic and belongs mostly to autistic savants. It is difficult to imagine a person with normal intelligence engaging in it.

The calendar is based on a small number of rules. There are a fixed number of days for 11 months, but to compensate for the fact that each year contains just under a quarter of a day more than 365 there is an extra day in one February every 4 years, except years at the beginning of three out of four centuries. Furthermore, the patterns of the calendar days within a year repeat themselves in 28-year cycles. Regularities in the Gregorian calendar include the existence of 14 different "year templates," which describe the possible configurations an annual calendar may have: the existence of "paired months" within calendars where certain months have an identical day—date structure; the advance of 1 weekday for a given date on successive (non-leap) years; and the periodic occurrence of leap years. Algorithms based on these rules permit the successful calculation of simple day—date problems (Howe & Smith, 1988; Miller, 1999; Spitz, 1994). If one need specify the correct day of the week for a given calendar date, one can find the corresponding tables or algorithms devised on the basis of calendar rules—they are published in various encyclopedias, handbooks, etc.

There is no evidence that autistic calendar calculators follow any of these commonly known methods. Most of them do not read well enough to follow written instructions and are unable to perform the necessary calculations. Furthermore, their responses are astonishingly quick, leaving no time for them to have made use of known rules. Instead, they are *self-taught*, and most do not have any idea how they do it.

---

[27] As discussed above, normatively the inner structure for calculation is a specific spatial schema, combining quantity and order; this schema is connected with the left parietal-occipital region, BA39,40.

Calendar is based on periodicity of natural phenomena, but it also a product of human culture, and, as such, exposure to calendar and learning are necessary. It is no surprise then that calendar calculation talent usually manifests later than other savant talents. Autistic calendar-calculators also received help and encouragement from other people. For example, after the twins' parents were largely unsuccessful in teaching them numerical and reading abilities, they noted that one of the boys, when he was 6 years old, would spend hours pouring over an almanac that contained a perpetual calendar. What is surprising was that he could decipher the almanac. The father later gave him a silver perpetual calendar, which he played with for hours. He made no errors from the beginning. A paternal aunt who was a legal secretary would even call to check dates of documents with him (Horwitz et al., 1965).

The opportunity to study a perpetual calendar is clearly important, but what exactly do calendar savants learn from their "studies" and how do they learn it? Horwitz and colleagues point out that the twins demonstrate their calendar ability in areas where memory, learning, and recall are not available. They operated in a range of calendar calculation far prior to and beyond our usual 200 or 400-year perpetual calendar. The longest known perpetual calendar extends to about the year 2400 and yet one of the twins could reach beyond the year 7000. They also calculated dates for which no formal calendar exists, such as the years in which a certain day in a certain month falls on a Sunday. Did they invent a general formula? This would be impossible because of their low intellectual level and inability to abstract. Furthermore, implementation of any formula would take time, and their responses were instantaneous.

6.9.1 *The Right Inferior Temporal Region (BA37), Autistic People's Memory, and Autistic Savant's Calendar Calculation*

How do they do it? Let's return to Sacks' observations of the twins performing their calendar calculations:

> [O]ne may observe, though this is not usually mentioned in reports, that their eyes move and fix in a peculiar way as they do this—as if they were unrolling, or scrutinizing, an inner landscape, a mental calendar. They have the look of 'seeing', of intense visualization... (Sacks, 1987, p. 197) (see footnote 20).

In most cases, the impressive speed of their responses could not be timed: they occurred "in a flash." Visualization, instantaneous response, and inability to explain how they do it all point to calendar calculation being a product of RH processing.

For explanatory purposes, I artificially separated BA37 of the RH temporal-occipital region and BA39,40 of the RH parietal-occipital region, but in life they are intimately integrated. Indeed, integrated functioning of different brain regions is much more potent in the RH as opposed to the LH.

In BA37, the VSS, saturated with affect, is perceived as a moment of experience of the outside world. The VSS is also stored just as it was perceived in the right

BA37. It is a global, indivisible entity that includes visual images, movement-action, emotion, etc. The moment of environmental time when the situation was perceived is internalized and represented as an equivalent of the situation (its temporal marker) in the right frontal lobe. "Marking" the constantly changing impressions from the outside world establishes an order in the continuity of one's memory images. In other words, temporal marks allow situations to be stored in their natural succession, going from early to later layers of memory. As such, these ordered situations represent one's past experience or autobiography. The temporal sequence of events per se, represented in the frontal lobe, makes one's life timeline. Thus, what is stored in the right BA37 is not just memory! What is stored is the experiential, unique-for-each-individual part of one's self.

The brain is organized hierarchically, and only the highest level presents what we are "conscious of," with all lower levels being assimilated as a background or "behind the scenes." The "self" I am speaking of here in regards to the right BA37 is a fundamental and deep (most ancient) part of self. However, there is no access to this level in the norm, and by no means do we "remember" every situation we experienced. This layer of self belongs to the sensory-motor level and is "covered" by a higher level in the RH, the symbolic level, where symbols are created by reducing a series of situations into compressed form. This latter, symbolic part of RH self will not be on the foreground of consciousness either. Instead, it will be "worked out" and re-interpreted by the LH, which is responsible for the one "conscious" self. Consider again what Sacks writes about the twins, in light of this idea about RH self of the sensory-motor level:

> They can tell one the weather, and the events, of any day in their lives—any day from about their fourth year old.... Give them a date, and their eyes roll for a moment, and then fixate, and in a flat, monotonous voice they tell you of the weather, the bare political events they would have heard of, and the events of their own lives—this last often including the painful or poignant anguish of childhood, the contempt, the jeers, the mortifications they endured, but all delivered in an even and unvarying tone, without the least hint of any personal inflection or emotion....[T]here is no personal reference, no personal relation, no living center whatever....[W]hat needs to be stressed...is the magnitude of the twins' memory, its apparently limitless extent, and with this the way in which memories are retrieved. If you ask them how they can hold so much in their minds—a three-hundred-figure digit, or the trillion events of four decades—they say, very simply, 'We see it.' And 'seeing'—'visualising'—of extraordinary intensity, limitless range, and perfect fidelity, seems to be the key to this....[T]here is no doubt, in my mind at least, that there is available to the twins a prodigious panorama, a sort of landscape or physiognomy, of all they have ever heard, or seen, or thought, or done, and in the blink of an eye, externally obvious as a brief rolling and fixation of the eyes, they are able (with the 'mind's eye') to retrieve and 'see' nearly anything that lies in this vast landscape (Sacks, 1987, pp. 198, 199) (see footnote 20).

This is not just "memory" for the twins, it is the exposed self of the RH sensory-motor level. How can this "exposure" happen? It is as though the sensory-motor level is split from higher levels (symbolic) in the RH and also freed from the dominating interpretations of the LH. Only then, it would seem, can the twins' self of the sensory-motor level, a qualitatively different mental awareness, emerge on the surface.

But what about Sacks' observation that the twins deliver their "stories" "in an even and unvarying tone, without the least hint of any personal inflection or emotion... there is no personal reference, no personal relation, no living center whatever."

Such a lack of "personal reference" would seem like a contradiction to what was just stated about their having an exposed *self* of the sensory-motor level. However, "personal reference" and "personal relation" refer to the type of self and its emotion that is connected with the LH[28]; its emotion is what is usually called emotion, as such (feeling happy, sad, angry, etc.). Such emotion is a subjective experience that reflects the attitude of an *individual* to valuable objects and to oneself. It is a kind of "free energy," when we experience feelings of happiness or sadness and are still able to distinguish them from the objects toward which they are directed. It is not so at the deeper layers of experience.

To better understand these deep, hidden layers of our psyche, we can again turn to the collective representations of primitive societies. Consider the following contrast: whereas a modern man might say, "I experience fear of a dead body," aborigine people would say: "Taboo (that is, fear) is sitting in the dead body" (Kretschmer, 1927). For the RH's sensory-motor level, emotion is bound up with the VSS, projected onto visual images in that situation, and indivisible from the situation itself. There is no one individual self at this level, there is no division of subject and object. Self at this level is only *subjectively felt* situations.

6.9.2 The Right Inferior Parietal Region (BA39 and 40), Archetypal Calendar, and Autistic Savant's Calendar Calculation

Now we will focus on the right parietal-occipital region's (BA39, 40) contribution to autistic savant calendar calculation. In this brain region, visual representations will be spatial-temporal arrangements of situations, perceived as gestalts and stored exactly as they were perceived. As mentioned before, temporal concepts (as well as quantitative concepts) originate from spatial representations. This can be illustrated by an example from the history of language. Levy-Bruhl (1930) indicates that aborigine languages in America and Australia described with exactness of detail all relations concerning location in space, a distance, singularity, repetitiveness, and *time* were indicated with words that initially designated *place.* There was no verb "*to be*"; the word used instead, was "*this*," but it had multiple meanings: to be here, to be in some place, to be in this or that moment. In general, everything related to time is expressed in these languages with the words that once designated spatial relations. Words that initially had meant distance in space, later came to be meant distance in time. Relative sequences such as *before*, *after*, *first*, *second* are the LH signs of temporal relations. In the RH, sequence of time is experienced through and not divisible from the sequence of the visual-spatial situations. In other words,

[28] We will discuss this idea in Chaps. 7–9, where we will talk about the very core of the autistic pattern.

sequence of time is embodied within the spatial images. But this is WHAT calendar is! Or at least some kind of archetypal image of calendar, and autistic calendar "calculation" teaches us that we, indeed, have this archetype.

The calendar represents recurrent patterned sequences based on the periodicity and repetitiousness of natural phenomena. A "pattern of sequences" is implicitly contained in the file of repetitious visual-spatial situations from the outside world. The visual-spatial situation, then, is the medium for autistic calendar "calculation."[29]

Howe and Smith's (1988) descriptions of Dave, the 14-year-old calendar calculator with an IQ of 50, are elucidative. They suggest Dave was able to call upon calendar information in some kind of visual-spatial image. He was asked to say which month in each year started on a Friday. The authors speculate this task would be easy for a person who retains some kind of representation of the structure of an entire month or is able to generate an image-based representation of the arrangements of days forming a month, where the configurations for adjoining months are not independent of one another. In fact, Dave quickly ran through the correct months in every year from 1970 to 1990, and he indicated that he could have continued further if asked to do so. Next, he was shown a list (which was also read out to him) of 7 months falling in particular years, as follows: January 1971, September 1972, June 1973, etc., 6 of the 7 months began on a Friday. However, Dave was not told this, but was simply asked, "Which one is the odd one out?" With little hesitation Dave gave the correct answer (February 1971). When asked why he chose that month he said that he did so because it began on a Monday. Note that the question itself did not hint at the kind of difference the examiners sought. The authors indicate that an image-structure of the entire month within which each date occurred is needed to solve this task.

What Dave did with calendar months might be similar to autistic children's operations with shapes on the Raven test: it is right hemispheric playing with holistic forms! When autistic children are finding a missing piece to complete the whole of visual patterns presented in the Raven test, they *instantaneously* "see" the pattern (design) as the *totality* of as well as *relations* of unique qualities. Autistic calendar calculators also "see" patterns in repeated sequences of month and year configurations. We do not know exactly what these configurations or "archetypal landscapes" entail. Dave, himself, was preoccupied with drawing an actual calendar. His drawing had idiosyncratic features, which turned out to be a literal reproduction of the calendar on the wall in his kitchen:

> He drew vague figures at the beginning of some of his months: when asked about them, he reported (correctly) that they represent the moon. When further asked why he drew them he replied: 'they always go there' and muttered something about a kitchen calendar on the wall.... [W]hen asked the day of the week for 6 April 1939, he said, 'Thursday, that's black.' Asked why he said this he replied: 'Thursdays are always black'. When further

[29] "Calculation" is in quotes above because what autistic calculators are doing is *not* calculation per se.

pressed, he said: 'they are black on the kitchen calendar'. Again, asked if he was sure about another of his answers he muttered, as if thinking aloud, 'Yes, it's on the top line' (Howe & Smith, 1988, p. 379).

I do not think that calendar "archetypal landscapes" represent actual calendar configurations. First of all, we are talking here about an innate ability to perceive patterns of periodically repetitious experience, while the actual calendar is the product of culture and human convention. Secondly, RH cognitive mechanism is not abstract, but neither is it concrete. The essence of RH cognitive mechanism is unity of the single and the whole, where relations between images are thought of as the images themselves. The RH innate ability was the source of the calendar when it first originated in human history. However, once again, the end result of a RH product was dictated by LH interpretations of RH experience, with its concomitant descriptions of rules and algorithms and its expression of RH gestalt in spatial schemes and numbers. After the calendar was invented, it then became part of one's daily environment, and thus a part of RH experience. The image of the actual calendar might be stored in both hemispheres, but its meaning will be different according to interhemispheric differences in information processing.[30]

The actual calendar image might be included into RH visual-spatial gestalt (as in Dave's case, where the kitchen calendar was a part of his visual-spatial situation). An actual calendar image might also be included into a continuous series of RH gestalts, reflecting a spatial-temporal organization of the world.

We remember that autistic calendar savants are *self-taught*. It seems they have discovered the initial source—the RH "archetype"—for our calendar.[31] This is what all autistic calendar savants have in common. As to what degree they use their LH, it varies in each individual. LH usage correlates with IQ, and it can give autists some partial insight into what they are doing, but it does not correlate with the degree or power of their talent.

Thus far, we can say that the autistic calendar calculation talent is based on the intimate interaction between the right inferior temporal region, BA37 and the right inferior parietal region, BA39,40. Experience of dates for a specific year (BA37) parallels a specific year's spatial-temporal configuration (BA39,40). But how do savants give the correct day for dates they have not experienced: the past and the future? Something is missing in this explanation. That which has been left out of our discussion thus far, of no small importance, is the brain area in which the *where* system culminates: the dorsal lateral prefrontal cortex (DLPC).

---

[30] Such interhemispheric differences were visited before in this chapter in regard to the perception of word sound and numbers.

[31] Recall Temple Grandin's bay window experience discussed earlier in this chapter. Our neurophenomenological analysis showed that her brain stayed on the beaten path of the human brain's historical development when her right inferior parietal area was used to discover grammatical constructions underlying the linguistic meaning of *relation*. In calendar calculators this same region is used! Indeed, the autist's brain is in some way a "walking history" of the human brain.

6.9.3 *Contribution of the Right DLPC to Autistic Savant Calendar Calculation*

So far the posterior part of the cerebral cortex has been in focus. The network *RH TPO-RH DLPC* reflects the posterior–anterior axis in the brain three-dimensional differentiation. As discussed in Chapter 1, the posterior brain perceives, processes, and stores information. The anterior brain (the frontal lobe), on the other hand, operates with the information presented to it by the posterior brain. It unfolds the posterior brain's images in time, creating representation of *action.* In the LH, representation of *action* includes temporal order, programming of movements, logical thinking, behavior. Our topic now involves the frontal lobe in the *right* hemisphere, so what is the representation of action there? An opinion exists that the RH gives a subjective view of the world, incorporating time and space and providing a framework for our *weltbild* (world perspective) or a spatial-temporal background for one's world image [Lange (1936); Cloning et al. (1968) cited from Cutting (1990)]. The right DLPC operations with spatial situations of BA39 and 40 might then be a continuous flow of spatial situations, where time is perceived through movement, i.e., "world movement." This makes sense given the holistic, implicit nature of RH cognitive mechanism. In autistic people, I have argued, RH representations are on the foreground of consciousness. Autists may actually in some way be "seeing" this space unfolding in time.

6.9.4 *Contribution of the Striatum to Autistic Savant Calendar Calculation*

Then, what do autistic calendar savants discover?

Let us return to the calendar rules. The rules are not obtained by sequential logical operations, nor are they mathematical formulas or any type of abstraction. They are based on empirical observation of repetitive sequences, e.g., the patterns of the calendar days within a year that repeat themselves in a 28-year cycle; the existence of "paired months" with identical day-date structures; the existence of 14 *year templates,* which describe the possible configurations an annual calendar may have, etc.

It seems that what autistic calendar savants discover is *stereotype*! They discern patterns that are repeated again and again and never change—albeit very complicated patterns, they are always constant.

Now we can grasp why autistic savants are so attracted to the calendar. It is the repetitiveness of its patterns-configurations (holistic forms). It is steadiness, "sameness," in the streaming away of the autists' world! The intrinsic nature of a calendar meets the autist's subjectively experienced need for stereotypical behavior. Recall that stereotypy is the excessive production of one type of behavior that necessarily results in repetition (discussed at length in Chap. 5). Also remember that "playing" the patterned sequences over and over again is connected with the *striatum* where patterned sequences (automatisms) are stored. It is the *striatum* that is responsible for an irresistible desire "to repeat," a kind of internal motivational state similar to addiction. Thus *striatum*, the subcortical center of the sensory-motor level, connected with the DLPC, must be included in the network that gives rise to the calendar savant's performance.

We are now led to the conclusion that calendar savants' astonishing feats are not goal-directed. Indeed, savants are engaging in automatic, repetitive activity. Some authors tend to explain "calendar talent" by a savant's high motivation, intense concentration, and extensive practice. All these terms are relevant only to the case of goal-directed behavior, implemented by higher cortical levels as the leading ones (in Bernstein terms). The leading level in the autistic savant's functioning, however, is the sensory-motor level, including its subcortical sublevel (striatum). As such, the savant's mental awareness and voluntary control are not complete. Instead of the term "motivation," I would use an "irresistible desire to repeat," instead of "extensive practice," "preoccupation with sameness." As to "intense concentration," it is rather a special state of consciousness (some kind of meditative state) in which they are not fully aware of the surrounding world.

From the responses of a few savants who are able in some way to say what they are doing, one gets a sense that there is no voluntary effort: "We just see it."

By its nature, stereotypical behavior is repetitive, reinforcing "in itself" and therefore "of itself."[32]

How then can we imagine autistic savant calendar calculation of past and future dates that do not exist in the actual calendar? Figuratively speaking, *the tape is automatically unfolding, and missing links (from past or future) jump into their place.*

Again, instantaneous recognition of a missing calendar piece seems analogous to how autistic children perform on the Raven test: seeing a continuous whole in the relations of its parts. For autistic children, performing on Raven test, rules and regulations governing progressive alterations in Raven's Matrices are self-evident in the designs themselves. Autistic calendar savants also do not extract calendar algorithm and then apply it to their task. They do not *do* anything. They *perceive* algorithm in the flowing spatial situations themselves.

6.9.5 Contribution of the Inner Rhythm to Autistic Savant Calendar Calculation

Here, we will explore further the question: why are autistic savants so attracted to calendar? And also, why do ability to phenomenal calculation, attraction to the prime numbers and calendar calculation often go together?

The tape is unfolding in the time of outside space, in moments of environmental time, while the spatial situations are saturated with the inner propriomotor rhythm, as a result of the integration of *I-space* and *non-I-space*.

[32] See Chap. 5.

To have a better understanding of this inner rhythm, we return to the LH and movement's rhythmical aspect provided by the thalamo-pallidum level.[2] The thalamic level receives kinesthetic information from muscles and joints of the extremities, which are multilinked pendulums with various oscillation frequencies. Due to the organization of proprioceptive sense being according to one's body spatial coordinates, specific to the thalamic level, oscillation patterns of multiple "pendulums" are united into a single *resultant* rhythm. Bernstein called it *propriomotor rhythm.* This is the rhythm of one's body kinesthetic space.

As a background level, Bernstein indicated, the thalamic level brings the inner propriomotor rhythm into all aperiodic movements of the higher levels. An example of this would be periodic alternation of movement of the extremities while stepping. As such, the leading level for walking is the sensory-motor level. Nevertheless, if the thalamic level in the brain is damaged, a person will not be able to walk. Interestingly, such a patient would not be able to take a step on an even surface, but might go up and down the stairs briskly. Here, visual perception of the staircase steps makes external space *periodic*, thereby compensating for the lost inner rhythm.

If the LH is responsible for movements and actions in time, the RH is responsible for subjective experience, including subjective experience of one's inner rhythm. It can be suggested that as the propriomotor rhythm brings harmony (coordination) in all movements in the external world, so the subjective experience of one's inner rhythm brings harmony to the spatial organization of the external world.

In my purported archetypal calendar, repetitive patterns of outside space would be integrated with the rhythm of internal space. Visual-spatial representations of calendar (whatever they may be) would be united into a continuous whole by one's internal rhythm, which is perceived as some kind of vital (perhaps mystical) power... the calendar is alive for autistic savants! In doing calendar calculation, they communicate, or, more precisely, they commune (as Sacks astutely indicated) and experience mystical belonging. Here we can see similarities with the autist's perception of numbers. I argued above that autistic savants perceive prime numbers as spatial (holistic) forms assigned with some kind of vital (mystical) power. Interestingly, Sacks (1987) compares the twins' "sense" in their numbers to a musician's sense of harmony. Sacks further suggests that the twins perceive the harmony and order of the world in their own way—in numbers.

Now we can answer the question why autistic savants are so attracted to calendar. Calendar includes emotionally saturated experience (right BA37), it represents repetitive pattern (right BA39,40, right DLPC and Striatum), and it is spiritual for them (right LP of the Thalamus, right BA39,40). Thus, we have delineated the brain network of autistic savants' calendar calculation: *RH BA37–RH DLPC–RH Striatum* and *RH LP–RH BA39,40–RH DLPC–RH Striatum.* The corresponding parts of the left hemisphere participate as well, however, the LH contribution is not specific to this talent and varies significantly in individual cases.

Calendar calculation, this bizarre talent, illustrates the autism pattern in its extreme. It includes features of autism already discussed: stereotypical behavior and prevalence of the right hemisphere and the sensory-motor level. Not only autistic savants, but autistic children, in general, are often preoccupied with calendar numbers and dates. We can see this in examples from Kanner's material of Donald, an 8-year-old autistic boy:

> [He] was inexhaustible in bringing up variations: 'How many days in a week, years in a century, hours in a day, hours in half a day, weeks in a century, centuries in half a millennium,' etc., etc.… Sometimes he asked, 'How many hours in a minute, how many days in an hour?' (Kanner, 1943, p. 222)

What has already been shown about the *where* system in the autist's brain? It is hyper-functional in the right hemisphere. As mentioned in Chap. 2, some authors have promoted a so-called right-hemispheric theory of autism. They suggest that the RH in autism prevails secondary to damage to the left hemisphere. It is known that if the dominant LH is damaged, the RH becomes "disinhibited." The interesting thing about autism is that the left hemisphere is *not* damaged. I have made it a point to show by detailed analysis that the left hemisphere in autism is not functioning properly, but it is not dysfunctional either (at least in regard to what we know about brain functions and their disorders in brain-damaged patients).

Furthermore, the fact that autistic savant calendar calculation is stereotypical activity does not remove its mystery. I would maintain that while performing calendar calculation autistic savants' consciousness is qualitatively different: RH immediate experience and lower levels in the brain are exposed. It was discussed in previous chapters that such consciousness is characterized by a lower degree of awareness and voluntary control, but also by a different quality of mental awareness. When the lower level gets exposed, it is its *space* that one becomes aware of. During autistic savant calendar calculation, we observe two types of RH experience that appear as disjointed phenomena, each of the two presents its own "consciousness" or awareness of *space* it represents: *subjectively felt VSSs* and *non-I-spaces* integrated with *I-space.* Each of these phenomena harks back to the deep history of our self-consciousness. However, for the autist, there is no personal center.

Thus, autistic calendar calculation is not just "activity," but the acting out of autists' selves, their way of living and communicating, their worldview and the meaning of their existence.

Patients with focal, left-hemispheric damage, often suffer from aphasia. Methods of rehabilitation include the use of the intact right hemisphere to compensate for their deficits. But results are quite limited. It is impossible to achieve re-construction of the language history in the individual brain. The mystery of the autistic brain is that it re-organizes itself at very deep levels. I will try to approach this mystery in the following chapters.

6.10 Endnotes

[1]In the third-century BC the Greek astronomer and mathematician, Eratosthenes, devised an algorithm to determine which numbers were primes. The method is reliable but very cumbersome and time consuming. So far, there is no mathematical formula or rule that can predict which numbers are primes. Their identification

remains dependent on trial and error methods, although shortcuts that suggest which numbers may be primes do exist (Hermelin & O'Connor, 1990). One of the tables for factorizing any number was devised in the early nineteenth century by Colburn. Colburn was one of the most celebrated calculating child prodigies who could, by the age of six, rapidly find the factors of any number up to a million or more. However, he could not tell people how he achieved this feat, and relates how he was sometimes reduced to tears by the persistence of his questioners. Hermelin and O'Connor rightly conclude that Colburn's special ability to carry out complex cognitive operations preceded by several years his capacity to account for the strategy he employed and that conscious access to the rules was not necessary for Colburn's effective use of them. I do not think that subconsciousness, or, in my preferred term, RH "flow of consciousness," has rules. Rules are abstractions, and they require logical operations. RH "flow of consciousness" on the other hand is immediate direct experience. The LH "rationalizes" and "interprets" this RH experience, "extracting" the rules from it. The rules will be "conscious," for they are the LH equivalent of RH experience, but they will only touch upon the "tip of the iceberg," not the depths of RH experience. Colburn is representative of phenomenal calculators who have a high IQ and thus highly developed LH abilities. His phenomenal ability is based on RH holistic cognitive mechanism, but his capacity to develop rules for factorizing derives from his LH. Colburn's tables simplify the time-consuming process of factorizing, but they cannot enable (or teach) someone to become a phenomenal calculator.

[2]In classical neurology, coordination of movements was thought to be due to the motor subcortical system, the so-called extrapyramidal system, specifically a formation within the extrapyramidal system, globus pallidum. It was assumed that the globus pallidum has a vast repertoire of motor schemes and, by some unknown ways, chooses the needed one. Bernstein was the first to show that the coordinational process is instead based on sensory corrections. The globus pallidum is able to organize simultaneous contractions of multiple muscle groups in an orderly, harmonious manner, Bernstein indicated, because it receives, as no other motor center in the brain, the fullest and detailed information about positions and movements of one's own body. The thalamus is the main proprioceptive center of the brain and feeds the globus pallidus with knowledge of the body's position. Proprioceptive information comes mainly from the extremities—kinematic chains whose links are united by joints. Each link is a pendulum, with its own spectrum of oscillation frequency. Bernstein noted that it is "not difficult to coordinate thirty muscles at once, but it is difficult to coordinate three joints of one chain at once." Difficulty in controlling dissenting kinematic chains increases with enlarging degrees of freedom: the larger the number of joints the more reactive forces threatening to pull the kinematic chain apart. Prompt signaling about all dynamic events happening in body positions allows the thallamic-pallidum level to reign in reactive forces and turn kinematic chains into a controllable system. One's inner rhythm is a complicated pattern as it is the total result of the oscillation patterns of multiple "pendulums" and dependent on the physical parameters of the constituent pendulums. Its contingency on physical parameters makes inner rhythm *individual-specific*. It manifests in all kinds of motions and behaviors of the particular individual. Although inner

rhythmical patterns are individual-specific, some authors made an attempt to classify them. The linguist, Sievers, and the musicologist Becking [cited from Jakobson (1970)] distinguish three basic kinesthetic types ("*Generallkurven*" or "*Personalkurven*"). They claimed that if the performer and the author of a literary or musical work belong to the same kinesthetic type, execution of the art would achieve the highest effect. It is thus important to recognize that the inner rhythm reflects the uniqueness of one's own body space (internal space).

Chapter 7
How Autistic Persons Feel (Cerebral Organization of Limbic Emotion and Autism)

7.1 The Amygdala and Its Connectivity Pattern

The amygdala, so named for its resemblance to an almond, is a subcortical region that is located within the medial wall of the temporal lobe. The amygdala marks the junction of information from the autonomic nervous system and its control center, the hypothalamus, about internal milieu and vital needs of the organism, and information coming from the *what*-system about external objects. The amygdala determines then biological significance of external objects as a potential source of satisfaction for the organism's internal needs. Evaluating external objects from the point of view of internal need satisfaction emerges as a qualitatively new phenomenon—subjective experience, emotion, and feeling. How neural activity translates attachment of significance to the object into subjective experience of certain feelings remains a mystery, but numerous experimental studies and clinical observations have demonstrated that this enigma is indeed linked to the amygdala.

Studies have shown that electrical stimulation of the amygdala can evoke any one of the full array of human affective states, with the most common response to stimulation being fear (ranging from mild anxiety to outright terror). Emotional distress such as sadness, guilt, loneliness, embarrassment, as well as anger, anticipatory excitement, and mirth can also result from amygdala stimulation. The sudden feelings of overwhelming fear, embarrassment, anger, and depression—without any connection to environmental stimuli—can occur just prior to seizures in patients with epilepsy (so-called emotional aura) or these feelings might represent a seizure as such. The amygdala is usually implicated in these seizures' location.

The amygdala is a complex structure composed of numerous nuclei, each of them having distinct connections and participating in discrete functions.[1] In addition to connections with extrinsic areas, there is an elaborate array of intrinsic connections among the amygdala nuclei.

[1] Our point of interest will mostly be the basolateral nuclear group.

DOI 10.1007/978-1-4614-4112-0_7,

It is said that in order to understand neural mechanisms of emotion, one must understand the amygdala. In order to understand the amygdala, we will have to understand its connections. This small subcortical region receives massive input from the highest cortical regions in the brain, and nowhere else in the brain do connections by themselves "speak" so eloquently as to their functional meaning.

We will start with the amygdala's relation to the *what*-system. Although the amygdala is reciprocally interconnected with higher level cortical regions representing all modalities, the input from the inferior temporal visual cortex (BA37) is by far the largest cortical input to the primate amygdala. There are two intriguing features regarding the amygdala–cortical visual system connectivity pattern:

1. The information from BA37 enters the amygdala via the lateral nucleus, which does not generate return projection to the cortex. Instead, a projection back to the visual cortex originates in the basal nucleus of the amygdala (Fig. 7.1). This setup means intrinsic processing within the amygdala (taking place in the connections between its lateral and basal nuclei) is required to "close the loop" (Amaral, Price, Pitkanen, & Carmichael, 1992). It is tempting to speculate about functional meaning of the lateral nucleus–basal nucleus connection. Is subjective experience of emotion born in this processing?
2. While visual information is processed in cortical regions in a hierarchical fashion, the amygdala receives input only from the highest level of this hierarchy—BA37.[2] The surprising and interesting aspect of the return projection is that the amygdala projects to much more of cortical visual system than that from which it receives input (Fig. 7.1). The amygdala's return projections extend to all levels of the visually related cortex in the temporal and occipital lobes, including secondary BA19 and 18, and primary BA17 (Amaral et al., 1992). The functional meaning of this pattern is that the amygdala influences visual processing at its early, preconscious stages (Halgren, 1992; Kling & Brothers, 1992).

Kling and Brothers (1992) indicate that the amygdala–cortical connections depend on the species-specific prevailing sensory modality. Primates (including humans) heavily rely on visual information. Correspondingly, the primate amygdala receives massive input from the visual temporal cortex. It also receives highly processed, but lesser, input from the auditory temporal cortex as well as the gustatory, somatosensory, and olfactory systems. These sensory regions project to a portion of the lateral nucleus distinct from the terminal field for visual input to the amygdala. Processing of nonvisual sensory information follows the same rules as that of visual information—requiring intrinsic interaction between the lateral and basal nuclei (Amaral et al., 1992). This suggests that the amygdala also evaluates emotional significance of auditory, gustatory, somatosensory, and olfactory stimuli.

The amygdala's major connection to the frontal lobe is its reciprocal connections with the posterior orbital and medial prefrontal cortical areas. The orbital frontal

[2] Recall that BA37 is involved in the higher order visual processing—object recognition. This field is also connected with supramodal functions: word meaning (concept formation) and semantic memory.

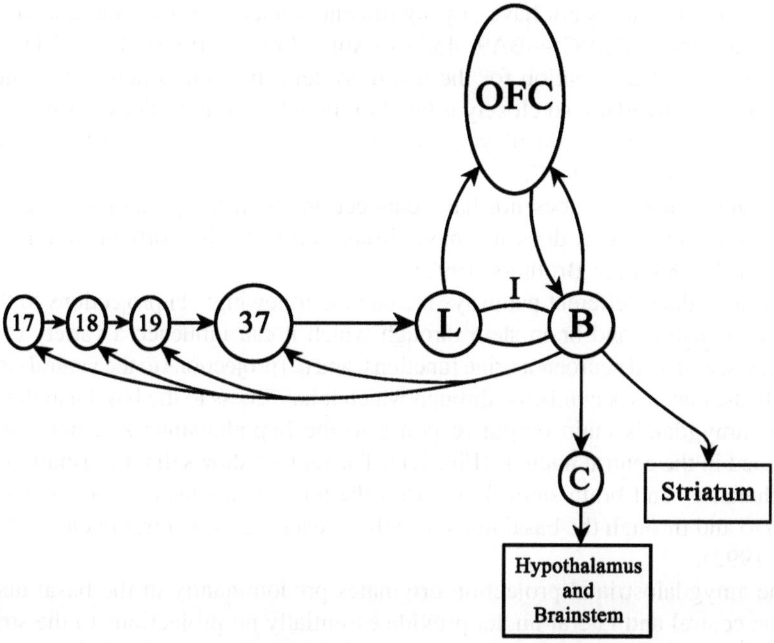

Fig. 7.1 The amygdala afferent and efferent connections. 17—the primary visual field, BA17; 18 and 19—the secondary visual fields, BA18 and BA19; 37—the tertiary field, BA37; *OFC*—the orbital frontal cortex; *L*—lateral nucleus; *B*—basal nucleus; *I*—intrinsic connection between L and B; *C*—central nucleus. Modified from D.G. Amaral et al. in: The amygdala: neurobiological aspects of emotion, memory, and mental dysfunction, 1992, New York: Wiley-Liss. Reprinted with the kind permission of John Wiley & Sons

cortex (OFC) is the region of the prefrontal cortex that possesses the strongest association with the amygdala (Zald & Kim, 1996a). The amygdala also has an indirect and alternative output to the OFC by way of its projection to the mediodorsal nucleus (MD) of the thalamus. While the MD projects to the vast territory of the prefrontal cortex, the amygdala fibers terminate in the part within the MD that projects to the same orbital and medial prefrontal cortical areas that receive direct projections from the amygdala (Amaral et al., 1992; Porrino, Crane, Goldman-Rakis, 1981). The amygdala's connection to the MD is unidirectional, from the amygdala to MD; The MD does not reciprocate projections back to the amygdala (Fig. 8.1).[3]

[3] This fact is very interesting in relation to the hypothesis that so-called *thalamic* emotion is connected with the MD (Glezerman, 1986; Glezerman & Balkoski, 1999). This may imply that integration of *limbic* and *thalamic* emotion is implemented through amygdala–MD–OFC pathway. *Thalamic* emotion would have no influence on the emotional evaluation of objects within the amygdala which would make sense in that limbic emotion could then remain in a "pure" form to serve its basic goal of individual and species survival. *Thalamic* emotion will be discussed at length later in this chapter and in Chap. 8.

The amygdala does not have any significant connections with the dorsal lateral prefrontal cortex (DLPC)—BA9, 45, 46 (Amaral et al., 1992). The DLPC is the ultimate point of destination for the *where*-system. It is important to take note of this fact: the amygdala, so closely related to the *what*-system, does not have direct connections with the main players of *where*-system, BA39 and 40, nor with its destination point, the DLPC.

The amygdala also does not have connections with the premotor cortex (BA6 and 8), meaning that it does not have direct access to the cortical motor system (Gloor, 1992; Kling & Brothers, 1992).

The amygdala's efferent pathways include the following: (1) projections to the lateral hypothalamus and brain stem through which it can influence a variety of autonomic, visceral, and neuroendocrine functions, and (2) projections to the ventral striatum (mostly the nucleus accumbens) through which it has access to the basal ganglia.

The amygdala's main output response to the hypothalamus and brain stem is generated in the central nucleus (Fig. 7.1). The picture shows that information to the hypothalamus and brain stem flows from the amygdala's lateral nucleus (sensory input) to and through the basal nucleus before reaching the central nucleus (Amaral et al., 1992).

The amygdalostriatal projection originates predominantly in the basal nucleus, and the central and lateral nuclei provide essentially no projections to the striatum (Amaral et al., 1992).

This connectivity pattern tells us that the amygdala's response through the efferent pathway to the autonomic and motor systems follows and is separate in time from intrinsic processing within the amygdala—processing that is, most likely, emotional evaluation of stimuli.

The amygdala projects substantially to the hippocampus, while the hippocampal projection back to the amygdala is rather meager. Amaral et al. (1992) indicate that, in this regard, the amygdala is poised to have significantly greater influence on the hippocampus than vice versa. It would seem that the amygdala's dominant relationship with the hippocampus is to ensure that emotionally important information will be remembered.

7.2 Limbic Emotion: Emotional Experience

Let's return to our formulation that the amygdala evaluates external objects as a means of internal need satisfaction, which is manifested by the emergence of subjective experience—emotion. I will refer to this evaluation as "limbic emotion"[4] (Glezerman & Balkoski, 1999) to distinguish amygdala-connected emotion from

---

[4] According to the traditional definition, the amygdala is a key component of the limbic system in the brain.

the emotion-like subjective experience connected with the thalamus ("thalamic emotion") that will be discussed later on.

Limbic emotion is basically what, in daily life, we call "emotion" or "feeling." Izard (1991) distinguished seven main types of emotion—joy, surprise, fear, sadness, anger, disgust/contempt, and interest. It has been emphasized that limbic emotion is object-related subjective experience (Glezerman & Balkoski, 1999). Distinguishing the separate object in consciousness is a function of the left hemisphere (see Chap. 2). Thus, limbic emotion must likely be the product of the left amygdala. What is meant by "object" in this context? I mean it to refer to people, objects, and oneself.

Connected with the left hemisphere, limbic emotion is conscious, can be categorized and named, and is part of the network that underlies goal-directed behavior. Having said this, one might postulate that amygdala-related subjective experience is conventional. In other words, do different people have the same feelings within each category of limbic emotion?

The evolutionary meaning of object-related subjective experience is to serve the species-specific goals of survival and reproductive success. We obviously cannot speak directly about inner emotional states in animals. However, extensive amounts of studies have examined how the primate brain processes the emotional significance of stimuli. Animals' attachment of appropriate significance to sensory stimuli is also presumably manifested by an internal state, namely affect. Such affects, according to Kling and Brothers (1992), would constitute critical inner signals to their possessor, immediately differentiating whether the approaching conspecifics' intent is to threaten, to initiate a friendly grooming session, or to copulate. Without affect, social interactions are meaningless and would appear to the observer as social indifference. Deficit of affect, Kling and Brothers (1992) indicate, may be why amygdalectomized monkeys fail to socialize. [Brothers (1990) cited from Kling & Brothers, 1992] speculates that primates' trend toward increasingly complex group life has created the condition for an expansion of the array of the innate affects the central nervous system can engender. Such affects would be the inner signals "differentiating the playful intention of an ally from those of a foe, or the solicitations of a dominant from those of a subordinate, etc." (Kling & Brothers, 1992).

In humans, when electrical stimulation of the amygdala evokes a feeling of fear or pleasure, patients' subjective experience often also includes the visual images of other individuals directing an action toward him, e.g., speaking to him, threatening him physically, having sexual intercourse with him [Gloor (1986) cited from Kling & Brothers (1992)]. These instances clearly illustrate that *object-relatedness* is the *sine qua non* of limbic emotion. Furthermore, the subjects' affects signal a particular aspect of objects' emotional significance—*the intentions of others.* Here are more instances of "social affects" elicited by electrical stimulation of the amygdala: a patient reported feeling "as if you are demanding" and "as if I did not belong here" (like being at a party and not being welcome) [Gloor (1986) cited from Kling & Brothers (1992)]. These are descriptions of feelings directly connected with a social situation and others' intent. Kling and Brothers suggest that, in human beings, evaluation of dispositions and intentions of others is mediated at least in part by amygdala activity.

We can conclude that amygdala-generated emotional experience, always object related, is intrinsically social.[5] We can further deduce that emotional experience is *conventional.* This conventional meaning can then be lost if the amygdala is damaged. For example, a patient G.R. underwent bilateral amygdalotomy for chronic self-mutilation and thereafter complained of difficulties describing her own affect, reporting most affects to be unfamiliar, strange, and vague. She consistently bemoaned that all her affects, except for fear and depression, could "not be pigeon-holed into any familiar category." Emotions felt qualitatively different and unfamiliar to her, but not dull. She conveyed: "It does not feel like it did before, nor is any word to describe it" (Jacobson, 1986). The interpretation of such statements is that the conventional meaning of emotion is lost to her. Feelings became idiosyncratic, and a connection with a categorical framework was lost.

One more aspect of limbic emotion needs to be emphasized. Congruent with the LH's discrete and analytical cognitive mechanism, limbic emotion is related to an object but is also *dissociable* from it. The above patient, examined by Jacobson (1986), reported that following the loss of her job or her room in her parents' home, she felt anger but without accompanying thoughts. She also experienced odd excited, either sexual or joyful, states of feeling which lacked associated fantasy, exhibited no observable expression, and occurred in the presence of her father and her therapist.

While stimulation of the amygdala in humans may evoke the subjective emotional experience (feeling), the reverse is also true: emotionally significant stimuli induce activation in the amygdala.

Studies of brain activity using event-related potentials (ERP) have shown that the human amygdala responds significantly to objects, words[6] and faces (Halgren, 1992). This correlates with the major afferent input to the amygdala being the inferotemporal cortex. Amygdala activity in response to words and faces is manifested by N2 and N4 potentials, which appear to come from the inferotemporal cortex; the N2 and N4 potentials also co-occur simultaneously in multiple supramodal cortical areas. This implies that the amygdala's emotional evaluation of events occurs synchronously with their cognitive evaluation in the cortex, and thus prior to any conclusions being obtained from the cognitive processing (Halgren, 1992). Halgren indicates: "From the onset of the amygdala N2 at 120 ms poststimulus onset to the completion of N4 about 500 ms later, there is ample time for many iterations of cortico-amygdala communication" (Halgren, 1992, p. 212). Halgren concludes that the amygdala potentials indicate "preConscious emotional processing of events en route to Consciousness" (p. 212).

This rapid response illustrates the amygdala's vital role in the realization of species-specific biological goals: emotional significance of the object is appreciated near-instantaneously.

[5] Indeed, the social situation is just a particular case within the broader definition of limbic emotion.

[6] We may consider *words* as object equivalents. As discussed in Chap. 2, word meaning is connected with the phylogenetically newer part, the "nucleus," of BA37 in the inferotemporal cortex, while the older part of this field is responsible for object recognition.

If we postulate that limbic emotion is conscious, then the question arises: when and where does conscious awareness of emotion occur? This is a difficult question to answer. Halgren (1992) suggests that degradation of ERP will tend to eliminate both cognitive and emotional processing. If we consider emotional experience to be the subjective equivalent of the amygdala's evaluation of external objects' biological significance, emotion must become conscious DURING this processing, when the amygdala is still activated. The amygdala is heavily interconnected with the OFC. The OFC receives information from the sensory input center of the amygdala—the lateral nucleus (Fig. 7.1)—as well as from the basal nucleus, the latter providing information that underwent intrinsic processing within the amygdala (see connection between the lateral and basal nucleus, Fig. 7.1). The OFC also sends projections back to the basal nucleus. Interaction between the amygdala and the OFC occurs in parallel to emotional processing in the amygdala and cognitive processing in the inferotemporal cortex. Does emotion reach conscious awareness during this interaction with the OFC? Even before emotion becomes conscious, the amygdala influences (modulates) visual processing at its early stages in response to emotionally significant stimuli (see Fig. 7.1 for the efferent connections from the basal nucleus to all components of *what*-system).

7.3 Limbic Emotion: Emotional Expression

There are two types of emotional expression: somato-motor and viscero-motor. Somato-motor expression of limbic emotion includes facial expression, body language, and intonational modulation of voice. It is triggered by the amygdala's efferent outputs to the motor system (basal nucleus—striatum, see Fig. 7.1). Viscero-motor expression represents autonomic responses (flushing, sweating, change in blood pressure and heart rate, change in gastric motility, etc.) triggered by the amygdala's efferent outputs to the autonomic/neuroendocrine centers (basal nucleus–central nucleus–hypothalamus and brain stem, see Fig. 7.1). The evolutionary meaning of emotional expression is the development of a signaling system for intra-species communication. Kling and Brothers (1992) indicate that, while aggressive and fear and "flight" behaviors are all elements of social interaction, the ability to signal one's disposition to fight or flight (by producing some elements of the behavioral pattern) may be more important in successfully negotiating a social situation than the fully executed behavior. These authors consider external state (affective display), internal state (affect) and emotional evaluation of a stimulus to be inseparable.

> Affect is no more and no less than the confluence and integration of sensory information in several modalities, combined with immediate coactivation of somatic effector systems (motor, autonomic, and endocrine)....An animal whose pupils are dilated, whose fur is erect, and which is hissing or growling also may be presumed to be feeling something (Kling & Brothers, 1992, pp. 371–372).

It is difficult to determine anything about the subjective experience of animals, but we should not forget that what we know about amygdala's connectivity pattern is mostly based on brain research of nonhuman primates. Recall the diagram of primate amygdala connections (Fig. 7.1) showing that information entering through the lateral nucleus undergoes intrinsic processing within the amygdala (presumably emotional evaluation of the stimulus) before reaching the output nuclei and efferent connections. If nonhuman primates' evaluation of the object's emotional significance (affect) is separate in time and neuronal activation from emotional expression, then it seems likely that their emotional experience (affect) is the *base* for emotional expression.

The impression that emotional evaluation of the object (subjective experience of emotion) and emotional expression are one and the same may have arisen because these events happen so fast.[7]

The amygdala is an amazing structure: on the brain's posterior–anterior axis, it combines both systems, afferent and efferent. Within the amygdala, the lateral nucleus can be considered as an input center—afferent system (Fig. 7.1); the basal nucleus is then the output center for the effector (somato-motor) system, while the central nucleus provides output to the autonomic (visceral-motor) system. Transition from the amygdala's afferent to efferent parts is carried out by intrinsic connections between the lateral and basal nuclei within the amygdala. Such proximity between its afferent–efferent systems allows rapid response, which, again, is required by the evolutionary goals of emotional expression.

As described earlier, the amygdala does not have direct access to the cortical motor system. The somato-motor response orchestrated by the amygdala is most likely implemented by the striatum. Recall that the striatum (in particular, the nucleus caudatus) is the repository of patterned motor sequences (see Chap. 5). These fixed "behavioral chunks" are automatically "released" when called into play, so that somato-motor response is fast, effortless, and does not involve a high level of conscious awareness.

There is some evidence that the motor cliché for facial expression of emotion are stored in the striatum. Neurophysiological investigations of single neuron recordings in monkeys have found a small population of neurons in the striatum with responses selective to faces [Rolls & Williams (1987) cited from Rolls, 1992].

[7] It may be that full separation of emotional experience from emotional expression is completed only in humans. This severance process would be connected with the formation of the human-specific cytoarchitectonic fields in the auditory and visual temporal cortex, and, correspondingly, with the further development of "world of sounds" and "world of meanings," respectively. In nonhuman primates, vocalizations communicate both affect and information. With the development of secondary BA22 and BA42, the emerging speech sounds serve communication of information but no longer emotional expression. Development of tertiary BA21 manifests the formation of language's phonological system that translates the expanding "world of meanings" into sound. Only intonation and inflection of voice remain as emotional expressions that are orchestrated by the amygdala's efferents. At the stage of separation of information *about* object and an *attitude toward* the object, emotional expressions became external signals communicating internal emotional state.

Facial expressions are rapid and emotion specific (Izard & Buechler, 1979). Do facial expressions present stereotypical motor responses fixed in evolution to a particular affect? Charles Darwin was the first to propose this idea. In 1872 he postulated that primates have universal facial expressions for basic emotions with a common evolutional origin and that the recognition and understanding of such emotional expressions is an innate ability. Indeed, facial expressions of basic emotions are conventional and universally recognized. Studying visual identification of the primary affective states across widely isolated cultural groups, Ekman (1982) found that irrespective of sex or culture (including preliterate cultures) facial expressions of the seven basic emotions—happiness, fear, contempt, anger, surprise, sadness, and interest have virtually the same meaning to all people. The conventional and universal character of facial expressions suggests their left-hemispheric nature. It is interesting in this regard to note that fMRI studies have consistently shown that the LEFT amygdala appears to be specifically activated in recognition and identification of facial emotional expressions.

7.4 Interhemispheric Differences in Amygdala-Related Emotion

So far, we have spoken about emotional expression as a conventional language—its social-communicative role. We also addressed the necessity of emotional expression's universality and conformity in order to accomplish its evolutionary tasks. In reaction to this seeming banality or the cliché nature of emotional expressions, one might protest: What about elusive, subtle, and individually specific expressions, or all the ambiguous, idiosyncratic, enigmatic expressions—the magic of a human face, the eyes that are "the mirror of the soul"? What about Mona Lisa's smile?

To quell such a revolt, we should return again to Bernstein. Recall that Bernstein, in describing the thalamic level in the brain as a background level for all complex movements, indicated that a few movements are a direct product of the thalamic level as a leading one. They include half-voluntary plastic-rhythmical movements, and pantomime: inborn emotional-expressive movements of the face, extremities, and the whole body. These movements are highly individually specific, because they depend upon the uniqueness of one's own body space.

Conventional emotional expressions are then superimposed upon individual-specific expressions. The latter reflect the state of the subject, the so-called *thalamic emotion* (as opposed to *limbic emotion*). The concept of *thalamic emotion* was introduced in previous works (Glezerman, 1986; Glezerman & Balkoski, 1999). It will be discussed in detail in Chap. 8. Here, it needs only mentioning that thalamic emotion, defined as an attitude to one's own bodily sensations, is closely connected to *right* hemispheric processing (see also footnote 3).

In discussing limbic emotion, we have mostly considered the object-related subjective experience generated by the left amygdala. We can turn now to the right amygdala. In the RH, the object is not distinguished out of the visual scene-situation, so that emotional significance is assigned to the whole situation (network: the right amygdala—the right BA37), and the affect is not separable from the situation. The emotionally saturated, subjectively felt situation has not just one feeling, but contains emotional complexes, including opposite emotional tones not differentiated from each other (Glezerman & Balkoski, 1999).

The LH with its analytic-sequential processing is unable to adequately evaluate the entirety of a situation (event). The RH limbic emotion then is complementary in that it is an attitude to the object, which is included into a situation. RH emotion is always situationally (contextually) bound. Because of this, the RH appreciates complexity and subtlety of emotional meaning. When patients with RH damage are read stories that convey emotional states of characters, they do not recognize the emotion implied in the situation. For example, when retelling a story that included the phrase: "The little girl began crying; her heart pounded as she crept in," a patient stated: "The little girl did not express any opinion or feeling except being excited. She did not wet her panties, she did not kiss anybody around, and she did not hug anybody" (Wapner, Hamby, & Gardner, 1981, p. 25). We can see the patient's intact LH makes inferences about emotion in a behavioral framework: matching up certain actions and behavioral displays with emotion.[8]

Recognizing the facial expression of emotions depends on two strategies (1) expression is perceived in terms of its properties alone (LH analytic processing) and (2) expression is perceived as a whole—a meaningful gestalt pattern (RH holistic processing). LH cognitive mechanism performs inferential decoding of facial expressions, while RH cognitive mechanism achieves direct, immediate knowledge of their emotional meaning. RH perceives all the shades, nuances, and subtleties of facial expressions. Moreover, the individual's face might be included into a symbolic system of meaning, identified with other faces by common affect and with the series of visual scene-situations behind the identified faces. All this allows RH emotion to be infinitely complex, but also ambiguous (emotional complexes may include opposite emotional tones).[9]

To better understand the difference between RH and LH emotion, recollect the comprehension of the metaphor "heavy heart" in patients with unilateral RH or LH damage (see Chap. 2). Patients were asked to explain this expression in two ways: by verbal explanation and by picture choice. Patients with RH damage gave adequate, although behavioral (from the point of view of an outside observer), explanation:

[8] Here we deal with the network Left Amygdala–Left OFC–Left DLPC–Left Striatum. The OFC places emotion into a behavioral context, and the DLPC is responsible for implementation of behavior per se. Repeated behavioral patterns that follow emotions become stereotypes, which are stored in the striatum.

[9] Due to identification and formation of symbolic systems, RH emotional meaning is polysemantic and gives rise to intuitive emotional sensitivity. On the other hand, RH gestalt of facial expression's emotional meaning (content) cannot be distinguished from the visual image of the face (form): they are united into a single integrated representation. Emotional meaning is always implicit, perceived through the visual image. Is this the "unconscious"? I do not think so.

he's got many troubles; in contrast, their picture choice was grossly inappropriate: *a person staggering under the weight of a large red heart.* As compared to other groups of patients (including patients with dementia), only patients with RH damage did not "see" this literal depiction to be amusing or absurd (Winner & Gardner, 1977). On the other hand, the "RH" (patients with LH damage) chose the correct picture (*a man crying*, reflecting an inner state) but gave a literal, physicality-based explanation: *It's heavy, the heart, a lot of weight.*

The brain mechanism of metaphor formation, discussed in Chap. 2, is based on right hemispheric identification of things from alien domains (usually physical and psychological). For the RH, the physical–sensual image of a heavy heart IS someone crying, which is why the patient with LH damage chose the picture of a crying person. The task of verbal explanation, however, is an impossible demand for RH holistic processing: the heavy heart and the crying person are the same, and they are not divisible and represent a new whole. Again, in the RH symbolic system, emotion (meaning) is expressed through the visual sensual image (form) but cannot be extracted from it thus remaining implicit. We really cannot say then that the physical–sensual explanation of the patient with the LH damage (*it's heavy, the heart, a lot of weight*) is incorrect: it *is* RH emotion of sadness.

7.4.1 FMRI Studies of Emotion Recognition by the Left and Right Amygdala

Now we turn our attention to facts based on fMRI studies of facial emotion recognition. Usually in these studies healthy volunteers are presented with a standard series of emotionally expressive faces. There are three types of experimental design (1) explicit, in which subjects are asked to attend and judge facial expression; (2) implicit: subjects are asked to attend and judge to the *gender* of each face. Critchley et al. (2000a) justly noted that in this design, facial stimuli were presented long enough for the subjects to process both the emotional and the gender attributes of the faces, so that covert, but conscious processing of facial expression is not excluded; (3) short and masked: subjects are presented with target stimulus (i.e., angry face) for less than 40 ms and then immediately shown a neutral face (mask), these subjects report seeing the mask but not the target.

In both explicit and implicit tasks, response was unilateral: only the left amygdala was activated (Critchley et al. 2000a)

We will discuss in detail the Morris et al. (1996) study since it raises several points of interest relevant to our topic. In this study, subjects were presented with pictures of angry and happy faces. The experiment's design was implicit: subjects were instructed to classify faces by gender. Only the left amygdala was activated. No activation in the right amygdala was observed even at low thresholds for either angry or happy faces. Activation in the left amygdala increased concomitantly with escalating emotional intensity of facial fear, but it decreased with the increasing emotional intensity of happiness. Indeed, it was a continuum of activation from the most happy to the most fearful faces. This finding corresponds with data that facial expressions of fear and happiness are psychologically at opposite ends of a spectrum (Adolphs, Tranel, Damasio, & Damasio, 1994). Also, a study of patients with unilateral amygdala damage found that ratings of emotional intensity in facial expressions were significantly lower in patients with left versus right amygdala damage (Adolphs, Tranel, Damasio, & Damasio, 1995). The above results illustrate LH processing—operating with continual combinations of discrete signs (Fig. 1.2) (in contrast to "all or nothing" holistic RH processing). To the left amygdala, basic emotions are clear and distinct qualities because left hemispheric mechanism allows for distinguishing opposite feelings with gradual quantitative differences between them.

Thus, in Morris et al. (1996) study, the left amygdala was activated as a response to salient features of facial expression. These features present external social signals: danger or threat signals in angry faces and safety signals in happy faces. Interestingly, in this study there was no activation in the prefrontal region. This indicates that in humans, as in nonhuman primates, receiving an expression of fear in a conspecific may trigger an automatic response to a potential danger without necessity of higher level processing.

In an fMRI study with a masked experiment design (Morris et al., 1998), the healthy volunteers were presented with pictures of angry faces in masked and unmasked conditions. The task was explicit: subjects were instructed to detect any occurrence of the angry faces. Subjects reported seeing none of the masked angry faces but all of the unmasked angry faces. A significant activation was found only in the *right* amygdala to masked presentation, while unmasked presentation of the same angry face showed activation of the left but not the right amygdala. The authors conclude that the lateralization of amygdala response is a function of the level of awareness. They further suggest that processes related to conscious awareness, such as engagement of language, may inhibit the right amygdala's activation. Finally, they conclude that the right amygdala is activated for emotional stimuli that subjects are unable explicitly to recall.

One would tend to assume then that the "dominant" LH with its language processing inhibits RH activation. However, the same segregation of conscious and unconscious processes is observed in split-brain patients. For example, increased heart rate responses were provoked by masked emotional visual stimuli when presented to the RH in split-brain patients, but the same stimuli shown unmasked produced greater increase in heart rate after presentation to the isolated LH (Ladavas, Cimatti, Del Pesce, & Tuozzi, 1993).

Moreover, the split-brain patients would deny awareness of any unmasked visual stimulus presented to their RH (recall that people with intact brains reported 100 % of unmasked stimuli) even if they had just verbally reported the same stimulus when it was presented to their LH (Gazzaniga, 1995). Furthermore, nonverbal tasks revealed that stimuli presented to the RH of the split-brain patients were detected and recognized. Thus, it seems more likely that the RH's peculiar performance in processing faces (masked vs. unmasked) is due to the intrinsic nature of RH processing rather than external factors such as interhemispheric interaction.

Recall that in Morris et al. (1998) study, the task was explicit: to detect any occurrence of angry faces under conditions of a very short, masked presentation. Timing for "LH consciousness" is longer than stimulus detection. Breaking down the whole by analysis of its features, then re-constructing a "synthetic" image from those features is required to distinguish an object in consciousness. Only once it has been distinguished in consciousness can the separate object be named. The process of conscious awareness is sequential. Meanwhile, in the RH, the "whole" is *registered.*

It was postulated before that the RH is constantly registering the continuously changing impressions from the outside world, which are subjectively felt visual scene-situations. This creates the alert state—attentiveness—and provides the basis for LH conscious awareness. This was called RH consciousness or RH "flow of consciousness" (in contrast to LH "focus of consciousness") (Glezerman & Balkoski, 1999). However, this RH consciousness is not a part of what we usually call "consciousness," which instead refers to LH conscious awareness.[10] RH consciousness is always immediate experience, with no potential to lead to conscious awareness. RH consciousness is ever implicit and always an indivisible whole, by definition. This provides a plausible explanation as to why the RH cannot name the stimulus it detected and recognized: RH content is implicit. On the other hand, the right amygdala "detects" threatening stimuli before the left and gives a "warning" by triggering changes in autonomic and endocrine activity, priming the organism for the appropriate behavioral response.

Finally, we should address the differential roles LH and RH limbic emotion play in communication. By distinguishing clear, explicit signals (social cues) LH processing allows a fast behavioral response. Further, the network Left amygdala–Left OFC interprets emotions in a behavioral context necessary for appropriate goal-directed behavior. While the LH is preoccupied with interpretation, the RH is preoccupied with identification. We have already spoken about identification as the main law of RH processing and the foundation for a sense of *interconnectedness.* We can now propose: the emotional understanding the RH contributes to communication is *empathic understanding.* According to Jaspers (1968), empathic understanding cannot be achieved by sense perception or logical thought but has to be grasped by transferring oneself, so to speak, into the other individual's psyche. In other words, empathic understanding is our inner representation of the other person's subjective experiences (Jaspers, 1997).[11] Empathy is a necessary underpinning for any communication. It modulates behavior in correspondence with the individual's situation.

Before applying the results of our exploration of the amygdala to autism, it is necessary to clarify the terms that will be used in this work.

---

[10] However, I would not call RH consciousness "unconscious" either.

[11] I use the term *empathic understanding* to emphasize that, in empathy, there is no free floating affect (that applies to LH object-related limbic emotion), but there is identification with the other's experience, the whole within which emotion is embedded.

The term *implicit* will be used exclusively to refer to right hemispheric representation, where the whole's parts do not lose their uniqueness but also cannot be "extracted" out of the whole by means of RH processing alone: they remain implicit. They cannot be explicitly defined. They are defined with something else that is identified with them.

I will avoid using the terms *conscious* and *unconscious*. Instead, I will use RH *consciousness* and LH *consciousness*. As already discussed, the *conscious/unconscious* dichotomy is not congruent with the phenomenon of RH *consciousness*/LH *consciousness*. The term *consciousness* will refer to the content of subjective experience, with the R/L *consciousness* dichotomy based on the qualitative differences in RH/LH experience. Also, instead of "all-or-none" *conscious/unconscious* dichotomy, the idea of a *degree of conscious awareness* or *degree of awareness* will be put forth. Finally, the term *automatic* will be used for subcortical processing (versus cortical voluntary processing). The term *automatic* is not synonymous with *unconscious*. I earlier illustrated the role of automatic processes in skill acquisition in the norm, and in stereotypical behavior in pathology when the prefrontal region is damaged (see Chap. 5).

Recent studies have shown that activity in prefrontal regions falls away as behavior becomes automatic. Nevertheless, we can easily revive our awareness of a skill, like driving, placing our attention to the action (Zeman, 2001).

Regarding the amygdala, automatic activity can either be completed without involvement of higher levels (prefrontal cortex) or emotional processing of events can be continued along the connection amygdala–OFC (and prefrontal cortex–inferotemporal cortex). In the former instance, amygdala processing can be called *preConscious*, en route to *Consciousness* (Halgren, 1992): this applies to the left amygdala (LH consciousness).

7.5 Do Autistic People Recognize Emotion?

Do autistic people recognize emotional expressions of others? There is an enormous amount of literature on this topic, but there is no definite answer. It looks as though both HFA and LFA are able to recognize basic emotions (happiness, anger, sadness, etc.) on facial expression tasks, but they make more errors than control groups. However, an interesting recurrent pattern emerges from these studies:

1. Double dissociation in the ability to recognize emotional expression compared to personal identity was observed between autistic and control groups. The performance of normal and nonautistic mentally retarded subjects was characterized by "facial expression of emotion superiority effect": these groups did better on the emotional recognition than on face identity tests. In contrast, autistic subjects performed worse on identifying facial expressions of emotion compared to facial identity (Celani, Battacchi, & Arcidiacono, 1999; Hobson, Ouston, & Lee, 1988).

   Studies of patients with brain lesions provide sufficient evidence that facial identification and ability to recognize and interpret facial expression are dissociable.

For example, Young, Newcombe, de Haan, Small, and Hay (1993) report selective impairment of facial expression recognition and face identification in patients with brain injury. In their material, disorder of familiar face recognition is observed only in patients with RH damage; interestingly, selective impairment of identifying emotion was found only in patients with LH damage.[12]

The double dissociation in autistic and control groups between facial identity and facial expression is evidence, the authors concluded, for the specificity of an emotional recognition deficit in autism.

2. When autistic, nonautistic mentally retarded and normal children were presented with photographs of human faces differing by both discrete emotional (i.e., happy–unhappy) and nonemotional features (absence–presence of hat or type of hat, i.e., floppy–woolen), and then, without any further explanation, they were asked to select pairs of photos that were most similar, only autistic children more often selected photos that were similar on nonemotional features. Both nonautistic mentally retarded children and normal children sorted by facial expression of emotion before they sorted by type of hat or other nonemotional features ("facial expression of emotion superiority effect" mentioned earlier). However, when subjects were given direct instruction to sort by emotional features, autistic and control groups did not differ (Beggeer, Rieffe, Terwogt, & Stockmann, 2006; Jennings, 1973; Weeks & Hobson, 1987).

What does this change in autistic children's response to direct versus indirect tasks mean? In searching for an answer, we can turn our attention to fMRI findings on facial expression recognition in autistic individuals.

Critchley et al. (2000a) found that during implicit processing of emotional facial expression in HFA, significant activation was observed only in two areas: the left middle temporal cortical region and the left DLPC. The authors emphasize that, in contrast to the control group, HFA did not activate the left amygdala during the implicit task. Apparently the visual "signs" of emotional expression were recognized (middle temporal area activation) by the autists, but emotional significance was not assigned.

---

[12] Within a hemisphere (here we speak about the axis of intrahemispheric difference as opposed to the axis of interhemispheric L–R differentiation), facial identity (as we already know, see Chap. 4) is connected with the fusiform gyrus, the so-called Face Area in the RH. The middle temporal gyrus (BA21) is a visual cortical area that is important in processing of facial features and is activated during explicit processing of facial expressions (Critchley et al., 2000b). These authors indicate that the middle temporal gyrus may represent a human homolog of the superior temporal sulcus (STS) region in nonhuman primates. In this book, area STS was mentioned as a part of facial processing network in humans, based on work of Puce et al. (1998), see Chap. 4.

What is the visual image of facial expression? It is a particular pattern of muscle contractions that make up emotional expression. What then might this representation be in the LH? It must be (distinguished out of a face's gestalt) the salient and typical features specific for each basic emotion. This visual image of a "moving" face, assigned with emotional significance by the amygdala, has its "dynamic" motor counterpart stored in the striatum (the amygdala and the striatum are both activated during explicit processing of face expression!).

In Baron-Cohen, Ashwin, Ashwin, Tavassoli, and Chakrabarti (2000) study, fMRIs were taken while HFA and controls were presented with a series of photographs of eyes and asked to indicate by button press which of two simultaneously presented words best described what the person in the photograph was feeling. Examples of word pairs include: unconcerned–concerned; sympathetic–unsympathetic. The control group demonstrated a significant activation in the left amygdala and left DLPC. The autism group did not activate the left amygdala at all but activated the left DLPC.

In this latter study, the task was more complicated and included cognitive components requiring frontal lobe involvement. As illustrated earlier, recognizing basic emotion (for example, danger signals from an angry face) may trigger an automatic response from the left amygdala without activating the prefrontal cortex. The autists (in the above studies), however, demonstrated no automatic left amygdala response to emotionally significant stimuli, but the left DLPC was activated. We are coming to a conclusion that the left DLPC is recruited in HFA for compensation. The left DLPC is the purely cognitive region responsible for verbal reasoning (verbal–logical thinking) and implementation of goal-directed behavior. Having a visual image of facial expression in their middle temporal cortex (activated during face processing in autistic individuals), autistic individuals "recognize" emotion by knowledge of behavioral consequences of the expression (for example, angry face—threat; happy face—safety).[13]

We can find support for this interpretation in a recent study where both LFA and HFA selected photos of faces based on nonemotional features (moustaches, glasses) in a sorting test, but turned their attention toward emotional features when asked to pair the photos based on their expectations of people's future actions (Beggeer et al., 2006).

Further of note is that autistic individuals' performance on an emotion recognition test did not differ from the control group, if matched by verbal mental age; however, when the groups were matched according to nonverbal mental age, autistic individuals performed significantly worse than control subjects on the same task (Celani et al., 1999). An explanation for this finding would be that autists in general have a specific profile on IQ tests: nonverbal IQ is higher than verbal IQ. Thus, when autistic and control group are matched by nonverbal IQ, autistic individuals' verbal reasoning must be much lower than in control group. This corresponds to the suggestion that autists compensate for their deficiency of emotion recognition by utilizing the left DLPC, the area responsible for verbal reasoning.

Additional support for this suggestion comes from Wang, Lee, Sigman, and Dapretto (2006) study. These authors performed fMRI while HFA listened to short scenarios and decided whether the speaker was sincere or ironic. Scenarios included knowledge of the event outcome and/or strong prosodic cues: sincere or sarcastic intonation. Here are examples: a scenario with both event knowledge and prosodic

[13] Routinized behavioral patterns associated with basic emotions are stored in the striatum. They are retrieved by the DLPC "command" in response to angry/happy faces.

cues—*Steve went to the barbershop. When Jen sees his BAD/NICE haircut, she says, 'You look great!'* (intonation is IRONIC/SINCERE). A scenario with event knowledge only: *Steve went to the barbershop. When Jen sees his BAD/NICE haircut, she says, 'You look great'* (intonation is NEUTRAL). Another scenario with prosodic cues only: *Steve went to the barbershop. When Jen sees his NEW haircut, she says, 'You look great!'* (intonation is IRONIC/SINCERE).[14]

Overall the results of this study showed that children with autism recruited their prefrontal region more strongly than the control group. The authors suggested that this increased activity may reflect a compensatory strategy involving the use of verbal reasoning skills in place of "instinct." An indirect proof of this theory was the correlation between degree of engagement of the prefrontal cortex and verbal IQ in children with autism. No such correlation was found in the control group.

7.6 Which Hemisphere Do Autistic People Use to Recognize Emotion?

It should be noted that the tests used to examine autistic individuals' ability to recognize emotional expressions mostly target the LH. In the typical sorting task, subjects are asked to pair pictures of people, with one or the other of two target pictures. As in the example earlier, the pictures could be grouped according to emotional expressions or according to age, sex, presence–absence of hat, or other nonemotional features. The typical matching task requires subjects to match a set of pictures with a target picture, e.g., matching faces according to emotional expression. The tasks performed during functional brain imaging are even more simplified: subjects are required to make choices between two discrete features, i.e., *happy–angry; happy–sad,* etc. These tests appeal to feature analysis, the LH domain of distinguishing discrete, salient, and typical signs (the species-specific external signals to communicate danger or safety). It is not surprising then that fMRI studies uniformly showed activation in the left amygdala in healthy volunteers during such emotion recognition tests. Equally, it is not surprising that when brain-damaged patients are presented with facial expression tests, only patients with damage to the LH show disorder of emotion recognition (although a significant role of the RH in the processing of facial expression is a well-known fact) (see Young et al., 1993).

There are few studies where autistic individuals are given emotion recognition tasks that call upon RH holistic processing. In Celani et al. (1999) study, autistic children and two mental age-matched control groups (nonautistic children with Down syndrome and normal children) were presented with two tasks: *facial expression condition* and *emotional situation condition.* In the *facial expression* task, subjects were presented with a set of pictures where the same individual exhibited

[14] As indicated at the beginning of this chapter, the human amygdala, though it relays heavily on visual information, receives input from all sensory modalities, including auditory.

neutral and happy faces. In the *emotional situation* task, subjects were presented with pictures of agreeable and disagreeable situations (examples of pleasant situations include a cloud on a blue sky, a snow-covered pick, a sailboat on water; some examples of unpleasant situations were garbage among sand and rocks, a duck in oil, pizza with roaches). In both tasks subjects were requested to pick the picture they liked best.

The three groups did not differ in their performance on the *emotional situation condition.* Indeed, all subjects gave congruent responses to the situational pictures. In contrast, on the *facial expression condition,* autistic subjects performed significantly worse than both control groups. While both nonautistic mentally retarded children and normal children showed a significant preference for happy faces (congruent choice), autistic children's choices on the task did not differ from that of chance selection.

The visual emotional situation task, on which autistic children performed well, apparently targets more the RH than does the facial expression task on which autistics failed.

Note that both tasks touch upon not only emotional expression, but also upon emotion experience. The authors indicate that autistic subjects are "not inclined to use happy facial expression as a meaningful index in judging the pleasantness of a face" (p. 64). I would further specify that there is a LH deficiency in assigning emotional significance to a face (object).

Heaton, Hermelin, and Pring (1999) showed that autistic children can perceive the basic feelings in music. Children were presented first with schematic pictures of happy and sad faces. When explicitly asked: "what does this face look like?," they were able to label the faces as "happy" and "sad." They were then presented with musical stimuli and asked to indicate whether the music was happy or sad. Children with autism performed correctly on this test as well.

We pick up on another's emotions by their facial expressions. Facial expression is a visible sign of emotion, it is a mediator between one who "feels" and one to whom emotion is communicated. Music is unmediated experience where emotion is communicated through sound and rhythmical movement, but not distinguishable from them.[15]

We can say then that in the above study autistic children demonstrated their ability to recognize emotion expressed in music.

In conclusion, there is evidence that autistic individuals use their RH (right amygdala) more in emotion recognition, and they use the LH (left DLPC) to compensate for their left amygdala deficit in emotion recognition.

---

[15] Note that musical ability is mostly connected with the RH. Luria described a distinguished Russian composer V.G. Shabalin who had a stroke in the LH, and as a result became aphasic. At the same time, however, his ability to compose was intact, even better than before his stroke.

7.7 Do Autistic People Express Emotions?

7.7.1 Facial Expression in Autistic Children

All observers indicate a severe, wide-ranging disorder of emotional expressiveness in autists. Indeed, this expressive disorder is included in the clinical pattern of autism. Below are a few examples from Asperger's material (Asperger, 1991):

> Fritz, 6 y.o.: "His gaze was strikingly odd. It was generally directed into the void. When somebody was talking to him he did not enter into the sort of eye contact which would normally be fundamental to conversation. He darted short 'peripheral' looks and glanced at both people and objects only fleetingly. It was 'as if he wasn't there.' The same impression could be gained of his voice, which was high and thin and sounded far away" (p. 42)
>
> Haro, 8 y.o. boy: "He had few facial expressions and gestures… He talked slowly and in a deadpan way without much modulation. He never looked at his interlocutor while talking. His gaze was far away… Often he did not respond to questions but let his talk run single-mindedly along his own tracks" (p. 52)
>
> Ernst, 7 y.o. boy: "His eye gaze was highly characteristic, far away and unfocused. The eye did not seem to grasp anything and was vaguely aimed into the distance. Mainly for this reason, the boy looked as if he had just 'fallen from the sky'. His voice too fitted with this. It was high, slightly nasal and drawn out, roughly like a caricature of a degenerated aristocrat" (p. 60)

As we can see from these vignettes, the disorder of emotional expressiveness in autism is strikingly uniform. Autists convey a peculiar absentmindedness without internal preoccupation. As Asperger put it "as if he wasn't there [But *where?*]." It induces feelings of unconnectedness without the bizarre quality observed in patients with schizophrenia. (A schizophrenic's bizarre quality is due to the "glimpses" we perceive from the schizophrenic patient's internal world—autistic thinking.) The gaze is highly characteristic: no direct eye contact with people or objects and use of peripheral vision. The specificity of the disorder of emotional expression in autism and its meaning will be addressed later in this book. For now, we can simply conclude that autists do not express emotions. Or can we?

Nadia, an autistic child with severe cognitive deficits and extraordinary drawing ability (Selfe, 1977), was discussed in Chap. 3. I argued earlier that Nadia's drawings illustrate RH "flow" of consciousness, with the projection of her sensory-motor level (physical world of the external spatial field) and thalamic level (kinesthetic body space). Selfe (1977) describes Nadia as clumsy, poorly coordinated, and excessively slow in her movements:

> She did not respond to command or instruction and it was extremely difficult to know whether she merely did not comprehend or whether she was refusing to cooperate. She would sit staring into space or wander slowly and aimlessly about the room…. She would sit immobile apparently taking no notice of her surroundings (pp. 3, 119).

This picture radically changed when Nadia was drawing:

> She generally drew swiftly and deftly, becoming animated…her motor control was highly developed…. She usually sat back to survey the effect, moving her head perhaps, to vary the viewing angle. This usually gave her great pleasure and after surveying intently what she had drawn she often smiled, babbled and shook her hands and knees in glee. However, it

> was at this point that she was most distractible and she could fall into a staring reverie perhaps lasting for several minutes before she continued drawing[16].... Nadia enjoyed drawing and this was when she was most animated and lively.... She drew quickly and laughed a great deal while working (pp. 8, 9, 13).

Nadia's smile and laughter are not produced for or from communication. They reflect her inner state of experience when the thalamic and sensory-motor levels in the RH are leading.

Switching to the tune of thalamic and sensory-motor levels makes Nadia's movements dexterous. In this connection, it is interesting to recollect Bernstein's remark that movements implemented at the thalamic level as a leading one (for example, half voluntary habitual mechanical-monotonous movements, stretching, hugging) are smooth and deft, even in clumsy people.

Another example that comes to mind is the secret smile of the twins who were exceptional calculators, as described by Oliver Sacks (see Chap. 6). Observing the twins "playing" with prime numbers, Oliver Sacks concluded that they did not communicate, they *communed.* As was hypothesized in this book, they communed via sharing "sacred objects"—numbers. It was thus suggested that the twins were projecting their subjective experience of inner rhythm into numbers, identifying with each other and with numbers (see Chap. 6). The twins' mystical experience is reflected in their enigmatic facial expressions, insightfully described by Oliver Sacks: "secret smile on their faces," "the strange pleasure and peace they now seem to have," "serene and meditative and almost holy intensity" (Sacks, 1987, pp. 201–202).

7.7.2 Prosody in Autistic Children

Prosody can be defined as nonverbal features of speech, provided by the speaker's voice, which modify speech meaning.[17] As an aspect of emotional expressiveness, prosody is abnormal in autism, illustrated in the above examples from Asperger's material.[18]

---

[16] Don't we recognize here LFA thalamic level's movements with their meditative quality?

[17] Articulation of a given language's speech sounds is independent of one's voice qualities.

[18] Although a detailed analysis of prosody is beyond the scope of this book, we will briefly outline the following components of prosody (Ross, 2003):

1. Linguistic prosody modulates and enhances linguistic meaning, for example, raising the intonation in the end of the sentence to indicate a question or using stress to highlight an element of information within a sentence.
2. Dialectical and idiosyncratic prosody refers to regional and individual differences in speech quality.
3. Affective prosody (emotion in voice) includes attitudinal and emotional prosody. Attitudinal prosody reflects feeling about the content of the utterance, as skepticism, doubt, enthusiasm, or boredom. An example of attitudinal prosody would be the emotional recognition task, given earlier in this chapter, where sincere or sarcastic intonation is the prosodic cue to decide whether the speaker was sincere or ironic. Emotional prosody injects emotions into speech, such as happiness, sadness, fear, and anger. It also conveys a speaker's general feeling state.

Affective prosody is a lateralized function of the RH, while linguistic prosody is likely bilateral (Heilman, Bowers, Speedie, & Coslett, 1984; Merewether & Alpert, 1990; Ross, 2003).

The acoustic means for prosody are pitch (fundamental frequency), intensity (amplitude), and duration (Paul et al. 2005). They are the base for the prosodic cues such as stress, intonation contour, rhythm, etc. Prosody shares significant acoustic features with music (Jarvinen-Pasley, Pasley, & Heaton, 2008), and they are both connected with the RH. There is a dissociation of these two right-hemispheric functions in autism: while prosody is characteristically disordered, musical pitch and pitch contour processing are enhanced.

Abnormal prosody, together with disorder of understanding humor, which autistic people share with RH-damaged patients, led to the hypothesis that these symptoms of autism are a result of RH dysfunction (Ozonoff & Miller, 1996). We will now compare aprosodias resulting from RH dysfunctions with abnormal prosody in autism.

Ross and Mesulam (1979) proposed that functional organization of affective prosody in the RH is analogous to the organization of language in the LH. This hypothesis was supported by MRI studies of patients with right unilateral brain damage. In the case of damage to the right inferior frontal region (counterpart of the Broca region in the LH), loss of spontaneous affective prosody was observed (analogous to aphasic nonfluency), whereas injury to the posteriosuperior temporal lobe in the RH (counterpart of Wernicke region in the LH) caused loss of affective-prosodic comprehension (analogous to aphasic loss of comprehension) (Ross, Orbelo, Burgard, & Hansel, 1998).

The patients with spontaneous affective aprosodia complained bitterly of their inability to insert affective variation into their speech; however, they did not have difficulty perceiving affective displays and insisted that they could feel and experience emotions inwardly (Ross & Mesulam, 1979).

In autism, we observe abnormality of spontaneous prosody (expressive disorder), but it is not loss of prosody, but rather distorted, inappropriate prosody. There are various deviances in rhythm, intonation, and pitch production: aberrant stress patterns; deficit in pitch and intensity control; inappropriate emphasis within and across words; incorrect intonation and inflection; both insufficient and excessive pitch rise for stress; excessive inflection but also contraindicated and stereotyped inflection; and both inappropriately loud and quiet volume (Jarvinen-Pasley, Pasley, & Heaton, 2008; Paul et al. 2005; Simmons & Baltaxe, 1975). These deviances make up the quality of autists' speech, such as robotic, pedantic, singsong, and stilted. As Paul, Augustyn, Klin, and Volkmar (2005) noted, it is this "vocal presentation of individuals with autism that most immediately creates an impression of oddness."

Furthermore, patients with spontaneous aprosodia resulting from damage to the right inferior frontal region have selective expressive prosodic disorder. Such patients' emotional experience of prosodic cues remains intact; this is not the case in autism. Also, there is no evidence of a motor basis for the abnormal prosody in autism.[19]

---

[19] Later in this chapter we will discuss the autistic savant, Stephen Wiltshire, who gives a perfect musical performance within the VSS, with exact intonation and gestures.

Disorder of affective-prosodic comprehension connected with lesions in the right posteriosuperior temporal region does not match the abnormal prosody of autism either. Jarvinen-Pasley et al. (2008) showed that autistic children were superior in perceptual interpretations of intonation patterns (pitch contour) of speech stimuli compared to normal controls. These authors conclude that superior processing of perceptual information in speech is specific to autism. We must clarify that this study refers to prosodic perceptual information, i.e., RH processing. Indeed, it is the amygdala that assigns auditory stimuli with emotional significance while the perceptual characteristics of affective prosody are provided by the right superioposterior region (counterpart of Wernicke zone in the LH). Thus, we have again encountered the prevalence of the RH sensory-motor level in autism. At the perceptual level then there is no dissociation between prosody and music processing in autism, which in turn indicates enhanced RH function.

We have compared so far disorders of prosody in autism with selective aprosodias in patients with circumscribed lesions in the RH. As to patients with more extensive RH damage, they do not match autism either. Flat prosody in these patients might be explained by their mental state of indifference, inappropriate emotional reactions, and verbal disinhibition.

In conclusion, the analogy of abnormal prosody in autism with aprosodias in patients with damage to RH does not work. As with faces, we cannot explain disorder of prosody in autism by dysfunction of the cortical regions that in the norm serve as prosody's brain basis.

7.7.3 Laughter in Autistic Children

Hudenko, Stone, and Bachrowski (2009) conducted an acoustic analysis of laughter as part of emotional expression in normal and autistic children. First they described the acoustic structure of laughter that includes two qualitatively different components: voiced and unvoiced. Voiced laughter is periodic and contains a fundamental frequency; perceptually, it has a tonal, song-like quality. In contrast, unvoiced laughter is aperiodic, it is largely atonal and noisier, produced with turbulence arising in the supralaryngeal cavities. Only voiced laughter is strongly associated with the experience of positive affect. Unvoiced laughter is used to negotiate the subtleties of social interaction, for example, "grunt-like" unvoiced laughs affirming others during conversation. These laughs are only loosely coupled with an individual's internal affective state.

This experimental study revealed a qualitative difference in the expression of laughter between children with and without autism. Children with autism did not produce unvoiced laughter, in striking contrast to nonautistic children whose laughs were 37–48 % unvoiced. The authors concluded that laughter in children with autism reflects their internal state. They indicated that children with autism do experience positive affect, and this state is signaled with voiced laughter. They also found individually specific rhythmical characteristics of voiced laughter in children with autism.

The acoustic properties of voiced laughter described in this study suggest its relation to the thalamic level's inner rhythm. Also note that voiced laughter is similar to the RH-associated affective prosody (emotion in voice).

7.8 Viscero-Motor Expression of Emotion in Autistic People

So far, we spoke about somato-motor expression of emotion in autism, but what about their viscero-motor expression?

Emotional expression as mentioned earlier seems to follow the amygdala's evaluation of the biological significance of objects with its subjective equivalent being emotional experience (likely via the intrinsic network: lateral nucleus–basal nucleus). Autonomic, neuroendocrine responses are also triggered by the amygdala: from its basal nucleus—through the central nucleus—to the hypothalamus and brain stem (Fig. 7.1). Several studies provide evidence that, in contrast to the somato-motor emotion expression, autonomic response to emotionally significant stimuli is relatively intact in autistic children throughout the spectrum, from LFA to HFA (Ben Shalom et al., 2006; Bernier, Dawson, Panagiotides, & Webb, 2005; Blair, 1999; Willemsen-Swinkels et al., 2000). Other studies showed atypical autonomic reactivity in autism, for example, less arousal when viewing sad pictures but higher arousal while processing neutral stimuli. No lucid picture was provided by these studies as to how altered autonomic response and affective report were related in autists (Bolte, Feineis-Matthews, & Poustka, 2008). Gaigg and Bowler (2008) found that autistic individuals, like normal, control participants, exhibited significantly increased skin conductance response (SCR) to emotionally charged as compared to neutral words. However, correlation between SCR and subjective rating was significantly higher for control than autistic group for whom this relationship was only marginally above chance. Hirstein, Iversen, and Ramachandran (2001) found highly abnormal electrodermal activity (SCR) in children with autism in response to eyes ("look at me") and during a wide variety of everyday behaviors. These authors emphasized that SCR reflects activity of the sympathetic branch of the autonomic nervous system with its primary neurotransmitter norepinephrine.[20]

Autistic children appear to have a higher baseline electrodermal level (Palkovitz & Wiesenfeld, 1980). It was also suggested that autism involves chronically high arousal level (Hutt & Hutt, 1965). Note that this is commensurate with our hypothesis that a hyper-alert state in autism is a result of unopposed RH flow of consciousness (see Chap. 3). In Hirstein et al. (2001) study, most children with autism had abnormally high electrodermal activity, which was dramatically decreased when autistic children were engaged in repetitive behavior. The authors also found

[20] The attention system that maintains the general alert state is dependent on norepinephrine pathways which are lateralized to the right hemisphere (Cutting, 1990; Harris, 1995; Heilman, Watson, & Valenstein, 1985; Posner & Peterson 1990).

autists' range of fluctuation of autonomic activity to be much greater than in normal children.

I think we should exercise much caution in interpreting the above data.

First of all, the OFC sends projections to the hypothalamus not only through the amygdala, but also directly. The latter may trigger autonomic response without amygdala involvement. Second, an increase in autonomic activity during tasks may be not specific, and may just reflect the hyperarousal state characteristic for autism (as suggested to be a manifestation of exposure of the RH sensory-motor level). Finally, there is a lack of correlation between affective report and autonomic response in autism, which might reflect right amygdala activity. The right amygdala is not involved in verbalizing emotion (affective report) and is more closely connected with the autonomic system.

7.9 Do Autistic People Feel?

Asperger warned that children with autism "cannot be understood simply in terms of the concept 'poverty of emotion', used in a quantitative sense. Rather, what characterizes these children is a qualitative difference… They are full of surprising contradictions…" (Asperger, 1991, p. 83).

What is this qualitative difference and how it is related to what is known about the brain? This is now our primary inquiry.

In looking for an answer, we will turn to Oliver Sacks' writings about Stephen, an autistic savant artist, who also had an outstanding musical ability. This is how Oliver Sacks described Stephen when they met: "He reminded me… of the autistic children I had seen before, with a head-nodding mannerism or tic, and some odd flapping movements of the hands. He never looked at me directly but seemed to glance at me, briefly, out of the corners of his eyes" (Sacks, 1995, p. 204). At an early age, Stephen "would not play with other children and tended to scream or hide in a corner if they approached. He would not make eye contact with his parents or anyone else. Sometimes he seemed deaf to people's voices, though his hearing was normal (and thunder terrified him)… he was virtually mute" (p. 197). His schoolteacher wrote:

> He had virtually no understanding of or interest in the use of language. Other people held no apparent meaning to him except to fulfill some immediate, unspoken need; he used them as objects. He could not tolerate frustration, nor change in routine or environment and he responded to any of these with desperate, angry roaring. He had no idea of play, no normal sense of danger and little motivation to undertake any activity except scribbling (Sacks, 1995, pp. 197–198).[21]

And this is how Sacks describes an episode with Stephen and his music teacher:

> Evie asked Stephen if he would sing 'What a Wonderful World'. His singing seemed to be full of genuine feeling, and his gestures while he sang were not his usual stilted, ticklike

[21] From AN ANTHROPOLOGIST ON MARS by Oliver Sacks, copyright © 1995 by Oliver Sacks. Used by permission of Alfred A. Knopf, a division of Random House, Inc.

> ones.... Then Evie gave him an exercise in 'interpretation'...playing a theme he had never heard before, Schumann's 'Traumerei'. Stephen listened intently and told us his 'associations' as he listened: 'It is about... air in the field, daffodils in springtime...a stream... sunshine...rose gardens...light breezes, fresh...children come out to play with their friends Sacks (1995, pp. 238–239) (see footnote 21).

Note that he gives his interpretations not in terms of feelings, but as visual scene-situations that convey emotions similar to those evoked by the music.

Here is another example from Sacks' brilliantly told anecdotes about Stephen—one where he performs the song "It's Not Unusual":

> He sang with the great enthusiasm, swinging his hips, dancing, gesticulating, miming clutching the imaginary microphone to his mouth, addressing himself in imagination to a vast arena. 'It's Not Unusual' has become the theme song of Tom Jones, and in his version, Stephen took on Jones's flamboyant physicality, adding to it a flavor of Stevie Wonder. He seemed completely at one with the music, completely possessed—and at this point there was none of the skewed neck posture that is habitual with him, none of the stiltedness, the ticcing, the aversion of gaze. His entire autistic persona, it seemed, had totally vanished, replaced by movements that were free, graceful, with emotional appropriateness and range. Very startled at this transformation, I wrote in my notebook, "AUTISM DISAPPEARS." But as soon as the music stopped, Stephen looked autistic once again Sacks (1995, p. 239) (see footnote 21).

This is really a puzzle:

Sacks observed Stephen "clutching the imaginary microphone to his mouth, addressing himself in imagination to a vast arena," but what is characteristic for autistic people is that they do not have imaginations!

Did Stephen *imitate*? We know that inability to imitate is also a particular problem specific to autism. Autistic children do not imitate adult's actions to acquire skills. They have a particular difficulty in mechanical learning—unable to be taught by adults in conventional ways. The same child who astounded others by solving complex math problems could not learn the simple calculation methods even children with mental retardation are able to acquire. Asperger's point on this was that autistic individuals are always original, and they need to create everything out of their own thought and experience. While describing Stephen's talented imitation of a popular singer, Sacks indicates Stephen was "completely at one with the music, completely possessed" and "full of genuine feeling." What exactly was going on: was it genuine feeling or emotional expression without emotional subjective experience?

WHAT WAS IT? Perhaps we have a clue: Stephen's miraculous transformation was taking place strictly within the boundaries of the visual-musical situation.

As mentioned before, music is unmediated experience—it is gestalt, a separate domain, as is the visual scene-situation. Using this analogy, we may speak about a "musical situation" in which emotion is embedded.

It seems to me that Stephen, for the time of his performance, did not imitate the visual-musical situation—he *was* the visual-musical situation, his being was the visual-musical situation, acting it out.

WHAT is this "acting out" in terms of brain mechanisms? What level is exposed and projected outside? Apparently, the right-hemispheric situational level (the visual-musical situation) where a visual scene and the sounds, feelings, and actions therein are not distinguishable, but represent a singular whole. Stephen "feels" but the feeling is in within the situation. Stephen's gracious movements "within a visual-musical situation" indicate that not only the sensory motor, but also the thalamic level is exposed.

Taking together the emotional expressiveness of Stephen and that previously described of Nadia and of the "twins"—what kind of communication does it represent? I think it is EMPATHY—empathic identification with the physical world.

7.10 Amygdala—Temporal Cortex Functional Network, Penfield's "Brain Archive" and Memory of Autistic People

John Hughlings Jackson in the latter part of the nineteenth century described so-called psychic or experiential phenomena activated by epileptic discharge arising in the temporal lobe (Jackson, 1880, 1958). He reported cases of epilepsy with "dreamy states" consisting of scenes remembered from the past. The experiential phenomena were always the same or essentially the same: with each seizure of the patient, a scene was stereotypically "replayed." It was involuntary reminiscence, sudden re-experiencing of the past. Patients indicated an astonishingly vivid immediacy of these experiences. However, patients also realized that the involuntarily experienced "flash-backs" were separate from present reality. This is why Jackson called this condition "doubling of consciousness." The patients' insight clearly distinguished these phenomena from psychotic hallucinations and illusions (Gloor, Oliver, Quesney, Andermann, & Horowitz, 1982).

In 1933 Wilder Penfield, while operating on a conscious patient with intractable seizures, for the first time, evoked a memory "flash-back" by electrical stimulation of the temporal cortex. This was much to his own and the patient's surprise: "I was incredulous. On each subsequent occasion I marveled… I was more astonished each time my electrode brought forth such a response. How could it be? This had to do with the mind!" (Penfield, 1975, pp. 21, 27).

Following this first discovery, Penfield for many years evoked and carefully recorded such experiential responses by electrical stimulation of the temporal lobe in numerous conscious patients during surgery for epilepsy. In the patients' minds, the situation or event from their personal past was not a typical recollection, it was more vivid and intrusive—it was a "re-experiencing" and had a compelling immediacy similar to or sometimes even more vivid than those occurring in real life. Penfield concluded that epileptic discharge and electrical stimulus may both be activating the same physiological mechanism. He also believed that evoked responses represented activation of a neural mechanism that keeps a record of past experience. He recounts:

> There is activation too of the emotional tone or feeling that belonged to the original experience. The responses have that basic element of reference to the past that one associates with memory. But their vividness or wealth of detail and the sense of immediacy that goes with them serves to set them apart from the ordinary process of recollection which rarely displays such qualities. Thus, with stimulation at point 11, J.V., case 15, said, 'There they go, yelling at me, stop them!'...with stimulation at point 14, T.S., case 37 said, ...'That music from the stage hit, Guys and Dolls.' When he was asked if he seemed to be there or was remembering it, he said, 'I seemed to be there'.... Although there is a strong sense of immediacy in these experiences, none of the patients has ever confused the hallucination with reality except perhaps for a moment. All have retained awareness of the operating room and the events occurring in it, even during the time of an experiential response. For example, G.L., case 22, at point 15b said, 'I see people in this world and in that world too at the same time' (W. Penfield & P. Perrot, The Brain's Record of Auditory and Visual Experience: A Final Summary and Discussion, *Brain*, 1963, 86, p. 679. By permission of Oxford University Press).

Again, we can see that patients know where they are, what is going on, and are able to report their experience. They know that what they are experiencing is from the past, but at the same time, they feel it to be real—"to be there."

Penfield believed that our brain keeps a remarkable record of the stream of each individual's consciousness. During a seizure or electrical stimulation of the temporal cortex, the patient is re-experiencing a particular situation from his past because the epileptic focus or place of electrode application happened to be where this particular experience is stored. Each time a seizure or electrical stimulation was repeated, the given segment of the brain's record involuntarily replayed. Penfield and Perot write:

> [T]he evoked experiential response is a random reproduction of whatever composed the stream of consciousness during some interval of the patient's past waking life.... It may have been a time of listening to music, a time of looking in at the door of a dance hall, a time of imagining the action of robbers from a comic strip, a time of waking from a vivid dream, a time of laughing, conversation with friends, a time of listening to a little son to make sure he was safe, a time of watching illuminated signs, a time of lying in the delivery room at childbirth, a time of being frightened by a menacing man, a time of watching people enter the room with snow on their clothes.... It may have been a time of hearing someone call your husband's name, a time of listening to your mother scold your brother, a time of watching a guy crawl through a hole in the fence at a baseball game, a time of standing on the corner of 'Jacob and Washington, South Bend, Indiana'...a time of grabbing a stick out of a dog's mouth, a time of listening (and watching) your mother speed the parting guests, a time of listening to the broadcast from Philadelphia...a time of hearing your father and mother singing Christmas carols (W. Penfield and P. Perrot,[22] The Brain's Record of Auditory and Visual Experience: A Final Summary and Discussion, *Brain*, 1963, 86, p. 687. By permission of Oxford University Press).

Penfield indicates that in past experiences recalled electrically, perceptions are largely auditory, visual, or both. That these experiential responses were evoked by

[22] These examples are from the records of different patients.

electrical stimulation of the temporal lobe only is very important. Furthermore, there were more of such responses from the right than from the left temporal lobe. The experiential response points were located between the primary auditory area of the temporal cortex and the visual occipital cortex. For visual experiential responses (scenes, persons, or objects), prevalence of the right temporal cortex was even more remarkable than for auditory responses. A considerable collection of points (visual responses) was found in the posterior temporal region, a few of which lie quite close to the zone of transition between the temporal cortex and the secondary visual cortex (BA19). As we know, this area corresponds to BA37. Penfield gives an eloquent example of a striking delimitation between the occipital visual area, BA19 and the posterior temporal area, BA37. In patient R.W., case 3, occipital cortex stimulation produced colored flashes of light, stimulation immediately anterior to this, just in front of temporal–occipital transition[23] produced a sharp change of response from crude sensation to experiential: the patient then saw robbers with guns and his brother aiming an air rifle at him.

Experiential responses to electrical stimulation included perceptual-cognitive and emotional components. Vividness, a strong sense of immediacy and "of being there" are due to re-activation of emotion that belonged to the original experience. Here we return to the amygdala—we now are dealing with the functional connection between the temporal cortex (mostly BA37) and the amygdala. It is the amygdala that assigns visual things with emotional significance. Indeed, Gloor (1990) obtained experiential responses by electrical stimulation of the amygdala, not the temporal cortex. In the responses Gloor obtained, the emotional component was the most vivid, while the perceptual was more vague. He gives the following example of a 22-year-old epileptic patient whose *right* amygdala was stimulated during stereotaxic depth electrode exploration:

> [T]he patient immediately said, 'Yes'. When asked what he felt, he found it difficult to describe; it was a feeling like falling into water. When asked to elaborate, he replied that it was as if something had covered his eyes, nose and mouth.... Stimulation was then repeated...but without warning. He immediately exclaimed, 'Now', and then excitedly added, 'Could you do it again, doctor?' when asked why he wanted the stimulation repeated, he explained that when it began he had the words on his lips to describe the feeling, but then later could not do so.... Stimulation was repeated once more without warning. The patient opened his mouth with an astonished look, sat up and said that he now knew what it was. It was the feeling of being at a picnic in Brewer Park in Ottawa: 'A kid was coming up to me to push me into the water. It was a certain time, a big special day during the summer holidays and the boy was going to push me into the water. I was pushed by somebody stronger than me. I had experienced that same feeling when I had 'petit mals' before.' He called it 'bad feeling'.... On questioning, he said this had been a real event in his life when he was about 8 yrs old (P. Gloor, Experiential Phenomena of Temporal Lobe Epilepsy: Facts and Hypothesis, *Brain*, 1990, 113, p. 1676. By permission of Oxford University Press).

[23] The peripheral part of BA37.

As we see, the patient described his "re-experience" from the vantage point of his feeling, rather than directly reporting the visual scene.

Gloor et al. (1982) indicate that their findings do not contradict Penfield's data, but illustrate close anatomical and functional connection of the amygdala and temporal cortex (in this book's terms, the amygdala—BA37). They also emphasize these areas' different contribution to individual experience. However, the totality of a given experience is recreated even if only one part of the network (here, the amygdala "end" of network) is stimulated.

Interestingly, Gloor et al. (1982) in their series of amygdala stimulation, observed a much smaller number of auditory versus visual experiential phenomena, which, the authors indicated, is evidence of the visual modality's privileged access to the primate amygdala.

Do these electrically evoked memory 'flashbacks' represent what we described as RH immediate experience—momentary visual scene-situations, stored as perceived?

Are these experiential recollections indeed the visual scene-situations that are on the foreground of consciousness in autistic people? Are they the visual scene-situations that represent memory of autistic savants?—YES and NO. *Is there more to learn about the visual scene-situations and autism from the memory 'flashbacks' described by Penfield?*—YES!

First of all, it is very significant that the events re-experienced were most often *insignificant*. This corresponds to our definition of RH flow of consciousness—every moment registered as a visual scene-situation, the "flow" of which is the base for maintaining an alert state. The visual scene-situation is stored as perceived, but even when it was perceived, it was not on the foreground of consciousness. Thus, the stored VSSs are not really "memory" as we understand the term—they cannot be recollected voluntarily because they have never been in consciousness! Penfield's terms "brain record" or "brain archive" then are very insightful.

Why does the brain keep a record of insignificant events that did not even go to the surface of consciousness when initially perceived? They are needed not for their content but as *moments* of experience in time that give rise to a continuity of one's life time line, an important part of self. Recall that the visual scene-situation is a single whole, not dividable into component parts and not connected with other situations. Namely because situations are *discrete*, their sequence makes up one's *continuous* time line.

Oliver Sacks cites some pieces from Proust's *Remembrance of Things Past* that are of one accord with this idea:

> ...Proust's image of life was as "a collection of moments," the memory of which are "not informed of everything that has ever happened since" and remain "hermetically sealed," like jars of preserves in the mind's larder.... Proust writes: "A great weakness, no doubt, for a person to consist entirely in a collection of moments; a great strength also; it is dependent upon memory, and our memory of a moment is not informed of everything that has happened since; this moment which it has registered endures still, lives still, and with it the person whose form is outlined in it" (from Sacks, 1995, p. 172).

What a beautiful metaphor for RH self at the sensory-motor level exposed in autism! And it is such moments that are evoked and re-experienced involuntarily when, as Penfield believed, the electrode "touches" the part of a patient's "brain record" where they were stored.

But here, the similarity between memory in autistic people and electrically evoked experiential recollections ends. Recall how Sacks (1987) described the experiential recollection of the twin-autistic savants:

> They can tell one the weather, and the events, of any day in their lives—any day from about their fourth year old.... Give them a date, and their eyes roll for a moment, and then fixate, and in a flat, monotonous voice they tell you of the weather, the bare political events they would have heard of, and the events of their own lives—this last often including the painful or poignant anguish of childhood, the contempt, the jeers, the mortifications they endured, but all delivered in an even and unvarying tone, without the least hint of any personal inflection or emotion...there is no personal reference, no personal relation... (Sacks, 1987, p. 198).[24]

Lack of emotional response here was explained (see Chap. 6) by the fact that, at the sensory-motor level in the RH, emotion is included *into* the visual scene-situation and not distinguishable from the visual images. On the contrary, Penfield indicated that electrically evoked experiential recollections are often accompanied by the emotion experienced originally. Penfield emphasizes that although most events re-experienced when the temporal cortex was stimulated were not significant ones, some of them *were* significant and charged with intense emotion. Penfield also notes that "significant" events are more prone to re-activation during electrical stimulation, and he thinks it is because these events are more heavily represented in the brain.

How can we make sense of all this in terms of brain regions? We must return to the *amygdala-BA37* network. As discussed before, BA37, the temporal cortex's higher order visual area, provides the visual content of experiential recollections, while the emotional component is a contribution of the amygdala. In the RH, with its holistic processing, these two components are so intimately integrated that they are not separable—emotion and visual image form a single integrated representation. In the LH, the separate object is singled out of a visual scene-situation, and the amygdala attaches emotional significance to the object. But, again, because of the specificity of LH cognitive mechanism, emotion is "detachable" from the object and can be experienced per se.

The RH flow of consciousness is a subject for LH analysis, which will distinguish "significant" events in the context of a person's current goals, needs, and motivations (LH focus of consciousness). Thus, the very idea of a "significant event" is a secondary process and the "work" of LH consciousness. RH moments of experience

---

are equipollent, and they cannot be compared because they are not connected with each other. "Significance" or "nonsignificance" is irrelevant to RH cognitive mechanism.[25] Emotionally significant events selected by the LH become conscious and are stored in the LH. They are by definition individually specific, personal. These events are what we recall from our past and can retrieve voluntarily, such as the separate pieces of vivid emotional scenes remembered from childhood.

We may think of a "brain archive" that is storing one's whole stream of consciousness as connected with the visual area in the right temporal cortical region, while voluntary emotional memory would be associated with the homologous area in the left temporal region. In general, we are talking here about an interaction between the hemispheres: left amygdala—left temporal region with right temporal region—right amygdala. The left amygdala has direct access to the content of our memory from its stores in the left hemisphere, which may explain why emotional memory does not stay the same, it may be modified, re-constructed, and its emotional emphasis changed, depending on one's current emotional condition, goals, and motivation. In other words, what is significant *now* may alter how significant events from the past are remembered. Emotional memory may then be changing. The "brain archive," however, can never be changed—it is a component part of the self. Emotional memory can be retrieved voluntarily, but one's "brain archive" is beyond the reach of voluntary effort.

We may think that, in the special case of autism, the 'brain archive' is on the surface of consciousness and available without any voluntary effort.

It was mentioned earlier that experiential responses can be evoked by stimulation of any "end" of the network, either the amygdala or the temporal cortical region. In other words, activation of any part of the network can turn on the whole network. In this context, Penfield's view that significant events have massive representation in the brain and can be more easily evoked is justified: they are part of brain record in the RH and part of emotional memory stored in the LH. Thus, when any "end" of the network left amygdala—left temporal cortex or right amygdala—right temporal cortex is stimulated, these events would be recalled.

We may conclude that experiential recollections due to epileptic discharge or electrical stimulation often involve both hemispheres, while subjectively felt visual scene-situations exposed in autism are of purely right hemispheric origin.

7.11 The Amygdala, Object Relatedness, and Autism

The amygdala's role in the generation of emotional experience is well known from clinical and experimental data. However, the *object-relatedness* of limbic emotion is rarely emphasized. The formulation of limbic emotion as object-related

[25] Implicit in the concept of significance is always the question "significance for *whom*?"—this "whom" is related to the left hemispheric self.

subjective experience, I believe, is crucial for understanding autism. Recall that in the network of the left amygdala—left inferior temporal cortical region (BA37), BA37 is responsible for identification of a separate object, while the amygdala is responsible for evaluation of the biological significance of the object with generation of subjective experience (emotion) *toward* the object. People with autism are able to identify objects and name them, and they do have a subjective (emotional) experience. Remember Elly's intense emotion evoked by certain numbers, or Anthony's[26] indirect expression of his fondness for his grandmother through number: "I like best 55" (grandmother's age). What then *is* missing in autism? Recall Stephen's example given earlier in this chapter. Stephen *feels*, but he feels *within the frame of the visual scene-situation*. It is RH limbic emotion which the autist is experiencing "consciously." In the RH, feeling is expressed through the visual image, and they form a single integrated representation. Thus, what we observe in autism is lack of emotional relation to the SEPARATE object (left-hemispheric deficit). Instead, the whole VSS or visual symbolic system (to which the target object belongs) is saturated with affect. In the case of Elly and Anthony, numbers are saturated with affect, symbolizing the VSS that is behind them.

It is a tribute to Kanner and Asperger, their powerful clinical thinking and great insight, that from the very beginning, they both (independently from each other) delineated the clinical pattern of autism and "extracted" the core of the disorder. This is how Kanner formulated it: "The outstanding, 'pathognomonic' fundamental disorder is the children's inability to relate themselves in the ordinary way to people and situations from the beginning of life" (Kanner, 1943, p. 242). He illustrated this formulation with a vignette about an autistic child: "…he was never angry at the interfering person. He angrily shoved away the hand that was in his way or the foot that stepped on his blocks, at one time referring to the foot that stepped on the block as 'umbrella'. Once the obstacle was removed, he forgot the whole affair" (Kanner, 1943, p. 220).

It directly follows from the object-related essence of LH limbic emotion that it is intrinsically social. In light of this claim, the ongoing intense debate in the current literature about whether the primary deficit in autism is emotional or social-cognitive (Corden, Chivers, & Skuse, 2008; Dawson, Webb, & McPartland, 2005; Gaigg & Bowler, 2008Schultz, 2005) seems artificial and superfluous.

The social nature of limbic emotion can be well illustrated by exploring its phylogenetic roots.

The amygdala has a broader role in nonhuman primates than in humans. In nonhuman primates, vocalization, used for communication, represents a syncretic whole of sound, affect, and information. Thus, vocalization in nonhuman primates is a message (communication) that includes knowledge about the object (cognitive part) and an attitude toward the object (emotion). Here the amygdala is responsible for generating affect and for recruiting the auditory temporal cortex (implementation of sound) and visual temporal cortex (*what*-system) through its efferent connections. Note that in humans the amygdala influences cognition through the same efferents!

[26] Anthony was the 5-year-old autistic boy from Kanner's material discussed in Chap. 6.

Neurophysiological studies show that N4 evoked potentials are specific to language processing: both phonological (N4 occurs to pronounceable nonwords but not to nonpronounceable nonwords) and semantic. N4 is also generated in the amygdala in response to words, objects, and faces (Halgren, 1992). Halgren suggests that since N4 appears to be generated in both the language cortex and the amygdala, in response to both words and to social signals such as faces, it (N4) might be an evolutionary link between primate social communication and human language.

In humans, damage to the amygdala does not disrupt language, and lesions in the language-specific areas do not cause emotional disorder. Autism is a unique pathology in that emotional deficit is intrinsically connected with language deficit, mostly its cognitive-communicative part, and there is no evidence that specific language zones are damaged in autism. The strikingly specific language atypicality in autism cannot be explained just by the difference in manifestation of acquired and developmental brain disorders. In Chap. 6 I hypothesized that the phylogenetic roots of human brain functions (long since "overgrown" historical pathways) might still be "alive" in autism. Is it the case here? We cannot help but see the phylogenetic link between emotion and social-communicative function in autism in pathological form.

7.12 The Orbital Frontal Cortex and Its Connectivity Pattern

The OFC represents the regions in the prefrontal cortex situated at the very anterior surface of the cerebral hemispheres, above the eye orbits (BA11 and BA47). As is the case with the amygdala, examining the OFC connectivity pattern is crucial for understanding its function.

The OFC is a region of cytoarchitectural transition between the agranular cortex (similar to the olfactory cortex) and the granular cortex (similar to the rest of the prefrontal cortex). Along this transitional continuum, two parts can be distinguished within the OFC: the posterior OFC (mostly agranular) and the anterior OFC (mostly granular). The posterior and anterior have different origins (paleocortex and archicortex, respectively) and distinct patterns of connectivity (Zald & Kim, 1996a). The posterior OFC has a close relationship with the amygdala; it is notably the only prefrontal cortical region that has direct connections with the amygdala. Figure 7.2 shows that the OFC receives afferents from both the amygdala's lateral nucleus (sensory input zone) and the basal nucleus, the latter sending information that has already flown through the intrinsic pathway between the lateral and basal nuclei. Interestingly, the OFC's connection with the basal nucleus is reciprocal, i.e., the OFC also sends direct efferents to the basal nucleus. Such a connectivity pattern suggests that the OFC may control the amygdala's influence on sensory processing of emotionally significant stimuli and events, perhaps even emotional experience itself.

The OFC also receives indirect input from the amygdala via the mediodorsal (MD) nucleus of the thalamus (Fig. 8.1). As discussed before, the MD nucleus projects onto the vast territory within the prefrontal cortex, but amygdala fibers terminate

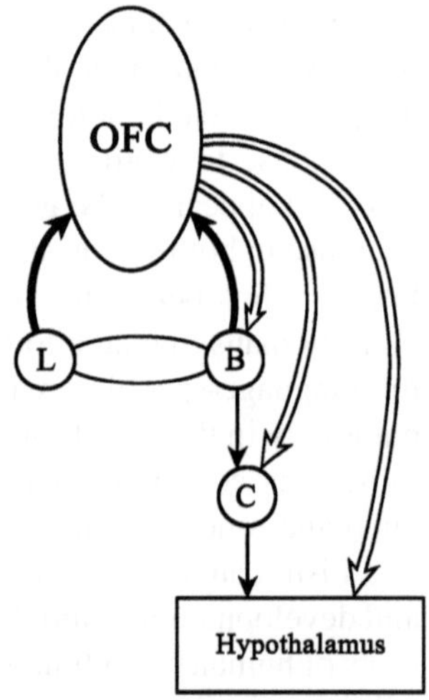

Fig. 7.2 OFC connections with the amygdala. L—the lateral nucleus of the amygdala; B—the basal nucleus of the amygdala; C—the central nucleus of the amygdala. Modified from *Anatomy and Function of the Orbital Frontal Cortex, I: Anatomy, Neurocircuity, and Obsessive–Compulsive Disorder*, by D. Zald and S. Kim, The journal of Neuropsychiatry and Clinical Neurosciences, 1996. Reprinted by permission

only in that part of the MD nucleus which projects to the OFC. Thus, the MD pathway to the OFC contains information about both limbic and thalamic emotion. I believe that human-specific emotion is a result of an integration of limbic and thalamic emotion (the last will be described and discussed in Chap. 8). It looks as though the place of integration of these two emotions is the OFC—most likely the right OFC, due to its holistic processing. It should be noted that limbic and thalamic emotions are not symmetrical. Limbic emotion is object related and as such is more connected with the LH. Thalamic emotion, subject related (based on bodily feelings), is more connected with the RH and its visual-symbolic thinking.

The OFC's efferent connections to the central nucleus (Fig. 7.2) may influence amygdala output to the autonomous and endocrine systems. Also important to note is that, of the prefrontal regions, the OFC alone projects directly to the hypothalamus.

The posterior OFC practically speaking connects with the same areas the amygdala connects to. For example, it also receives highly processed sensory information from the inferior temporal cortex (BA37), visual being its main sensory input. However, the OFC has connections that the amygdala does not have, especially with the DLPC. The DLPC occupies most of the convexital surface of the frontal lobe, and, as discussed before (see Chap. 1), the left DLPC plays a major role in operations with concepts, verbal (semantic) associations, abstract thinking, verbal–logical thinking. The area in the prefrontal cortex between the OFC and the DLPC has recently been distinguished as the ventral lateral prefrontal cortex (VLPC). The meaning behind the OFC–VLPC–DLPC network is illustrated by the following studies. PET brain imaging in patients with familial pure depressive disease showed increased activity in the left amygdala, left OFC and left VLPC (Drevets et al., 1992). In subsequent experiments, during contemplation of sad thoughts in healthy subjects, increased blood flow (brain activity) was found in the left OFC and the left VLPC, but not in the amygdala (Pardo, Pardo, & Raichle, 1993). Finally, during performance by healthy subjects on a word fluency task (a purely cognitive task), increased brain activity was observed in the left VLPC and the left DLPC, but not in the left OFC (Drevets & Raichle, 1995). The authors suggest that the left OFC and VLPC are responsible for bridging emotions and concepts, with the left VLPC more connected with making verbal associations as such, playing the major role in the production of negative ruminative thoughts in patients with depression.

7.13 Role of the Orbital Frontal Cortex in Human-Specific Emotions and Social Behavior

Although the amygdala may be the seat of basic emotions, it is the amygdala–OFC connection upon which more complex, specifically human emotions and their social and cognitive aspects rely. The social aspect of human emotion derives from object-related limbic emotion: the evaluation of objects as to their potential for biological need satisfaction. Derivative of left hemispheric limbic emotion are *social* emotions connected with the left OFC such as guilt, shame, embarrassment, pity, pride, jealousy, envy, etc. It seems that the left OFC also modifies emotional expressions in accordance with the norms of the given social group (perhaps the OFC's efferents to the amygdala's basal nucleus are involved here?). The following example illustrates the above statement. Spontaneous facial expressions and upper body gestures were compared in left and right brain-damaged patients, Parkinson patients, healthy age-matched adults, and preschool children when shown slides depicting either familiar people, scenery, unusual scenes, or unpleasant situations (Buck & Duffy, 1980). In children and adult controls, familiar slides generated the most pronounced and identifiable responses, unpleasant situations generated few if any emotional responses, and scenic and unusual slides generated intermediate responses. Although right brain-damaged and Parkinson patients showed significantly diminished expressivity, they still maintained the same profile. In contrast, left brain-damaged patients showed substantial responses to all slides without any particular profile.

In most formal situations, positive emotional expressions (such as cheerfulness and attentiveness) are expected to be in keeping with social "display rules" (Ekman & Friesen, 1975). In the above-cited experiment, only left brain-damaged patients did not follow display rules, i.e., did not suppress the expression of negative emotions.

In the left orbital–frontal cortex, emotional significance of the object is integrated into a social-behavioral context; evaluation of the object is translated into foundation of interpersonal relationships, the roots of what is judged "good" or "bad," and the rules and beliefs of the group.

Judgments and beliefs and not purely intellectual categories, they represent integration of concepts and an individual's emotions within the context of his or her given social group's value system. Left hemispheric limbic emotion then evolves in the prefrontal cortex by its union with cognition, becoming *beliefs*.

How can we apply the "action" metaphor (see Chap. 1) to the frontal lobe's function here? The OFC is situated at the crossroads of connections involving the amygdala, the medial prefrontal cortex (MPC), and the DLPC—the TPO (BA39,40) network.

Correspondingly, the OFC receives information about emotional significance of the stimuli from the amygdala, information about the motivational state of the organism from the MPC, and knowledge about territorial environment from the parietal cortex (mostly indirectly via the DLPC). The current understanding of the OFC function is, thus, that the OFC integrates emotional-motivational information and knowledge of the territorial environment (estimation of risk) into a behavioral context, with the OFC efferent connections to the DLPC triggering the DLPC's purely cognitive component of motivated goal-directed behavior, namely realization of inner programming, unfolding action in time.

It can be concluded that although the posterior OFC resembles the amygdala in its pattern of connections, and the OFC appears at least to a certain degree able to process information in parallel to the amygdala, in the network *amygdala–OFC*, each component contributes differently to the control of social behavior. The amygdala's role is to detect object or event significance, to produce clear and distinct threat or safety signals, and to trigger the autonomic–neuroendocrine changes necessary to prepare the organism for a behavioral response. The OFC makes use of this information to guide goal-directed behavior and to adjust behavior in accordance with changing conditions (Bachevalier & Loveland, 2006; Bechara, Damasio, Damasio, & Lee, 1999; Zald & Kim, 1996a), including monitoring autonomic response (via the OFC's direct projections to the hypothalamus).

Damage to the OFC results in inability to experience complex, human-specific emotions, emotion disregulation, and inappropriate social conduct. Patients are coarse, tactless, and irresponsible, lacking social restraints and concern for other people. They are irritable and moody, with shallow, unstable emotional expression and sexual disinhibition. Such a patient is aptly described in the following vignette:

> A patient makes profane remarks, he is facetious and boastful, but soon becomes apathetic and indifferent; at some later time, he may show a sudden burst of short-lived anger. Soon after, an unstable state of exuberance may be observed occasionally interrupted by a flare-up of irritation (Damasio & Anderson, 1993).

OFC-damaged patients are unable to enjoy pleasurable stimulation if it involves social, intellectual, or aesthetic rewards. On the other hand, they are excessively involved in pleasurable activities, with apparently decreased sensitivity to negative risks. For example, there are numerous reports in the literature of OFC-damaged patients making disastrous business decisions. Such patients might intellectually process their behavior as morally wrong or risky, but they are unable to experience human-specific social emotions, like guilt and shame. Mere intellectual knowledge of an action's consequences has limited influence on behavior (Zald & Kim, 1996b).

Although literature presents examples of "acquired antisocial personality disorder" in patients with unilateral left OFC lesions (Meyers, Berman, Scheibel, & Hayman, 1992), there is an opinion that RH or bilateral damage to the OFC has a greater effect on social conduct, decision-making, and emotional processing (Manes et al., 2002).

As done in prior discussions about interhemispheric differences, we will remain within the paradigm that qualitative differences of the LH and RH syndromes are based on the difference between LH versus RH modes of processing (not "what" but "how" they process information).

One may suppose that this social incompetence in the LH OFC syndrome is at its core a nonintegration of emotions and concepts, impaired "surveillance" of the social-territorial environment, and a disorder of integrating emotions into a behavioral context.

In the RH OFC syndrome, one might expect more general problems with emotional sensitivity and empathy, which would manifest as a profound dullness to social stimuli and situational nuances.

Obsessive-compulsive disorder (OCD) is a psychiatric disorder that is characterized by a pathological increase in OFC functional activity, as revealed by the brain imaging studies. It was suggested that brain pathology in OCD involves the anterior part of OFC whose connections are with higher cortical association areas (not with the amygdala) (Zald & Kim, 1996b).

Patients with OCD suffer from disturbing, intrusive, and repetitive thoughts (obsessions). The themes of obsessions relate to social-territorial concerns about danger, violence, sex, hygiene, and order. The danger is not real: input from the temporal and parietal cortex about a patient's environment does not signal any concern. It is the inappropriate hyperactivation of the OFC and its hypervigilant surveillance of the social-territorial environment that produces exaggerated harm concerns (hyperestimation of risk). Interestingly, many of the social-affiliative characteristics that are so deficient in OFC-damaged patients are accentuated in OCD patients (Zald & Kim, 1996b). While OFC-damaged patients show lack of concern and irresponsibility toward others, OCD patients often appear to be excessively concerned with how their actions will affect others. Furthermore, feelings of guilt, shame, anxiety, and concern for social norms, lacking in OFC-damaged patients, are pathologically excessive in OCD.

The OFC's commanding contribution to social behavior is illustrated by the above examples of its damage as well as its hyperfunctioning. Now, I will place an emphasis on the *temporal cortex-amygdala-OFC* network's role in generating human-specific emotions. To do so we will look at examples of experiential seizures in which such emotion is intensified, in particular the experiential seizures of two extraordinary individuals with "premorbidly" unusual development of their temporal and frontal brain regions. The first example given will be the case of Franco Magnani, a "memory artist" described by Oliver Sacks (1995).

At the age of 31, Franco began having sudden episodes of involuntary recollections that were very similar to the so-called experiential seizures described by Jackson in patients with the temporal lobe epilepsy. The scenes of his involuntary recollections were flashbacks, and they were always of the little Tuscan village where he was born: Pontito. Such consistency of "topic" in Franco's recollections corresponds with Jackson's observations that the experiential phenomena of temporal lobe seizures are either always the same or essentially the same for a patient.

In Franco's case he always experienced the scenes of Pontito's "streets and houses, the masonry, the stones…with the most microscopic, veridical detail, a detail beyond anything he could consciously remember." (Sacks, 1995, p. 159). The scenes were vivid ("like I was there") and emotionally overwhelming, charged with personal meaning and affect. Furthermore, Franco could

> …feel the churchyard wall; and, above all, he [could] smell what he sees—the ivy on the church wall, the mingled smells of incense, must, and damp, and, admixed with these, the faint smell of the nut and olive groves that grew around the Pontito of his youth. Sight, sound, touch, smell, at such times [were] almost inseparable for Franco… (Sacks, 1995, p. 162).

Of note, Franco commented that he had never seen images of such intensity before. This last aspect of Franco's recollections—their intensity—is consistent with Penfield's observation that his patients often described involuntary "re-experiencing" as more vivid than their original experiences. Penfield suggests that both epileptic discharge and electrical stimulation of the temporal lobe result in a more profound experience, with unnaturally amplified vividness and emotion, than that of normal phenomenon created by physiological activation in the brain. Finally, Franco also had a "doubling of consciousness" that is typical for experiential seizures. Sacks reports: "…when [Franco] is seized by a vision, a waking dream, a reminiscence of Pontito, he is transported—he is, in a sense, there" (Sacks, 1995, p. 165). Sacks further describes Franco's "doubling of consciousness": "he is, perhaps, talking with you quietly, and suddenly he leans forward, his eyes fixed and staring, in a rapture: an apparition of Pontito is rising before him" (Sacks, 1995, p. 162). As he began having more of these experiences, with ever deepening and intensifying of feeling, Franco started to experience an overwhelming urge, a "mission" even, to paint his visions. Before these experiences, Franco had never thought of being a painter. However, so strong was his urge to express these powerful experiences, he taught himself to paint and produced hundreds of depictions of his visions of Pontito. Sacks, reporting on the nature of Franco's painting, notes that when Franco had the "flash" of a scene, "he often feels a great urgency to get the scene down on paper immediately…" (p. 162).

Franco's life story is pertinent to uncovering the *amygdala-temporal cortex* network's role in his visions. Sacks writes that Franco had an idyllic early childhood, during which he was known for a remarkable "photographic" memory. However, his veritable childhood paradise ended with the outbreak of war: his father died in 1942 and a year later the Nazis occupied Pontito (after which the village never recovered). Franco lived in Pontito until he was twelve and then left for school. On completion of school, he did not return to his beloved village but searched elsewhere for work. When Franco was away from Pontito, he was markedly homesick. He would experience vivid involuntary recollections of the buildings and streets of Pontito. In 1965, under the stress of his decision to emigrate to America, Franco had full-blown episodes of what seems to have been experiential seizures accompanied by an irresistible urge to paint his "visions."

From this history, we learn that Franco had an intense nostalgic feeling for Pontito. He also had a great desire to recollect, to reminisce about Pontito—even once writing an autobiography. Thus, it is quite evident that what he was re-experiencing during his involuntary visions was highly significant for him personally. Given what is known about the amygdala, one can conclude that both right and left (but mostly left) *amygdala-temporal cortical* networks were activated to create Franco's visions. As proposed earlier, the right-hemispheric temporal cortex keeps a "brain record" while the left amygdala selects "significant events" for emotional memory, which are then stored in the left temporal cortex. Intriguingly, Franco's involuntary recollections were about the streets and buildings only, there were no people in these scenes. Sacks notes "he dreamed of Pontito, not of his family, not of activities or events, but of the streets, the houses, the masonry, the stones…" (p. 159). Franco's "visions" of Pontito were accompanied by nostalgic feelings as he experienced them in real life, yet such feelings were greatly deepened and intensified: "[a]n intense, strange excitement possessed him in these dreams: a sense that something had just happened, or was about to; a sense of immense, portentous, yet enigmatic significance, accompanied by an insatiable, yearning, bittersweet nostalgia" (p. 159). As noted earlier, epileptic discharge stimulates the temporal cortex beyond the physiological range, which can result in intense, ecstatic feelings that are very rarely experienced in real life.

Turning to Franco's paintings themselves (Fig. 7.3), they invoke emotions similar to those he experienced with his visions! What strikes the onlooker is the intensity and power of the paintings' emotional force. It's almost as if a condensed emotion is projected into the viewer—something dark and ominous, inexplicable and inevitable, portentous. What is overall seen in these paintings is NOT representation of the right hemispheric visual scene-situation, which would be a momentary snapshot of experience, a gestalt. Instead, Franco's paintings depict objects *selected* from the visual scene-situations, representing what is emotionally most significant for him—identity of Pontito (lost after the war). These scenes are thus a product of secondary, left hemispheric processing by which the left amygdala selected out images to represent *nostalgia* (a feeling that is very personal to Franco, to his life experience, but which was magnified by the epileptic discharge).

These visual images do not represent RH symbol, but rather LH sign. RH process of symbol formation is identification of the visual scene-situations by affect, so that feeling (affect) is not exposed, but is the very core for uniting the visual scene-situations, different in content, into a symbolic system of meaning. RH symbol is thus polysemantic and idiosyncratic. LH sign, on the other hand, is monosemantic. Franco's buildings are a means to express a powerful and complex, but clearly only one, feeling (nostalgia), and feeling *is* separable from the visual image.

The connection between Franco's left and right temporal cortex is intact. The right hemisphere contributes to these visual phenomena in that its "brain record" of the visual scene-situations is the source for his paintings' rich "experiential" details and their vivid, sensual quality.

Fig. 7.3 Franco's paintings. From *The Landscape of his Dreams*, in: An Anthropologist on Mars, by Oliver Sacks, 1995, Alfred A. Knopf, New York. Courtesy of http://www.francomagnani.com

The question then remains: "Why are there no people in Franco's recollections?" Franco is not autistic. Indeed, Sacks describes that when Franco showed him his paintings, he would "fill" his explanations with the situational details, all of which related to his early childhood in Pontito:

> "Look at this wall here—that's where the priest, he caught me climbing into the garden behind the church. He chased me all the way down the street. Oh, he always chased all kids away from there." Each reminiscence triggered others, and these still others, so that within minutes we were engulfed in a flood, without any clear direction or center, but all relating to his early life—to Pontito as he had experienced it as a child. He leapt from one story to another, without any connection that I could discern.... By midmorning, I had been enthralled again by Franco's paintings but had had enough of his reminiscences (p. 155) (see footnote 21).

Franco was verbose, circumstantial, and concrete—to the point of boredom. He calls to mind descriptions of patients with hyperactivated (disinhibited) left hemisphere (due to the temporary inhibition of the right hemisphere) with their verbosity and narrations with excessive details.

Although Franco was emotionally intense and nostalgic when telling stories about Pontito, he failed miserably in invoking commensurate emotion in the listener. Use of language was definitely not his artistic medium. Instead, the intensified feelings of his epileptic discharge both found expression in and revealed his talent for the medium of visual art. It must be remembered, as Sacks indicated, that "premorbidly," i.e., before he started to have his supposed experiential seizures, Franco was remarkable for his extraordinary visual memory and imagination, as were his mother and one of his sisters to a lesser degree. These exceptional abilities may thus be hereditary and connected with some peculiarities of the temporal region.

Finally, there is the task of describing how the last player in the *amygdala–temporal cortex* network, namely the left OFC contributes to Franco's images. It is helpful in this endeavor to first consider certain questions. Is nostalgia a human-specific emotion? If so, what exactly is nostalgia? Is it sadness, loss, or longing? Each of these three feelings contributes to the complex feeling of nostalgia, but as a sum they will not fully "make" nostalgia. Nostalgia is connected, on the one hand, with very personal life experience, and on the other hand with the cultural roots that play a role in the formation of each individual's beliefs and value system. Because nostalgia includes the junction of emotion and concept, it must be tied to the left prefrontal cortex. Franco's case well illustrates these assertions as to the complex emotion of nostalgia, for his nostalgic feeling includes not only personal loss, but also the loss of the culture he once belonged to. Sacks recounts that Franco's "family returned to the house in Pontito after the Germans had occupied it, and found things defaced and changed. Franco's mother was deeply upset, as was Franco. This 10-year-old fatherless child said to his mother, 'I shall make Pontito again for you, I shall create it again for you.'" (Sacks, 1995, p. 167).

Furthermore, imprints of the left prefrontal cortex's role are found in Franco's mission-like convictions surrounding his visions. His feelings associated with the personally significant events ("chosen" by the left amygdala) were magnified by the epileptic discharge. However, with continued bombardment of the temporal region

Fig. 7.4 Franco's paintings. From *The Landscape of his Dreams*, in: An Anthropologist on Mars, by Oliver Sacks, 1995, Alfred A. Knopf, New York. Courtesy of http://www.francomagnani.com

with more and more seizures, his experiences became increasingly intense, even ecstatic, which was then "translated" by the prefrontal cortex (union of emotion and concept) into mystical feeling-beliefs of "mission," "revelation," "of being called." Thus, we can conclude it likely that Franco's experiential seizures involved mostly the *left temporal cortex–left amygdala–left OFC* network.

Repeated hyperactivation of the temporal region in patients with temporal lobe epilepsy can result in chronic behavioral and personality changes. Patients can become religiously preoccupied or increasingly and overly involved in moral and philosophical issues (Geschwind, 1986). Geschwind indicates that preoccupations of such patients often have a cosmic or global flavor. This well describes what is seen in Franco, who increasingly became more and more "philosophically" oriented. There was also an evolution in his paintings toward cosmic motives with moralistic, instructive, and didactic tendencies. One such painting was titled "Pontito Preserved for Eternity in Infinite Space" (Fig. 7.4). Although those of his paintings that were characterized by expression of his beliefs and more "conceptualized" feelings are interesting, they loose the "natural" feel as well as the intense emotional and sensual appeal of his early paintings. Instead of drawing the onlooker in and evoking rich emotions, these more conceptualistic paintings are eerie and distancing. With such paintings, the involvement of his left VLPC is likely pronounced. Recall that the VLPC borders the DLPC and is concerned with verbal–logical and conceptual thinking. Conceptual thinking (as well as language, as noted earlier) was not the medium for which Franco was gifted.

However, there was a man with experiential seizures for whom language and conceptual thinking were definitely expressive media of which he was a master: Fedor Mikhailovitch Dostoiewski. Noted French neuropsychologist and neurologist Alajouanine gathered a rich collection of descriptions of Dostoiewski's seizures from letters and diaries, from what was said by his friends and relatives, and from descriptions of fits of his novels' characters which were in actuality a transcription of the attacks sustained by the author himself. This material allowed Alajouanine to diagnose Dostoiewski with temporal lobe epilepsy.

Alajouanine writes in the beginning of his fascinating study:

> To know that a man called Fedor Mikhailovitch Dostoiewski who lived in Russia in the second half of the nineteenth century was an epileptic subject, should not of itself arouse much interest, as this condition is not uncommon in any country of the world. But the interest of such an event appears at once when one recalls that this man has been one of the greatest Russian writers; that the experience of his fits has been shrewdly analyzed by himself, and furthermore was projected in certain characters within his attractive novels; and finally seems to have had an important role in his conception of life and of the world in general (T. Alajouanine, Dostoiewski's Epilepsy, *Brain*, 1963, 86, p. 209. By permission of Oxford University Press).

Here is an example of one of the emotional–ecstatic auras experienced by Dostoiewski:

> "I felt the heaven was going down upon the earth and that it had engulfed me. I have really touched God. He came into me myself, yes God exists, I cried, and I don't remember anything else" (p. 212).

From a neurological vantage point, one sees a rich psycho-sensorial experience here; a conglomerate of feeling from several integrated modalities—visual, tactile, kinesthetic. Dostoiewski's experience can thus be attributed to the right temporal cortex-right amygdala network. Furthermore, this experience is immediately felt as a metaphor, which is the working of his right prefrontal cortex. This felt metaphor is then 'rationalized' in the context of the cultural-religious concepts of Dostoiewski's given social group—the idea of God and heaven—which indicates involvement of the left prefrontal cortex. Finally, this extraordinary and powerful feeling, the result of epileptic discharge in his temporal lobe (amygdala), is translated by his left prefrontal cortex into belief—'God exists'

The following examples illustrate that epileptic discharge is truly "challenging" the temporal lobe, stimulating it to a degree beyond physiological range and resulting in such an intense, magnified feeling that it becomes ecstatic mystical experience.

> During a few moments I feel such a happiness that it is impossible to realize at other times, and other people cannot imagine it. I feel a complete harmony within myself and in the world, and this feeling is so strong and so sweet that for a few seconds of this enjoyment one would readily exchange ten years of one's life—perhaps even one's whole life (T. Alajouanine, Dostoiewski's Epilepsy, *Brain*, 1963, 86, p. 212. By permission of Oxford University Press).
>
> He (Prince Muichkine, the principal character in the novel "The Idiot") was thinking of the phase of the onset of his epileptic fits when they came upon him while awake. In a complete crisis of agony, stupidity and oppression, it seemed suddenly as if his brain was set on fire, and his vitality was prodigiously accelerated. During these moments as rapid as lightning, the impression of the life and the consciousness were in himself ten times more

intense. His spirit and his heart were illuminated by an immense sense of light; all his emotions, all doubts, all his anxiety calmed together to be changed in a sovereign serenity made up of lighted joy, harmony and hope; then his reason was raised up to the understanding of the final cause….These instants, to define them in one word, were characterized by a fulguration of the consciousness and by a supreme exaltation of emotional subjectivity (T. Alajouanine, Dostoiewski's Epilepsy, *Brain*, 1963, 86, p. 213. By permission of Oxford University Press)

At this moment, I (the Prince) have foreseen the meaning of this singular expression within the tenth chapter of the Revelation of St. John, THAT THERE SHOULD BE TIME NO LONGER (T. Alajouanine, Dostoiewski's Epilepsy, *Brain*, 1963, 86, p. 213. By permission of Oxford University Press).

Kirilov (a personage from the novel "The Demons"): there are some instants, they last 5 or 6 seconds, when you suddenly feel the presence of the eternal harmony, then you have reached it…it is clear indisputable, absolute feeling. You suddenly embrace the entire creation and you say, well, it is like that, it is true…It is not emotion, it is something else; it is felicity; you don't forgive anything because there is nothing more to be forgiven. It is not even love. The most terrible thing is that it is so frightfully clear and such an immense joy at the same time. If it were to last more than these five seconds, the soul would not be able to bear it and would disappear… (T. Alajouanine, Dostoiewski's Epilepsy, *Brain*, 1963, 86, p. 214. By permission of Oxford University Press).

Analyzing these experiences one can clearly see the different contributions by both hemispheres.

As for the right hemisphere, one notes there is a holistic ecstatic experience expressed in a symbolic (metaphorical) form, e.g., the person could "feel the presence of the eternal harmony" and an "immense sense of light." There is also a change in his subjective experience of time, speed, movement—"his vitality was prodigiously accelerated," "the impression of life and the consciousness were in himself ten times more intense." These phenomena are evidence of the right temporal lobe and probably also right prefrontal cortex's involvement in his seizure activity.

The left hemisphere's role in his experiences was to distinguish limbic emotion (joy, hope, happiness) out of the right hemispheric holistic experiences. The left hemisphere also gave verbal expression to his strong emotions. Finally, the transformation of sensual-emotional ecstatic experience into mystical thinking—"then his reason was raised up to the understanding of the final cause," and the interpretation of such experience within the cultural-religious context—"at this moment, I have foreseen the meaning of this singular expression with the tenth chapter of the Revelation of St. John" is the work of the left's amygdala and the prefrontal cortical areas (OFC and VLPC) which, as noted earlier, are responsible for bridging emotions and concepts. Interestingly, Dostoiewski's experience of "all his emotions, all doubts, all his anxiety calmed together to be changed in a sovereign serenity made up of lighted joy, harmony and hope," is opposite to the negative ruminative thoughts suffered by patients with depression and attributed by functional brain imaging to the left VLPC. Diametrically opposite experiences are often connected with the same area in the brain.

Alajouanine discussed how Dostoiewski's exceptional experiences, due to his seizures, influenced his work and his attitude toward life and the world, namely his ethics and philosophy. Figuratively speaking, it proved to be a "doubling of consciousness": Dostoiewski was a rationalist and a mystic at the same time.

In giving these detailed illustrations of temporal lobe epilepsy's manifestations, the intent was to show the contrast of this condition with autism—to show that different imbalances within the same system (*amygdala–OFC* network) result in profoundly dissimilar conditions. The examples of extraordinary individuals with "premorbidly" unusual development of the temporal and frontal brain regions were given to make the contrast more strikingly evident.

7.14 Amygdala—OFC Functional Network and Autism

7.14.1 Human-Specific Emotions in Autistic People

Temple Grandin in conversation with Oliver Sacks said that she was bewildered by Romeo and Juliet, reporting she never knew what they were up to. "She said that she could understand simple, strong, universal emotions but was stumped by more complex emotions and the games people play" (Sacks, 1995, p. 259). However, from Grandin's book *Thinking in Pictures* and from her dialogues with Sacks (*An Anthropologist on Mars*), it is clear she has strong beliefs and emotionally charged concepts that are attributable to the frontal end of the amygdala–frontal cortex network. As are people with temporal lobe epilepsy, Grandin is deeply involved in cosmic, philosophical issues. Yet there is a cardinal difference. In temporal lobe epilepsy, extreme limbic emotion, secondary to hyperactivation of amygdala, paves the way for its expression or justification in the frontal end of the network, so that personal feeling is translated in the frontal lobe as an ecstatic mystical experience (revelation, mission, etc.) with further rationalization in the context of the cultural-religious concepts of a given social group.

Let see what Temple Grandin says about her religious beliefs:

> It was quantum physics that finally helped me to believe again, as it provided a plausible scientific basis for belief in a soul and the supernatural. The idea in Eastern religion of karma and interconnectedness of everything gets support from quantum theory. Subatomic particles that originate from the same source can become entangled, and the vibrations of a subatomic particle that is far away can affect another particle that is nearby. Scientists in the lab study subatomic particles that become entangled in beams of laser light. In nature, particles are entangled with millions of other particles, all interacting with each other. One could speculate that entanglement of these particles could cause a kind of consciousness for the universe. This is my current concept of God (Grandin, 1995, p. 200).[27]

I read and re-read this passage trying to understand Grandin's mind—brain. She repeatedly tells her readers: "I am a completely logical and scientific person." She also frequently emphasizes that she always thinks and feels in visual pictures. "I THINK IN PICTURES. Words are like a second language to me" (p. 19). When

[27] From THINKING IN PICTURES by Temple Grandin, copyright © 1995, 2006 by Temple Grandin. Used by permission of Doubleday, a division of Random House, Inc.

she speaks, she indeed translates visual pictures into words. The same happens when she listens to others. "When somebody speaks to me, his words are instantly translated into pictures…[g]rowing up, I learned to convert abstract ideas into pictures as a way to understand them…" (p. 19, 33). Here is an example she gives: "WILL is a hard concept to visualize. When I think about it, I imagine God throwing a lightning bolt" (p. 33). The visual pictures Grandin speaks about are not complex symbols but literal visual scenes (visual scene-situations in our terms). I thought that whether Grandin conveys her thoughts to others or "extracts" meaning from listening to others, her MEANING must be different from ours. It must be literal and physical. I realized I would not be able to decode Grandin's emotional concepts and religious beliefs if I did not take this fact into consideration.

I then re-read the above passage, and an analogy came to my mind: isn't this similar to what aborigine people experienced…a continuity of things united by mystical forces, circulating in objects and creatures—the continuity of vital principle?! The aborigine's experience was interpreted before (Glezerman, 1986; Glezerman & Balkoski, 1999, see also Chap. 6) as an exposure of multiple "selves," a contribution of right hemispheric inferior parietal regions (BA39,40), where supposedly multiple *non-I-spaces* are integrated with *I-space.* And it is one's own body-space rhythmical pattern, or the inner rhythm of the thalamic level, that is projected into *non-I-space* and perceived as a mystical power uniting "*non-I-spaces*" into a continuous whole.[28] It was suggested that exposure of this part of self—multiple selves, not divisible from other people or from the environment—provides a profound mystical experience of "belonging." In the normal people, this ancient part of self is buried in the depths of consciousness and intimately integrated into the higher levels of self, so that it is very rarely experienced in a "pure" form.

In Grandin's case, a mystical experience is "muffled" by rationalizing, using scientific terms from her knowledge base to explain her experience. What ends up on the surface is "logical and scientific," a trick that is the work of LH interpretations. To "make" beliefs and moral concepts, Grandin's brain, instead of using left hemispheric *inferior temporal cortex–amygdala–OFC* network, employs the right hemispheric *thalamus–parietal cortex* network. The parietal–occipital region belongs not to the ventral-visual pathway (VVP), but to the dorsal-visual pathway (DVP), which terminates in the DLPC—not in the OFC where the VVP ends. Thus, we see here diversion from the VVP (*what* system) to the DVP (*where* system) and from the LH to the RH. The networks involved in this process of engendering Grandin's beliefs would then be *thalamus–RH TPO–RH DLPC–LH DLPC.* Would activation in her left DLPC then spread to the left VLPC and OFC, in turn paving the way to the left amygdala? It is difficult to say, but Grandin's emotional concepts and beliefs are quite intense.

In his story about Temple Grandin, Sacks says:

> Temple is an intensely moral creature. She has a passionate sense of right and wrong, for example, in regard to the treatment of animals; and law, for her, is clearly not just the law

[28] Recall that Levy-Bruhl indicated: "[a]ll animate and inanimate objects are imbued with continuous vitality resembling that vital power which they felt in themselves" (see Chap. 6).

> of the land but, in some far deeper sense, a divine or cosmic law, whose violation can have disastrous effects—seeming break-downs in the course of nature itself (Sacks, 1995, p. 296) (see footnote 21).

Sacks mentions her saying: "I have a feeling that if I do anything really bad, God will punish me, the steering linkage will go out on the way to the airport" (p. 296).

For us, the idea of God and some kind of mechanical device ("steering linkage") have nothing in common; they are from different categorical domains and different levels of abstraction. Thus, our emotional-conceptual left frontal lobe immediately registers irrelevance when they are put together…unless, perhaps this is not a metaphor. For Grandin "steering linkage" and God are neither concepts nor metaphor. She, herself, emphasizes the "literalness" of her mind many times, her experiential visual scene-situation being, as she remarked, "the only tangible reality or understanding of the world" (p. 37).

Sacks writes: "I was startled by the association of divine retribution with a broken steering linkage" (p. 296).

We can find a clue to her statement in her book:

> Doing something bad, like mistreating an animal, could have dire consequences. An entangled subatomic particle could get me. I would never even know it, but the steering linkage in my car could break if it contained the mate to a particle I disturbed by doing something bad. To many people this belief may be irrational, but to my logical mind it supplies an idea of order and justice to the world (Grandin, 1995, p. 200) (see footnote 27).

What is the connection here between physical order and moral justice?

Let's return to the network involved (*RH TPO–RH DLPC–LH DLPC*) and look at the input made by the separate links within this system. Although the RH TPO and the RH DLPC function as two closely integrated units, we can still distinguish their specific contributions. The RH parietal–occipital region is responsible for integration of *I-space* and *non-I-space*. In the RH DLPC, spatial situations are unfolded in time. In other words, there is a flow of spatial situations (which are *non-I-spaces* integrated with *I-space*) as moments of environmental time, or "world movement." We have already discussed in Chap. 6 the view that the right hemisphere and, in particular, the right frontal lobe, gives a subjective view of the world that incorporates time and space, or rather a spatial–temporal background for one's world image. Curiously, Grandin often speaks about her experience of "world order": "I hate the second law of thermodynamics because I believed that the universe should be orderly" (Grandin, 1995, p. 193). She also states her belief that "there is some ultimate ordering force for good in the universe—not a personal thing, not Buddha or Jesus, maybe something like order out of disorder…" (from Temple Grandin's conversation with Oliver Sacks, 1995, p. 296).

It was suggested before that in the RH, the inner propriomotor rhythm of the thalamic level provides harmony for one's subjective experience of the spatial–temporal organization of the external world. Harmony! This is what's associated with ideas of good and justice, and these are typical RH associations in that they have an emotional-experiential base. The physical experience of rhythm, harmony, and order may be some kind of exaltation, a strong RH experience of the deep thalamic level that can be exposed in autistic people and, in Grandin's case, "rationalized" in her LH DLPC as moral concepts. We can see that, as are people with temporal lobe epilepsy, Grandin is deeply involved in moral concepts and global cosmic issues, all of which are connected with the left frontal lobe. However, they (moral and cosmic concepts) come from different sources in Grandin's case versus TLE patients. While the left frontal lobe is their common endpoint, the brain networks that get them there are totally different. For Grandin, there is no personal emotion or feeling-concept of values of a social group, i.e., moral norms (connected with *left amygdala–left OFC* network). In her case, it is the right hemispheric experience of continuity, belonging, and harmony that give rise to left hemispheric moral concepts.

Finally, comparative neurophenomenological analysis of the experiential phenomena in temporal lobe epilepsy and autism shows us that ecstatic, mystical, and religious experience is heterogeneous and that qualitatively different types of mystical experience exist depending upon the different brain networks involved.

7.14.2 The Role of the LH DLPC in Compensation for the Deficient Left Amygdala—Left OFC Network in Autism

Recalling her school years, Grandin relayed to Sacks:

> "I couldn't figure out what I was doing wrong. I had an odd lack of awareness that I was different. I thought other kids were different…." Something was going on between other kids, something swift, subtle, constantly changing—an exchange of meanings, a negotiation, a swiftness of understanding so remarkable that sometimes she wondered if they are telepathic. She is now aware of the existence of these social signals. She can infer them, she says, but she herself cannot participate in this magical communication directly, or conceive the many-leveled kaleidoscopic states of mind behind it. Knowing it intellectually, she does her best to compensate… (Sacks, 1995, p. 272) (see footnote 21).

From this description we can see that Grandin is aware of facial expressions as visual-motor images, even more cognizant of facial gestures (as well as body language) than typically developed individuals. But she does not experience the internal meaning of these signals. There is a clear failure of her left amygdala's rapid automatic response by which emotional meaning is recognized immediately. This process is "preConscious on the route to Consciousness" (Halgren, 1992, p. 212), and in the case of basic emotion distinct signals of danger, safety, etc., automatic responses from the left amygdala, can be triggered without necessity of prefrontal cortex activation. How does Temple Grandin compensate for her lack of firsthand experience with the personal emotions behind facial expressions? This is what she described:

> She had built up a vast library of experiences over the years…. They were like a library of videotapes, which she could play in her mind and inspect at any time—'videos' of how people behaved in different circumstances. She would play these over and over again and learn, by degrees, to correlate what she saw, so that she could then predict how people in similar circumstances might act… (Sacks, 1995, p. 260) (see footnote 21).

What does her brain do here? It was suggested earlier in this chapter that the RH's brain record of experiences is available to autists without any voluntary effort. Can we say then that Temple Grandin manipulates with her visual-experiential memory?

She is intentionally, consciously, and voluntarily selecting (and classifying) parts out of the whole visual scene-situations as needed in concordance with her needs and goals, thereby making "collections" of explicit visual pictures of different types of social interactions. This is LH processing! More specifically, it can be attributed to the left DLPC. The left DLPC provides the inner programming for the execution of motivated, goal-directed behavior. While Grandin's "library of videotapes" is stored in her temporal cortex, the "acting out" of such tapes requires her frontal lobe. Grandin gives us then another clue: "My mind is like a CD-ROM in a computer—like a quick-access videotape. But once I get there, I have to play that whole part" (Sacks, 1995, p. 282). She cannot focus on the scene she needed in order to achieve her current goal, but must play in memory the "whole part" in steady progress (no "fast-forward"). This tells us that the process is not completely voluntary, but also in part automatic. The fact that the scenes in the "videotape" can be re-played only together as a "chunk," tells us it is a complex automatism—patterned behavioral sequences connected with the striatum.

We know that as skills are acquired, automatized behavior recedes from awareness. Furthermore, the prefrontal cortex activation necessary for attentive, effortful acquisition of a new skill falls away as the behavior becomes automatic (Zeman, 2001). This is not happening in Grandin's case, for her attention is directed toward finding the target scene within the "chunk."

We can easily revive our awareness of a skill (for example, driving), but then, the "chunk" comes apart, and performance becomes a sequential process—attentive, effortful, and slow. This is not the case for Grandin either.[29] For Grandin, automatisms remain automatisms, yet they are also the content of her consciousness. From the point of view of Bernstein's theories, it is the leading functional level that determines the goal of a task and that is conscious. Some technical components of a task might be switched to the lower levels and then become automatic. If the higher cortical level is damaged, the "next-in-command" lower level becomes the leading one (and, therefore, the conscious one). However, the degree and content of awareness will be different than that of the higher cortical level. Autists' higher cortical levels are not damaged, but, as has been argued in this book, the leading level is the sensory-motor level.

Recall also that the striatum is the subcortical center of the sensory-motor level. Thus, we are dealing here with the sensory-motor level, its awareness and voluntary control is not equivalent to higher levels. Grandin then has access to automatisms but her voluntary control in manipulating them is not complete.

However, Grandin does possess higher cortical level functioning, namely, her left DLPC as described before. Grandin "extracts" behavioral responses to social-emotional situations out of the context of visual situations and then puts them "on line" and plays them out. Zeman (2001) indicates that "the essence of willed or voluntary acts is that they have aims of which we are conscious and are, usually, willing to acknowledge" (p. 1278). We cannot deny that Grandin attains this "essence."

---

[29] Here again we are faced with the enigma of autism in that it does not fit any pattern known to neuropsychology or psychopathology.

How else can we explain Grandin's outstanding introspective aptitude, her ability to describe her way of thinking, or her intellectual insight into what she is missing? These are prefrontal cortex functions. Amazingly, however, the purely cognitive part of her left prefrontal cortex is functioning fully dissociated from her emotional brain, the *amygdala–OFC* network. Grandin's "library of videotapes," these visual stereotyped behavioral responses to social-emotional situations are only outward actions and do not provide her the emotional experience they should imply: "When she was younger, she was hardly able to interpret even the simplest expressions of emotion; she learned to 'decode' them later, without necessarily feeling them" (Sacks, 1995, p. 269). Sacks further elucidates this phenomenon:

> What is it then, I pressed her further, that goes on between normal people, from which she feels herself excluded? It has to do, she has inferred, with an implicit knowledge of social conventions and codes, of cultural presuppositions of every sort. This implicit knowledge, which every normal person accumulates and generates throughout life on the basis of experience and encounters with others, Temple seems to be largely devoid.... She herself, she infers, may never have had the normal social experiences from which a normal social knowledge is constructed (Sacks, 1995, p. 270) (see footnote 21).

Alas, there is no compensation for internal emotional experience.

We can clearly see here in Grandin a deficiency of the OFC's contribution to social-emotional behavior.

Let's trace the differences between the normal and autistic brain and then analyze Grandin's compensation for her emotional deficit.

Normal people do not have access to the "brain record." It is emotional memory that is conscious and can be retrieved voluntarily in the norm. The left amygdala "selects" significant objects and events (*amygdala-temporal cortex* network), and these emotionally significant perceptions have superiority for their encoding in memory (*amygdala-hippocampus* network). In the case of social interactions, emotional memory provides rapid access to appropriate emotional-motivational states when complex social situations or particular individuals are re-encountered.

In autism, the "brain record" of the RH experiences is accessible, but there is a deficiency of LH emotional memory. As a result, autists may have a memory of behavioral sequences, pertinent to social interactions, that is rich in meticulous exact detail, unbiased by emotion. Because such "factual" behavioral patterns are not connected with the emotional experience from which they are supposed to be originated (reactions split from experience), they present themselves as automatisms. Despite this "factual exactness" (never achieved by normal people) autists cannot respond fully appropriately to changing social situations, even at the behavioral level.

In conclusion, Grandin's network *left DLPC–left striatum* is recruited to compensate for her deficient *left amygdala–left OFC* network.

Analyzing the literature about autists' performance on various emotion recognition tasks, there is surprisingly uniform data suggesting involvement of the left DLPC in compensation for emotional deficit. I will summarize a few of such studies again. Beggeer et al. (2006) found a significant performance improvement in the facial expression task when autistic children were given instructions that emphasized the explicit knowledge of behavioral consequences of such expressions. The attention of autistic individuals to emotional words correlated with their verbal IQ, as opposed to the control group (Gaigg & Bowler, 2009). Furthermore, the performance of autistic individuals on facial emotion recognition test did not differ from the control group when the groups were matched by verbal IQ (Celani et al., 1999). A more direct and potent proof comes from fMRI studies of autistic individuals during emotion recognition tasks. Autists did not activate the left amygdala, but activated the left DLPC (Baron-Cohen et al., 2000; Critchley et al. (2000a)). In Wang et al. (2006) fMRI study of emotional recognition, the autistic group recruited their prefrontal region more strongly than control group. Also, a correlation was observed between the degree of engagement of prefrontal cortex and verbal IQ in autistic but not the control group.

While the above "objective" data clearly indicate recruitment of the left DLPC in compensation for the emotional deficit in autism, only Grandin's "subjective" report of her experience gives us a clue regarding its mechanism—the peculiar interaction between the left DLPC and the striatum.

Chapter 8
Autistic Persons' Sense of Self (Cerebral Organization of Self and Autism)

Whatever the ultimate solution to the problem of consciousness, it seems likely that it will be multifaceted, perhaps breaking apart in ways we have not yet begun to think about

M. Gazzaniga (ed.) (2000), p. 1355.

To understand the cerebral organization of self we must start with how one's own body is represented in the brain.

Ever since studies of split-brain patients revealed "two separate cognitive systems in one person each acting out of consciousness of the other" (Gazzaniga, 2000, p. 1355), the idea of "one body—two selves" continues to be an enigma. However, each hemisphere processes the body differently (Cutting, 1990), which *can be* the basis for two different selves.

There are numerous different mappings of the body in the brain. These body maps are not redundant but rather reflect the functional specificity of their underlying brain regions. Examples of such mappings include the somatosensory homunculus in the postcentral cortex, the viscerosensory homunculus in the insular cortex, and the body schema in the right parietal cortex.

Not every body map in the brain is relevant for an examination of the self's cerebral organization. For example, the somatosensory homunculus has a somatotopic organization, i.e., it represents body parts proportionally to their sensory innervation. Thus, this body image reflects the relative sensibility of body parts to *external stimuli* rather than the body itself.

A phenomenological approach to the self requires a "'transcendental reduction' [that] leads to a disclosure of the original underpinning of our experience…[i]t follows the constitution of self and reality down to the basic structures of corporality, spatiality, temporality, and intersubjectivity" (Fuchs, 2002, p. 321). Bernstein's ideas about the brain's organization fit these criteria perfectly and are therefore germane to a neurophenomenological analysis of the self. Based on his conceptual framework,

T.B. Glezerman, *Autism and the Brain: Neurophenomenological Interpretation*, DOI 10.1007/978-1-4614-4112-0_8, © Springer Science+Business Media New York 2013

models of a LH self and RH self have been developed (Glezerman & Balkoski, 1999; Glezerman, Balkoski, & Miller, 1996).[1] These models will be described below and applied to autism.

8.1 The Left-Hemispheric Self and Autism

8.1.1 The Multilevel Model of the LH Self

The model of the LH self is based on a multilevel hierarchy where one's body acquires a different meaning at each successive level.

The **paleokinetic level of tonus regulation** is concerned with keeping one's body in the vertical position. The afferentation for this level includes the totality of "static" proprioceptive and vestibular sense. The "space" of this level is the body's vertical position in the gravitational field (Bernstein, 1947, 1990; see also Table 1.1). Thus, *the body image at this level in the brain is a weight-object.* An internal sense of body mass is thought to be this level's contribution to the self.

At the **gnostic-praxic level**, "space" is that of an object (the LH's secondary model of the world). It results from distinguishing the functional signs of an object and re-constructing a so-called topological schema that is the meaning of the object's usefulness. *At this level the body is a functional object in the external world.*[2]

At the **symbolic level**, re-organization of "space" entails transformation from visual to categorical representations or "semantic" space. At this level, equivalent to the gnostic-praxic level's visual image of the body as a *functional* object is *I-concept* as an *agent.* The *I-concept* is an abstraction, resulting from a sequence of categorical signs, the most specific being *agentivity*: object → animated object → person → agent. The categorical sign is the common denominator for a given group of objects that sets it apart from other groups. Inherent in the categorical classification of the world is characterization of the object as such, as well as its connections to other objects. An object preserves its distinctive autonomy while sharing common categorical signs with other objects. Thus, self at the categorical level is not just a separate

[1] Bernstein's organizational model has already been expanded to include the LH–RH dimension and the symbolic (supra-modal) level (Glezerman & Balkoski, 1999).

[2] I do not include here a description of the thalamic level of one's internal space and the sensory-motor level of the external spatial field's contributions to LH self. Analysis of patients with unilateral brain lesions suggest that the RH represents the sensory body-space, while the LH is responsible for knowing the essential features of body parts rather than their spatial image (Cutting, 1990). Thus, it seems the thalamic level does not play a significant role in the LH self. As to the sensory-motor level, the somatosensory homunculus is the body image of this level; the meaning of this grossly distorted body image is not the body itself but its sensibility to external stimuli (recall that this level is completely extraverted).

me-object but *one, individual* self and at the same time a member of the categorical group (e.g., social group).[3]

One can see that there are multiple meanings of one's own body in the LH, providing a continuous line for self construction, but *I-concept* is the conscious one and incorporates the body representations of lower levels in an integrated fashion.

This model of LH self is based on the postulate that one's own body is represented in the LH as an object, which, just as any other object in the environment, is evaluated by the amygdala for its biological significance. It follows that relatedness to oneself and relatedness to any other object derived from the same source: amygdala-born limbic emotion, conscious and distinguishable from the object itself. LH self at the symbolic level then can be understood as an integration of limbic emotion and *I-concept*, the network being *LH amygdala–LH OFC–LH DLPC.*[1] *In a very distilled form, the LH self is the union of limbic emotion and I-concept, making a separate I-agent.* Another facet of the LH self, however, is being a member of a group, calling for integration of personal feelings[4] with the concepts of a given group or social–cultural context. This will make the social and moral self. Thus, at the symbolic level of the LH self, social communication is based on common beliefs and judgments—a value system.

In Chap. 7, the integration of emotion and concept was traced using the model of temporal lobe epilepsy (TLE) in two extraordinary individuals. Intensified and magnified personal feeling was shown to be transformed into belief or translated into a cultural-religious context presumably via the network of the left amygdala and left OFC.

8.1.2 Homeostatic System and Its Relation to the LH Self

There is an opinion in recent literature (Craig, 2002, 2003; Damasio, 2003) that cortical images of the physiological (homeostatic) state of the body is the foundation for self-awareness—equivalent to this book's hypothesized conscious self-concept. We will now take a close look at the homeostatic processing which these authors correlate with the "feeling self."

Homeostasis is defined as the autonomic, hormonal, and behavioral neural mechanisms that maintain optimal physiological conditions in the body. The afferent arm of the homeostatic system relays information from the baro-, chemo-, mechano-, osmo-, and metaboreceptors of internal organs and all other tissues of the body.

[3] At the gnostic-praxic level, functional signs characterize the object itself but do not relate the object to other objects. Instead, the functional signs are indirectly connected with situations in which the function (action) is performed. This means that social relation at the gnostic-praxic level is to act similarly (together) in certain situations, but it is not based on an internal connection (sharing common categorical signs).

[4] One's memory of emotionally significant objects and events.

Parasympathetic visceral afferents enter the central nervous system (CNS) through the cranial nerves and terminate in the nucleus of the solitary tract (NTS). They then pass through the parabrachial nucleus (PB) after which the visceral information goes to the visceral sensory thalamus and then, through the thalamo-cortical pathway, terminates in the insular cortex.

The sympathetic visceral afferents enter the CNS via the spinal nerves, and remain a component of the spinothalamic pathway until reaching the thalamic visceral sensory relay nucleus after which they pass on to the insular cortex. A key feature of this ascending sympathetic pathway is that it provides collaterals that converge with the parasympathetic visceral sensory pathway at every level of the homeostatic system: brain stem, mesencephalon, and diencephalon (hypothalamus). In another words, some sympathetic and parasympathetic fibers from the main pathway end at each level, the "stepped stations" of the hierarchical homeostatic system, such that only a modest number of the autonomic afferents reach the insular cortex. Only the cortical-terminating autonomic afferents give rise to conscious appreciation of visceral sensations. Most vital homeostatic corrections thus happen at lower levels without any involvement of conscious awareness. Autonomic afferents are not only projected to each lower level, but they are also connected with the efferent arm of the homeostatic system at "integration sites" or "pattern generators" in each level. These "stations" initiate specialized and complex but highly stereotyped patterned autonomic and hormonal responses (Saper, 2002).[5]

The hypothalamus, the autonomic control center, determines the overall "composition" of the response for general homeostatic functions such as maintenance of energy metabolism, body temperature, and fluid and electrolyte balance while the lower levels "may each produce a series of 'chords,' or more elementary response patterns, which are in turn composed of 'keys' consisting of individual autonomic actions" (Saper, 2002, p. 460).

Within the hypothalamic–spinal homeostatic system, there are specific subsets of neurons responsible for distinct physiological processes. "Command" neurons in the hypothalamus can contact a wide range of functionally related neurons at lower levels that can produce patterns of autonomic response in various tissues (Saper, 2002). Saper's conclusion is very important for our task: anatomical organization for this homeostatic processing is along the lines of functionally related cell groups rather than a body map. What then is the homeostatic image of the *body*? These are

[5] An illustration of one such patterned response involves neurons in the ventrolateral medulla that defend against low blood pressure (BP). These neurons receive information from baroreceptors that monitor vascular wall stretch. If BP falls, the reduced input from baroreceptors causes cardiovagal motor neurons to slow vagal firing, thereby increasing heart rate (parasympathetic response). There is also reduced excitation of inhibitory interneurons connected to neurons that regulate vasoconstriction tone. As a result, there is an increase in vasoconstriction tone (sympathetic response). Still other neurons in the ventrolateral medulla respond to a fall in baroreceptors' input by conveying this information to the brain area related to vasopressin secretion. Release of vasopressin increases fluid retention and causes vasoconstriction (endocrine response). Saper (2002) emphasizes that this pattern of response is intrinsic to the connections of the baroresponsive neurons in the ventrolateral medulla.

simply physiological and biochemical processes keeping the organism alive. Instead of *body* I would call it an *organism* and its tissues. As to representations of the visceral sensations in the insular cortex, they are organotopically organized. Thus, as in the case of the somatosensory "homunculus," the visceral sensory "homunculus" falls far short of an integrative image of the whole body.

I believe a phenomenological requirement for self-experience is *embodiment.* From a neurophenomenological perspective then, self-experience is "grown" from the body image (body experience): body image as a closed internal space (RH) and body image as an object functioning in the external space (LH).

According to Craig (2002, 2003), a cortical visceral sensory "homunculus" in the dorsal insula is re-represented in the right anterior insula for subjective evaluation of the body's physiological state. He then proposes that the right anterior insula serves as a neuroanatomical substrate for emotions and feelings. Craig also believes that the interoceptive re-representation in the right insular cortex corresponds to the ability to perceive the self as a physical and separate entity, that is, self-awareness. He bases his conclusions on functional imaging data of right anterior insula activation during emotional processing: "the functional imaging data strongly support the integral role of the right anterior insula in the feelings we perceive that are the basis of our perceptions of our selves, and therefore of consciousness" (Craig, 2002, p. 663). This assertion makes reductionistic leaps, and Craig himself admits that the amygdala, OFC, and medial prefrontal cortex are co-activated with the anterior insula in the imaging studies of emotion.

Saper (2002) justly remarks that when Penfield performed electrical stimulation of the insular cortex, his subjects reported a variety of descriptions about their visceral sensory experiences, but none of them felt a complete emotional response. Contrastingly, in Gloor (1990) study on electrical stimulation of the medial temporal lobe, patients reported emotional reactions or complex experiential phenomena with attendant emotions. Saper astutely concludes:

> [A]lthough the visceral sensory cortex does relay information to medial temporal lobe structures where it may be integrated with emotional experience, it is unlikely that the insular cortex itself functions as an emotional integrator, or that visceral sensation is equated by the brain with the emotional experience that often accompanies it (Saper, 2002, p. 436).

At each level (brain stem, mesencephalon, and hypothalamus) where sympathetic (spinal) and parasympathetic (cranial) visceral sensory afferents converge, they not only innervate the motor visceral system, but also project to the amygdala (Saper, 2002). The amygdala, in my understanding, receives information from the homeostatic system about the internal milieu and vital needs of the organism and then evaluates the biological significance of external objects as a potential source of internal need satisfaction. What emerges is subjective experience or object-related limbic emotion. In this indirect and non-specific way, the homeostatic system participates in limbic emotion and the LH self.

The cortical homeostatic system's role then is modification of conscious motivated behavior. The insular cortex's connection to the OFC may modulate the emotional state (mood), while its relationship to the medial prefrontal cortex may influence the motivational component of goal-directed behavior in line with current homeostatic needs.

8.1.3 *The LH Self in Autism*

Let us return to the example from Kanner's material of the 4-year-old autistic boy, Jay, who referred to himself as "Blum" whenever his parents questioned his truth-telling (see Chap. 2). The code to the idiosyncratic meaning of his odd referent was unveiled when he pointed to the advertisement of a furniture firm in the newspaper, written in large letters: "Blum tells the truth." J. Louise Despert speculates about the meaning behind this story.

> [W]hile 4-year-old Jay refers to himself as Blum, he does not say, and probably cannot say, 'I am Blum' or 'I am Blum because'. In so far as human relationships are concerned, the autistic child lives in an emotional vacuum; language symbols have emerged with overwhelming affect charges which have seemingly blocked the emergence of other symbols. Such experiences are possible because the binding power of free flowing affect characteristic of the normal child is lacking (from discussion in Kanner, 1946, p. 244).

What Despert refers to as the "binding power of free-flowing affect" corresponds to the proposed concept of object-related limbic emotion. Despert further emphasizes "[i]t is highly significant that the 'I not I' distinction is not established in the autistic child, as it is early in the development of language in the normal child…." (p. 245). On point is Kanner's observation that pronominal reversal was a typical sign in autism, one example being a child who "once told by his mother, 'Now I will give you your milk,' expresses the desire for milk in exactly the same words. Consequently, he comes to speak of himself always as 'you,' and of the person addressed as 'I'" (p. 244). Appearance of the first-person pronoun in language shortly follows the developmental stage that corresponds to the child's consciousness of being one, and apart from others.

It can be proposed that in autism, the symbolic level of the LH self (*I-concept*), an awareness of self as *one* separate individual, an agent (and at the same time a member of social group), is not developed or poorly developed.

This book's prevailing hypothesis is that the sensory-motor level leads in autism, with higher levels, gnostic-praxic and symbolic (categorical), functioning insufficiently. From this statement alone, one would suspect the LH self to be disordered in autism, since the gnostic-praxic and symbolic levels contribute most to the "construction" of the LH self in norm.

The LH self (mostly network *LH amygdala-LH OFC*) is responsible for complex human-specific emotions and personal feelings (see Chap. 7). A striking feature in autism is the lack of personal feelings (shame, embarrassment, pride, jealousy, pity, envy, guilt, etc.). Hence, we cannot make moral judgments about autists. They are incapable of felt altruism, yet also unable to experience selfishness. Such phenomena must reflect a deficiency in self as *one* and separate from others.

8.1.3.1 Face, Voice, Social Identity, and Autism

As all complex movements are "multistory constructions," facial movements (facial expressions) also consist of several layers. Let us consider emotional expressiveness at the various planes in the brain.

At the thalamic level, emotion is connected with bodily sensations (mostly kinesthetic, muscle sense, and from the extremities). Thalamic emotion is reflected in automatic, involuntary as well as half-voluntary, movements such as inborn and non-demonstrative facial expressions (smile, frown, and grin); expressive poses and gestures (a hand stretched before the face with the palm turned toward the danger, a general "unfolding" of the body, relaxation, calm); movements of the whole body (a high leap for joy; rolling into a ball from fear); and non-intonational cries (Maximova, 2008). One can see this level's emotional expressiveness in a LFA whom Kanner observed: "He ran around in circles emitting phrases in an ecstatic-like fashion. He took a small blanket and kept shaking it, delightedly shouting: 'Ee! Ee!'"; "He wandered about smiling, making stereotyped movements with his fingers, crossing them about in the air" (Kanner, 1943 pp. 228, 219).

Autistic savant-calculators, on the other hand, have enigmatic and meditative facial expressions that signify, as suggested in Chap. 7, mystical experience from the projection of their internal rhythms onto *non-I-space* (RH thalamo-parietal–prefrontal system).

At the sensory-motor level are the basic emotions.[6] These emotions are immediate responses to the negative (dangerous) and positive stimuli in the external spatial field. They are "*in-the-moment*" reactions. Emotional expressions at this level are always sincere and demonstrative. Although, innate, emotional expressions are acquired by imitation (Maximova, 2008). At this level, facial expressions, gestures, and intonational sounds are social signals. Given this description, it is evident that autistic persons lack this type of emotional expression. This very deficit is what Temple Grandin was referring to when she told Sacks how unavailable to her was "something was going on between other kids, something swift, subtle, constantly changing—an exchange of meanings, a negotiation, a swiftness of understanding so remarkable that sometimes she wondered if they are telepathic" (Sacks, 1995, p. 272, see Chap. 7). And yet, how does one explain autists' lacking emotional expression in light of this hypothesis that the sensory-motor level is the leading one in autism? The paradox is elucidated with the point that two dimensions have been emphasized in this book, not only the sensory-motor level but also RH processing dominance in autism. Remember Steven, described by Sacks (1995), who, unable to imitate (see Chap. 7), "plays" the visual–musical situation as a whole, in exact detail reproducing a famous singer's performances. From Stephen's spectacles, we can see that the neurological basis for facial expressions, intonations and gestures is intact in autists.

Transition to the gnostic-praxic level and to the symbolic level is characterized by emotion being ever more connected with concept and by higher awareness and voluntary control of emotional expressions ("space" and "time" at these levels are not "in the moment" anymore, but increasingly abstract).

At the gnostic-praxis level, emotions can be controlled (and the experience thereof repressed) with expressions modifiable or suppressible in line with social–cultural standards and one's ability to "fake it." All such modulating abilities are missing in autism.

---

[6] Described in detail in Chap. 7.

Finally, human-specific social emotions and beliefs emerge at the symbolic level. Autistic persons' symbolic level of the LH self is so disordered in this arena, it is striking. They are devoid of personal feelings, and if they have beliefs (as in case of Temple Grandin, see Chap. 7), the latter are phenomenologically different and are based on brain networks not involving *LH amygdala-LH OFC.* This disorder is highly specific to autism; it is not seen in any other kind of psychopathology, not even syndromes that involve the LH OFC. For example, in patients with damage to the LH OFC, human-specific emotion is impaired, but there is a concomitant disinhibition of lower-level emotions along with excessive involvement in pleasurable activities. Although shallow, unstable, and inappropriate, emotional expression is not lacking in such patients. Patients with LH OFC lesions also are severely inept at integrating emotion into a behavioral context (see Chap. 7). Autism contrastingly presents a peculiar situation where impaired emotion exists alongside with a superb aptitude for behavioral stereotypes.

In autism, deficient *basic* emotion is "compensated" for by routinized behavioral patterns connected with emotion—letting the autist know its behavioral consequences.

No such compensation is possible for more multifaceted human emotions (at the symbolic level), resulting in a *void* of complex emotional experience and emotional expression.

To understand this deficit of LH self in autism better, we will now turn to examples from brain pathology and fMRI data about the normal brain that illustrate the relationship between a "face" and *I-concept.*

Gallois (1988), cited from Cutting (2002), describes a patient who suffered damage to her left occipital lobe and the posterior third of her corpus callosum. She had color agnosia (dysfunction of BA19), object agnosia (dysfunction of the posterior part of BA37), and difficulty in recognizing her own face in the mirror. This last problem is our point of interest. Recall that the recognition of an individual human face is the function of the right fusiform gyrus—the "face area"—which was in this patient disconnected from the LH due to corpus callosum damage. With an intact RH, the patient could recognize her face, but she could not convey this knowledge to her *speaking* LH. She would report recognizing the face, but could not say: "It is my face." Instead, her answer reflected what her LH knew about the face *it* saw. The patient "referred to it [face] as that of the wife of her second husband (which was true of her), or that of a retired proprietor of a café (which was also true of her), or that of a hospital inpatient (which was true of her too)" (Cutting, p. 228). What does this tell us? The patient's isolated LH knows (represents) the essential function of her face, i.e., her social identity. Note that this patient speaks about her self-face as about the object. Due to lack, in this case, of the RH sensory-experiential contribution to self-face, "objective" nature of the LH contribution to self is exposed in a pure form.

A second example is the case of a patient with a circumscribed lesion in the LH area of the temporal–occipital junction (the counterpart of the "face area" in the RH), discussed in Chap. 4. In this patient, exaggerated function of the "face area" unopposed by its LH counterpart resulted in a peculiar deficit. Otherwise cognitively intact, the patient experienced erroneous familiarity for unknown peoples' faces. Her judgment about people was intact, and she was greatly embarrassed by the inappropriate behavior her errors caused. Intellectually, she knew "it is not correct" (the face was not familiar), but at the same time she experienced a compelling feeling of knowing, momentarily induced by the visual image of the face: "my eyes are caught by someone's face; I have strong feeling of knowing him or her without being able to place him or her…I have the feeling we have met in some place or talked together, but I cannot figure out where or when, or what we talked about." In contrast, she was confident in her correct recognitions, which were associated with immediate retrieval of a specific identity. "When I truly recognize someone, I have no doubt: I just know who the person is" (Vuilleumier, Mohr, Valenza, Wetzel, & Landis, 2003, pp. 892–893).

These two cases with their opposing lesions illustrate the drastic differences between RH sense of familiarity and LH knowing a face. The RH "sense of knowing" refers to one's subjective, idiosyncratic experience, while the LH "knowing" is abstract knowledge of a particular individual and his or her social identity (this knowledge includes one's own face as the previous example shows).

Finally, an fMRI study of the normal brain by Kircher et al. (2001) further elucidates the role of face in social identity. In this study, subjects viewed morphed versions of their own face.[7]

Because the interest of this study was not in mere familiarity, but rather self-recognition, the self-face was matched with the partner's face in order to control for emotional salience and familiarity. Comparing to controls (unknown faces, partner's faces), in addition to activation in the RH, it was *only* viewing one's own face that induced activation in the left prefrontal cortex. The authors suggest that visual recognition of one's own face includes self-recognition or self-awareness and that self-awareness is particularly connected with the left prefrontal cortex. While the right hemisphere recognizes the familiar face, the authors affirm "…only transcallosal transfer of information enables the sensory experience to reach awareness. The onset of self-recognition in human infancy correlates with the myelination of fibers in the frontal lobe" (Kircher et al., 2001, p. 7). This study shows that social and personal identity of self-face goes beyond the face processing network (discussed in Chap. 4). It involves *self-awareness,* connected with the left prefrontal cortex.

Kircher et al.'s (2001) conclusion that the left prefrontal cortex is responsible for self-awareness fully corresponds to our formulation of *I-concept*—awareness of self as *one* and separate from others, which is missing (undeveloped) in autism. "Social" self is also part of the LH self. We saw "social" self-face in a pure form in the first patient illustrated here (Cutting, 2002). Autist does not have "social" face. Disorder of social–emotional communication through face (and voice) pervades all levels higher than the thalamic: from social signals of danger/safety (sensory-motor level), to emotion modification in correspondence with social–cultural group norms (gnostic-praxic level), to personal feelings (symbolic level).

[7] A computerized morphing procedure was used to merge the target face with an unknown control face, so that the presented visual stimuli were sufficiently novel to prevent habituation yet were easily recognizable (Kircher et al., 2001).

We can now address the unanswered questions from Chaps. 4 and 7. Problems with face processing and lack of "social" face in autism do not originate in the FFA. They come from disorder of the LH self, for the autist is not fully aware of being a *separate* self and, correspondingly, of separate others. As a consequence, autists have unnatural and peculiar expressiveness (there is no individual to thereby characterize), and their voices lack the emotional tone that would normally mark out an individual as himself and unlike others.[8] In Chap. 7, we could not explain disorder of prosody in autism by the disorder of any link in the brain network responsible for prosody in the norm. In autism, it is not implementation of prosody that is disordered. Prosody is emotion in speaker's voice, but in autism there is no *speaker*, no agent to communicate his/her opinions, attitudes, feelings through voice.

8.2 The Right-Hemispheric Self and Autism

8.2.1 The Multilevel Model of the RH Self

Practically all aspects of the RH self were discussed earlier in this book when we considered different "layers" of consciousness. Here I will summarize them again to give a coherent picture of each level's contributions to the RH self.[9]

8.2.1.1 The Thalamic Level—Subjectively Felt Body

Of all the contributions to the RH self, the most fundamental is that of the thalamic level. It provides a physical (spatial–temporal) foundation. Proprioception (afferentation of the thalamic level) is not externally directed here. Instead, proprioceptive sensations are directed inward—they are *bodily feelings.* In the LH, proprioceptive sensations from the extremities are integrated according to one's body spatial coordinates ("space" of this level) providing information necessary for movements. In the RH, the same afferentation provides, it seems, subjectively felt kinesthetic body-space where bodily sensations (content) are inseparable from the space (form) itself. *I-space* has a unique-for-the-individual rhythmical pattern ("time" of this level). *I-space* is not part of conscious awareness, but it is intimately incorporated into the higher levels of RH self. The thalamic level's participation in the making of

[8] Both face and voice expressions in autism have individual features connected with the inner rhythm of the thalamic level. Recall the savants' enigmatic expressions during drawing or manipulating with numbers (Sacks, 1995; Selfe, 1977) and the data about individual-specific rhythmical characteristics of voiced laughter in autism (Hudenko, Stone, & Bachrowski, 2009).

[9] For the original description of this neuropsychiatric model of RH self, see Cerebral Organization of the Self and Schizophrenia, in: Glezerman, T. and Balkoski, V. Language, Thought and the Brain 1999, 250–259.

self is the subjective experience of the boundaries and continuity of one's own body, which is translated at higher levels into a sense of wholeness of self.

8.2.1.2 The Thalamic–OFC Network and Thalamic Emotion

In certain neurological (thalamic lesions) and psychiatric (schizophrenia) disorders, patients may experience unusual bodily sensations, which they describe as "new," queer, vague, or incomprehensible. These sensations are often unpleasant but cannot be characterized as pain. Although they cannot be attributed to any specific modality, they seem mostly related to kinesthetic experience. Patients feel certain body parts "got narrow," "became thicker," "swell," are "displaced," "compressed with some bonds," "fixed," "squeezed," or "shrunk." The above pathological bodily sensations are called *coenesthesia.* Most authors emphasize that *coenesthesia* is not just a disturbance of bodily sensation but, as a subjective experience, involves affect (Anufriyev, 1979; Dobrochotova, 1974). H. Head [cited from Kretschmer (1927), p. 27] described examples of peculiar unilateral changes of emotional tone in patients with focal thalamic lesions. One patient reported that he could not go to church because horrible feelings emerged on his affected side when the chorus started singing; another patient said that the right side of his body became more gentle—"my right hand needs consolation," "my right hand feels more artistic"; and a third patient shared that his soul on the left side of his body was different from the soul on his right. Huber (1957, 1992) who first described coenesthesia in patients with schizophrenia noted that it shows a phenomenal similarity to spontaneous sensations of the thalamus.

The syndrome of thalamic hyperpathy most often develops from lesions in the lateral thalamus—the major somatosensory relay nuclei (VP). It is characterized by agonizing and poignant but also diffuse and obscure sensations on the affected side. Most patients do not compare hyperpathic pain to "natural" pain, nor can they describe it (Cesaro, 1987). It is hypothesized that lesions in the VP lead to pathological hyperactivity of the medial dorsal nucleus (MD),[10] which would then be the direct cause of thalamic hyperpathy (Cesaro, 1987). There is some evidence supporting this hypothesis. Functional brain imaging has revealed hyperactivation of the thalamus in patients experiencing "thalamic pain." While pain evoked by electrical stimulation of the lateral thalamus (specific relay nuclei) is described as sharp and well-localized, electrical stimulation of the medial thalamus evokes pain described as diffuse and ill-defined. The latter effect is observed only in patients with thalamic hyperpathy syndrome, while electrical stimulation of the medial thalamus in patients

[10] The MD is the associative thalamic nucleus that so heavily and distinctly projects to the prefrontal cortex, the latter being defined as the cortical territory of MD projection (Fuster, 1985). It may be the most important subcortical nucleus of the human brain, yet little is known about its function (Jones, 2008). It has been suggested that in this nucleus somatic impulses are blended with feeling tone (Malcolm, Carpenter, & Sutin, 1983).

without this syndrome does not evoke any conscious sensation. Electrical stimulation of the MD nucleus of the intact thalamus has been observed to evoke unusual, inexpressible bodily sensations accompanied by emotional tone, i.e., coenesthesias (Smirnov, 1976). Coenesthesia is somewhat similar phenomenologically to thalamic hyperpathy, but less intense and more localized. Coenesthesia is also more incomprehensible, and less able than even hyperpathy to be described as pain. Occasionally reported by patients with focal damage to the thalamus, coenesthesia might be observed both in combination with hyperpathy but also separately (Anufriyev, 1979; Dobrochotova, 1974; Schmaryan, 1949).

As hypothesized elsewhere (Glezerman & Balkoski, 1999) subjective experience of *I-space* is transformed into a new subjective sense of bodily feeling, replete with emotional–sensual tone and timbre, within the MD nucleus. It was called "thalamic emotion" as distinguished from "limbic emotion." Thalamic emotion then is a subjective attitude to bodily sensations. This "emotion" is qualitatively different from the usual positive or negatively colored feelings, such as joy or sorrow, in which somatic sensations play no role. In contrast, thalamic emotion is "localized" in the body, and its sensational and emotional components are inseparable. In contrast to limbic emotion that is conscious, one is not aware of one's own body sensational feelings. We learn about thalamic emotion from pathology where it comes to the surface in a distorted and distressing form, namely, coenesthesia.

Now it is time to return to the MD's connections with the amygdala and the OFC. The pathway between the MD and the amygdala is interesting: it is unidirectional. The amygdala projects to the MD but does not receive return projections. One way to interpret this is to suggest (as argued in the previous chapter) that the amygdala influences thalamic emotion, but thalamic emotion does not weigh in on the amygdala's evaluation of external objects' emotional significance. "Objective" limbic emotion is needed for survival in the external world and should be in a pure form to execute its evolutionary tasks. "Afferentation" for limbic emotion is the homeostatic system (reflecting an organism's internal needs), not bodily sensations. Is it the amygdala that assigns affective tone to bodily sensations, or is there integration of limbic and thalamic emotion in the MD? We can only speculate.

While the amygdala projects mostly to the MPC and OFC, the MD, on the other hand, projects vastly throughout the prefrontal cortex. However, the part of the MD that receives input from the amygdala projects only to the same area within the OFC that receives direct projections from the amygdala. What does this tell us? In light of the OFC's connection with human-specific emotions, it seems as though limbic emotion conveyed through the amygdala's projection to the OFC is re-assessed or integrated in the OFC with information that the OFC receives from the MD about limbic emotion (from amygdala's indirect input via the MD) and about thalamic emotion (presumably formed in the MD).

8.2.1.3 The Sensory-Motor Level—Subjectively Felt *Non-I-Space*

The sensory-motor level is that of *non-I-space*. In the RH, visual objects are perceived and stored within a situational context. Visual scene-situation (VSS) is the term given for the unchangeable whole of visual picture (objects, events), action, time, and emotion that is a RH global entity. With emotion included into the situation, VSSs are not simply *non-I-spaces* but are subjectively experienced and felt—the RH self at this level.

Representing continuously changing impressions of the external world, VSS are momentary segments of real, environmental time. They are also internalized as an ordered succession of moments of experience of the external world (the time of this level). Because VSS are discrete, their sequence as moments of experience makes for a continuous timeline of one's life.

Thus, subjectively felt VSS and a sense of continuity to one's life is the contribution of the RH's sensory-motor level to self. This level of self is as fundamental as the thalamic level's contribution to RH self, and we are not consciously aware of it either; it is also probably the most ancient part of self.

8.2.1.4 The Thalamo-Parietal System and Integration *of I-Space* and *Non-I-space*

The thalamo-parietal system is responsible for integration of *I-space* and *non-I-space.* This RH identification is based on projection of the subjectively experienced inner rhythm of one's body-space into *non-I-spaces. I-space* and *non-I-space* are thereby united into a continuous whole. Recall that in the right inferior-parietal region, *non-I-space* is represented by singular visual *spatial*-situations. Imparted with the rhythm of *I-space, non-I-spaces* are now subjectively felt, multiple "selves," not divisible from the outside world.[11]

If experienced in pure form, this RH parietal "version" of self would give a profound feeling of mystical belonging to and unity with the world.

8.2.1.5 The RH Symbolic Self

Each lower component of RH self is integrated by the symbolic level. Figure 8.1 illustrates the distributed network, which, I propose, produces a new RH *gestalt*—the symbolic self.

The flow of VSSs saturated with affect (*right amygdala–right BA37–right DLPC*) along with spatial situations imparted with the individual inner rhythm (*right thalamus–right BA39,40–right DLPC)* is re-organized: from temporal sequence (self at the situational level) to semantic sequence. The same units participate in parallel (simultaneously) at both levels (situational and symbolic self). At the symbolic level, visual situations suffused with processed limbic and thalamic emotion (*right MD–right OFC–right DLPC*) are identified. United into a symbolic system by emotions they share in common, different visual situations become multiple

[11] Hence it is clearly not the RH that gives us a sense of being *one*, individual "I."

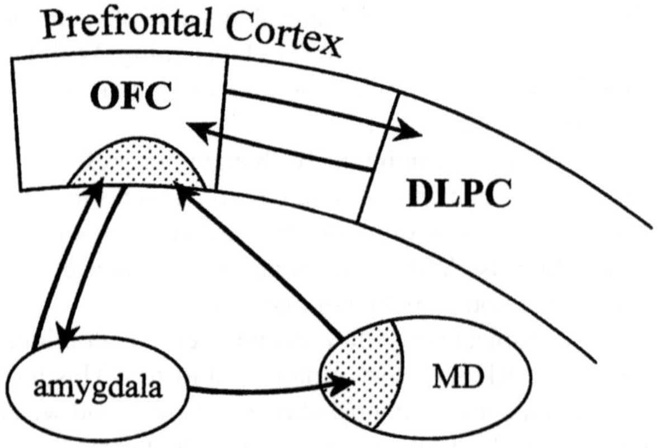

Fig. 8.1 Connections of the orbital frontal cortex with the amygdala and the medial dorsal nucleus of the thalamus (amygdala–OFC–MD triangle)

aspects of a continuous and indivisible whole—all situations are identical in their meaning yet unchanged from the instant they were formed. Visual images are identified by affect but also according to resemblance of the holistic forms within situations. As the RH self develops (symbolic level in the brain), it becomes determined less by external events and real time and more subjectively defined. Here it is important to point out a crucial role of thalamic emotion in the formation of a symbolic system. Otherwise inexpressible bodily feelings are "decoded" by a symbolic system. In other words, thalamic emotion is directly incorporated into the symbolic system, imbuing one's worldview. This role of thalamic emotion can be "exposed" and hence illustrated in pathology. It was proposed (Glezerman & Balkoski, 1999) that in schizophrenia continuity of *I-space* is disordered and bodily feelings are exposed. This disruption, along with a peculiar "compensation" whereby distorted bodily feelings are re-integrated into symbolic systems, would then be the basis for pathological symbolic (autistic) thinking. Here is an example from a schizophrenic patient of mine that illustrates such thinking:

> I have a bug in my back biting me all the time...and a lot of snakes inside.... I'm eating human bodies, snakes, and bugs on trays alive, and they're giving stiffness and pain...that's why we eat cats, rats, dogs, snakes and human bodies. Human bodies are a bunch of pigs. Why all that flesh and blood in the market.... Cats, rats, dogs, snakes, and human bodies are all the same faces, and they are pigs...[patient produced series of drawings of the human face and above animals transforming into each other].... On top of it, there [sic] eating them alive, and there [sic] up higher than pain. The world stinks with that (from Glezerman & Balkoski, 1999, p. 261).

We can see that bodily feelings are transformed into a visual metaphor of snakes and bugs biting inside, and immediately incorporated into symbolic system. The patient's feelings are projected onto the outside world with a flow of images identified and transformed (behind these images are VSSs—the patient was being interviewed in a cafeteria, after lunch, where pork chops remained on trays): snakes–dogs–cats–human bodies becoming the same as a pig; eating a pig is identical to eating human bodies alive; finally his worldview: "On top of it, there [sic] eating them alive, and there [sic] up higher than pain. The world stinks with that" (Glezerman & Balkoski, 1999, p. 259).

The worldview connected with the RH symbolic self is very different from LH beliefs and value systems in that the former is primary experience. While LH beliefs and values are based on integration of emotion and concept, RH values are based on identification with others, i.e., empathy. Undoubtedly, god as the RH symbol of value emerged in the history of humanity earlier than LH ideas of god and religion as a social institution. Adequate illustration of this point is found in experiential seizures of the mystical religious type in patients with TLE where RH holistic sensory-emotional experience is transformed by LH interpretation of such experience within a cultural-religious context. In Chap. 7, the LH component of mystical experience was emphasized in the context of discussing the LH *amygdala–OFC* network. Now it should be clarified that mystical religious experiences in patients with TLE most often involve *both* hemispheres. The RH contribution to such phenomena is inexpressible in ordinary language, a sense of the presence of some external being—a great spiritual figure—and an identification of and with that essence. Ecstasy and miracle are descriptions for this phenomenon (Trimble, 2007). We can see now that beliefs in normal people are more complicated than described in Chap. 7. Beliefs must include a RH sense of *value* as a primary experience in addition to the LH integration of emotion and concept. However, we must not forget that complex RH symbolic feelings always remain implicit. There is no opposition of *right* and *wrong*, no place for explicit rules or moral[12] codes in RH cognitive mechanism. Beliefs are realized by an interhemispheric connection, ultimately through LH interpretation.

8.2.2 The RH Self in Autism

8.2.2.1 *I-space* and Thalamic Emotion in Autism

As discussed before (see Chaps. 5 and 6), autistic persons' thalamic level is intact. It is even used for compensation. Autists at all levels of functioning, from LFA to HFA, instinctively and unknowingly explore their thalamic level space by "seeking pressure" to their bodies. Pressure increases the kinesthetic flow to the thalamus and thereby results in increased awareness of *I-space.* There are even examples of HFA who made special devices for applying pressure to their bodies, including the previously mentioned example of a "pressure suit" and Temple Grandin's "squeeze machine."

For Grandin, using her squeeze machine on a daily basis decreased her anxiety. Even more intriguing is Grandin's insistence that using the squeeze machine enabled her to have feelings for other people:

> From the time I started using my squeeze machine, I understood that the feeling it gave me was one that I need to cultivate toward other people. It was clear that the pleasurable feelings were associated with love for other people.... I built a machine that would apply the soothing, comforting contact that I craved as well as physical affection I couldn't tolerate when I was young. I would have been as hard and as unfeeling as a rock if I had not built

[12] The word *moral* in Webster's Dictionary is defined as "of or relating to principles of right or wrong in behavior; capable of right or wrong action." *Moral* then implies conformity to established sanctional codes or accepted notions of right and wrong. These definitions relate to the LH processing. I do not think the word *moral* can be used when speaking about RH processes, including *value as a primary experience.*

my squeeze machine and followed through with its use.... [W]ithout the machine I would have had no kind feelings toward [my mother] (Grandin, 1995, pp. 82, 83, 90).[13]

How striking! Increased awareness of *I-space* in Grandin somehow improved her *object-related* limbic emotion. Perhaps intense focus upon *I-space* stimulated her thalamic emotion via the *LP–MD* network within the thalamus. We should here consider again the MD's connections with the amygdala and OFC (Fig. 8.1). The amygdala projects to the MD and potentially influences thalamic emotion. However, thalamic emotion cannot influence limbic emotion, because there are not any return projections from the MD to the amygdala. Thus, it is not likely that subcortical connections are involved in the kind of compensation Grandin describes.

Again, the MD area that receives information from the amygdala projects to that part of the OFC which also receives a direct input from the amygdala. As to thalamic emotion, we do not know to which area in the prefrontal cortex it is projected, but the OFC is the likely candidate. The hypothesis that human-specific emotion connected with the OFC requires integration of limbic and thalamic emotion was mentioned above. Did thalamic emotion pave the way for Grandin, through its historically wired connections with limbic emotion? Originating in the amygdala, limbic emotion is united with concept—idea of emotion—in the LH OFC. This process is basically what Grandin describes when she tells about her experience of kindness and love. Her compensation is undoubtedly cortical and left-hemispheric:

[T]o have feelings of gentleness, one must experience gentle bodily comfort. As my nervous system learned to tolerate the soothing pressure from my squeeze machine, I discovered that the comforting feeling made me a kinder and gentler person. It was difficult for me to understand the idea of kindness until I had been soothed myself (Grandin, 1995, p. 82) (see footnote 13).

We see here some ersatz of limbic emotion. There is still no direct relation to the object. Grandin obtains the feeling-concept of kindness and love toward her mother not from embracing her mother, but from "embracing" the squeeze machine (increasing her awareness of her physical self, i.e., relation to the subject). The predominant emotion in Grandin is the thalamic one—a sensational feeling of one's own body:

In developing many varied, complex ways to operate the squeeze machine on myself, I keep discovering that slight changes in the way I manipulate the control lever, affects how it feels. When I slowly increase pressure, I make very small variations in the rate and timing of the increase. It is like a language of pressure, and I keep finding new variations with slightly different sensations. For me, it is the tactile[14] equivalent of a complex emotion and this has helped me to understand complexity of feelings (Grandin, 1995, p. 90) (see footnote 13).

The peculiarity of Grandin's thalamic emotion is that it remains physical—feeling-tone and timbre of bodily sensations—and is *not* incorporated into a symbolic system. Recall that thalamic emotion is not part of conscious awareness in the norm, precisely because it is immediately incorporated into one's symbolic system. In pathology, namely in schizophrenia, bodily feelings rise to consciousness in distorted form

---

[13] From THINKING IN PICTURES by Temple Grandin, copyright © 1995, 2006 by Temple Grandin. Used by permission of Doubleday, a division of Random House, Inc.

[14] Actually, she is referring here to mostly kinesthetic, not tactile, sensation.

(coenesthesia), and they also tend to be re-integrated into a symbolic system, which makes a pathological symbolic system with subsequent delusional interpretation (Glezerman & Balkoski, 1999).[15] In autism, thalamic emotion emerges on the surface because the thalamic level is split from the higher levels.

Kretschmer (1927, 1936/1999) proposed that human emotions are differentiated along two dimensions: mood and sensitivity. The mood dimension is represented on a continuum from joy to sadness, while the sensitivity dimension is represented on a continuum from sensitive to dull.[16] These dimensions are strikingly similar to our formulation of two types of emotions (limbic and thalamic), although Kretschmer's dimensions are purely phenomenological, without consideration of their underlying brain mechanisms.

Describing hypersensitive variant (within sensitivity scale), Kretschmer notes sensitivity to nature and art, taste in personal style, and hypersensitivity and vulnerability with regard to the daily irritations of life.

We are fascinated by the autistic savant artist's reproducing the physical world with high fidelity and exhaustive detail yet at the same time always maintaining his or her own unique style. The sensitivity dimension in Kretschmer's terms, or "thalamic emotion" in my definition, contributes to each savant artist's uniqueness of style. This "flavor" from the thalamic level is what Asperger empathically recognized when he emphasized the originality of autistic children's thought and experience as well as their natural appreciation for art.

> Another distinctive trait one finds in some autistic children is a rare maturity of taste in art.... Autistic children...can have a surprisingly sophisticated understanding, being able to distinguish between art and kitsch with great confidence. They may have a special understanding of works of art which is difficult even for many adults, for instance Romanesque sculpture or paintings by Rembrandt (Asperger, 1991, p. 73).

8.2.2.2 Subjectively felt *Non-I-space* and Integration of *I-Space* and *Non-I-Space* in Autism

Evidence that the sensory-motor level of *non-I-space* dominates the autistic person's mind and, correspondingly, that *subjectively felt VSSs* are the most important part of the autistic person's self has been presented throughout this book. This level of self,

---

[15] I once had a schizophrenic patient, D.A., who told me of a snake inside the wall of her apartment. When asked how, if she does not see it, she knows it is a snake, the patient replied: "You can hear them moving, turning around inside the wall [showed with the spiral movements of her hand], you hear crunching and biting through the cement...movement started in the wall...I could hear the wall shifting...noise shuffling." When asked whether she actually saw the wall moving, she denied it: "It's within the wall, as though a large living thing in the wall moving, turning." She went on: "[i]t is a white anaconda, it can swallow a cow... I don't want to be swallowed by a snake, come out of a snake...it is a certain look when you are eaten by the snake, you don't want to come out of a snake, very small bones." We can see the patient's metaphor conveys a physical, kinesthetic experience. She projects such feelings from the 'wall' of her body to the wall of her building (this patient was a real estate agent before she got ill). The patient concluded our conversation with a delusional interpretation: "The building should be inspected... What I am saying, every building should be inspected, it is just safer."

[16] Kretschmer's two-dimensional theory of emotion is described in more detail in Chap. 10.

exposed on the surface, gives rise to the special state of consciousness in autism. I will not revisit these themes further here.

At the next layer of self, wholeness and continuity of physical self is translated into wholeness and continuity of the outside, or *non-I-space*. It was shown in Chap. 6 that this thalamo-parietal "version" of self (*I-space* integrated with *non-I-space*) is also intact in autism, "acting out" in autistic savant-calculators and calendar calculators.

8.2.2.3 RH Symbolic Self in Autism

In autism, there is some identification of situations, but the last stage of polysemantic symbol formation (identification of different objects by affect) is missing. Symbolic meaning in autism is always inseparable from the single experiential situation. This is why the autists' symbols are so idiosyncratic. The parents of 4-year-old Jay (by now well known to the reader) would never have known why he referred to himself as "Blum," if he (being a fluent reader) had not pointed out the advertisement "Blum tells the truth" (Kanner material, see Chap. 2).

Grandin also illustrates this in how she describes using doors and gates to symbolize each transition in her life. A door as symbol is not unusual, but the difference between a non-autistic person and Grandin is that she literally identifies a particular step in her life with a visual-action-situation—going through a particular *physical door*.

> In order to deal with a major change such as leaving high school, I needed to rehearse it, acting out each phase of my life by walking through an actual door, window, or gate. When I was graduating from high school, I would go and sit on the roof of my dormitory and look up at the stars and think about how I would cope with leaving. It was there I discovered a little door that led to a bigger roof while my dormitory was being remodeled. While I was still living in this old New England house, a much larger building was being constructed over it. One day the carpenters tore out a section of the old roof next to my room. When I walked out, I was now able to look up into the partially finished new building. High on one side was a small wooden door that led to the new roof. The building was changing, and it was now time for me to change too. I could relate to it. I had found the symbolic key. When I was in college, I found another door to symbolize getting ready for graduation. It was a small metal trap door that went out onto the flat roof of the dormitory. I had actually to practice going through this door many times. When I finally graduated from Franklin Pierce, I walked through a third, very important door, on the library roof (Grandin, 1995, p. 34) (see footnote 13).

We can see that each step forward in life is connected with a singular physical door bound to a particular visual situation: "a little door that led to a bigger roof" (high school graduation); "a small metal trap door" (graduation from college); and the door "on the library roof" (graduation from Franklin Pierce).

We may then suppose that RH symbolic level in autistic people is not fully functioning since, as these two examples indicate, there seems to be only the initial stages of symbolic-*situational thinking*.

The LH symbolic (categorical) level is not used in autistic people either (see Chap. 2). This comes as no surprise in light of the previous suggestion that the two highest functional levels in the brain (LH categorical and RH symbolic) are closely interdependent in their phylogenetic and ontogenetic development (Glezerman & Balkoski, 1999).

8.2.2.4 Summary of Hypotheses About Autism and the Self Thus Far

"Anatomizing" self into LH and RH self is needed for the purpose of examination. However, in life LH and RH self function in an integrated fashion.

In autism, LH self is impaired such that a personal center is lost, and the self is fragmented.

The RH self at the symbolic level is not fully present either. However *I-space*, RH self as *subjectively felt situations,* and RH self as *non-I-spaces* integrated with *I-space* are all intact in autism. Thus, in autism we meet a peculiar situation where conscious LH self is impaired—the umbrella under which all parts of self are united is lost. Left to themselves, the splintered components of RH self rise to the surface of consciousness.

The most generic contribution of the RH to self is an experience of *whole-ness,* while the LH contribution to self is awareness of *one-ness*. In autism *wholeness* is intact, *oneness* is impaired. It might be that experience of *wholeness*—not opposed by the LH *oneness*—is exaggerated in autism. As a result the harmony of the spatial–temporal organization of the outside world is more accessible to autistic persons than in the norm.

8.3 Autist: Who Is He—Introverted Extravert or Extraverted Introvert?

Carl Jung distinguished two basic types of personality that he called extraverted and introverted. He defines them as attitude-types. The extraverted personality is oriented toward the outer "objective" world, while the introvert is oriented toward his inner "subjective" world.[17]

---

[17] Jung argued for the deep intrinsic nature of his personality types: "The two types are so different and present such a striking contrast that their existence becomes quite obvious even to the layman once it has been pointed out. Everyone knows those reserved, inscrutable, rather shy people who form the strongest possible contrast to the open, sociable, jovial, or at least friendly and approachable characters who are on good terms with everybody, or quarrel with everybody, but always relate to them in some way and in turn are affected by them. One is naturally inclined, at first, to regard such differences as mere idiosyncrasies of character peculiar to individuals. But anyone with a thorough knowledge of human nature will soon discover that the contrast is by no means a matter of isolated individual instances but of typical attitudes which are far more common than one with limited psychological experience would assume…it is a fundamental contrast, sometimes quite clear, sometimes obscured, but always apparent when one is dealing with individuals whose personality is in any way pronounced. Such people are found not merely among the educated, but in all ranks of society…. Such a widespread distribution could hardly have come about if it were merely a question of a conscious and deliberate choice of attitude. In that case, one would surely find one particular attitude in one particular class of people linked together by a common education and background and localized accordingly. But that is not so at all; on the contrary, the types seem to be distributed quite at random. In the same family one child is introverted, the other extraverted" (Jung, 1971, p. 330).

Jung did not speculate about the brain mechanisms behind introversion and extraversion, but he indicated that, as a general psychological phenomenon, the type antithesis must have some kind of biological foundation. He assumed that although each individual possesses the mechanisms that enable the development of both introversion and extraversion, the relative predominance of one of these innate capacities will ultimately determine personality trends.

Nicolai Bernstein emphasized the "introverted" nature of the thalamic level and the "extraverted" nature of the striatal–cortical, sensory-motor level when delineating brain's hierarchical function levels. Bernstein operated with concepts of "space" and "time," specific for each level, as a framework for "construction" of movements in the brain.

According to Bernstein, the space of the thalamic level is the spatial coordinates of one's own body. Time of this level is a complicated individually specific rhythmical pattern.

On the contrary, Bernstein suggested the space of the sensory-motor level is an external spatial field, and time here is presented by the moment, speed, and duration of environmental time.

What if theoretically we "extract" each of these levels separately?

The thalamic level is fully introverted, it knows nothing about the outside world. Its movements are directed at the body only: they are cyclical, organized according to a rhythmical pattern of one's own body.

The sensory-motor level is fully extraverted. It knows nothing about the body-space. Its movements are oriented toward the external spatial field—progressive linear movement.

This opposition—inward-looking thalamic level and outward-looking sensory-motor level—was extrapolated to psychological functions and personality trends (Glezerman, 1986; Glezerman & Balkoski, 1999).

These two contrary, denying each other levels are discrete, but they do not operate separately. They are both background levels, integrated and assimilated by the cortical symbolic level. It was proposed that the relative development of the thalamic versus sensory-motor level determines personality trend. In other words, predominance of the thalamic level will result in a more introverted personality, whereas predominance of the sensory-motor level will lead to greater extraversion. In general, the relative development of the thalamic and sensory-motor level gives a continuum of personality types from theoretically pure introversion to pure extraversion (Glezerman, 1986; Glezerman & Balkoski, 1999).

If this is the case and if, in autism, the sensory-motor level in the right hemisphere is the leading one, must autists then be extraverts? Jung indicated that in extraverts there is movement of interest toward the object. However, disorder of relation to object is the main deficit in autism! So then, is the autist an introvert? If autists do not have an "outward" extraverted attitude, do they look "inward"? If so, to where?

All researchers underscore the lack of fantasy, imagination—basically a lack of an internal world—in autistic individuals. The meaning we assign things comes from further internalization of information and processing it at higher levels: LH abstraction and RH symbolization. There is no such further working with information by autists. They do not abstract and do not "make" polysemantic RH symbol. Without the "distorting" involvement of higher levels and with the "eyes" of the sensory-motor level, autists see the outside world most "objectively," but do not relate to it. Indeed, the outside physical world *is* their psychical content, their internal world. Recall a peculiar statement of the high-functioning autistic individual, Nelson (Hurlburt, Happe, & Frith, 1994), that his thoughts took the shape of bricks. He made this statement while describing his inner visual experience of the brick wall of a house near his home:

> Nelson was feeling a 'little bit sentimental' about the bricks; he had seen them many times before, had grown up with them. He could not describe how this sentimental feeling presented itself, except that he was somehow paying particular attention to the colour of the bricks and this seemed to be a factor in the recognition of the sentiment.... During his description of many of his images, Nelson referred to the image as being 'the shape of my thoughts' (Hurlburt et al., 1994, p. 390).

So, is the autist then an introverted extravert?!

In the norm, reflection of the physical world in the brain undergoes transformations at higher levels with the goal, particularly in the LH, of using the outside world. In autists, the physical world is their psychic content, as a goal in and of itself. Poor connection of the sensory-motor level and its content with higher levels makes this so. Autists "swallow" the world but are disconnected from it. They are introverts in this sense.

Could the autist be an extraverted introvert?

Trying to explore this issue, we will examine the drawings of the autistic savant Stephen Wiltshire, so eloquently described by Sacks (1995).

One is first impressed by the exactness of Stephen's drawings. Then one forgets about their exquisite detail and is simply charmed by them. Compare, for example, the photo of Moscow's St. Basil's Cathedral with Stephen's drawing of it from life (Fig. 8.2). What does this allure derive from? Do his drawings directly project emotion? Not really. They are distant but alive in themselves. They are felt like a light, refreshing breeze. What makes them alive? What is the force behind his drawings? What kind of pleasure do they give us? Indeed they induce feelings of comfort, of harmony. Almost as if they sound to us...like music! Yes, music comes to mind. As such, his drawings reminded me of the expression: "Architecture is music in stone."[18] I think his harmony is composed of a periodic repetition of lines and forms, as well as recurring patterns (see, for example, Stephen drawings of the Doge's Palace in Venice, and the old houses on the Herengracht in Amsterdam—all these buildings were drawn from life, Figs. 8.3 and 8.4). One might think about projection of the thalamic level in Stephen's drawings. It was argued before that while in the "objective"

[18] As noted before, Stephen was found to have outstanding musical abilities after he was already accomplished in drawing. We may now guess why from an early age he was attracted to drawing buildings: he lived in London, and so architecture was readily available to him.

Fig. 8.2 Stephen's drawing of St. Basil's, in Moscow Red Square. St. Basil's photograph is in the upper left corner for the comparison. Modification from *Floating Cities,* Stephen Wiltshire line drawings, p. 103, Summit Books,1991. Copyright © Stephen Wiltshire Trust, Floating Cities, 1991. Reproduced with the kind permission of Johnson & Alcock Ltd

Fig. 8.3 Stephen's drawing of the Doge's Palace in Venice. From *Floating Cities*, Stephen Wiltshire line drawings, p. 25, Summit Books, 1991. Copyright © Stephen Wiltshire Trust, Floating Cities, 1991. Reproduced with the kind permission of Johnson & Alcock Ltd

Fig. 8.4 Stephen's drawing of the old houses on the Herengracht in Amsterdam. From *Floating Cities*, Stephen Wiltshire line drawings, p. 75, Summit Books, 1991. Copyright © Stephen Wiltshire Trust, Floating Cities, 1991. Reproduced with the kind permission of Johnson & Alcock Ltd

left hemisphere the propriomotor rhythm[19] brings harmony to one's complex movements, so, in the right hemisphere, subjective experience of the propriomotor rhythm, i.e., one's own body inner rhythm, brings harmony into the perceived visual spatial–temporal organization of the external world. I believe it is Stephen's subjective experience of his inner rhythm that is projected and gets conveyed to the viewer. When Nadia's drawings were discussed (see Chap. 3), the influence of the thalamic level was noted; however, the feeling induced in the viewer by her drawings was quite different. Musicality, lightness and airiness, together with natural orderliness is felt in Stephen's drawings; in Nadia's drawings, the horse, in full gallop, charges from the page (Fig. 3.2) or legs at rest are paradoxically alive with movement (Fig. 3.12). Nadia's drawings give the impression of tension, of movement, of muscular contraction—a kinesthetic experience. It was argued (see Chap. 3) that in Nadia's drawings, the kinesthetic sense that makes up the spatial boundaries of one's own body is projected. Thus, in Nadia and Stephen's drawings two different aspects of the thalamic level are projected: its space (kinesthetic space of one's own body) and its time (rhythmical pattern of one's own body kinesthetic space), respectively.

In Stephen's drawings there is clearly "subjectivization" of the external world. Metricity and geometricity of the outside world (objective factor) is united with subjective elements—his experiential situation and internal rhythm. Is Stephen then an extraverted introvert? One may ask: "Isn't this what is present in any work of art?" Yes and no. There *is* a difference in Stephen's case. To elucidate the distinction, one must consider whether Stephen's drawings are an expression of his internal life. Here, again, Sacks' observation is helpful:

> He portrayed…St. Mark's, the Doge's Palace, the great monuments of Venetian culture, and obviously enjoyed drawing them. But when asked what he thought of Venice, after a week in this high point of European civilization, he could only say, 'I prefer Chicago' (and this not because of its buildings but its American cars—Stephen had a passion for these and can identify, name, and draw every postwar model ever made in America). A few weeks later, plans were made for his next trip, to Amsterdam. Stephen approved of the trip for a very specific reason: he had seen photographs of the city, and said, 'I prefer Amsterdam to Venice because it has cars' (Sacks, 1995, p. 209).[20]

What do we have here? Stephen is obsessed with and enjoys drawing grand buildings in London from his childhood, but he does not relish *buildings*. His delight is in cars. This discrepancy reveals that Stephen's drawings are not the product of goal-directed, fully conscious activity. Instead they are expressions or, better yet, the acting out of RH fragments of self not subordinated by a personal center—the LH conscious, individual self. As to his enjoyment of cars, one cannot help but explain it as a manifestation of fascination with moving objects typical in autism (see Chap. 3).

Does Stephen have esthetic taste, esthetic feeling? Sensitivity and a strong feeling of measure and harmony are contained within his drawings—exactly what Asperger

---

[19] See Chap. 1 and endnote in Chap. 6 for detailed description of the propriomotor rhythm.

[20] From AN ANTHROPOLOGIST ON MARS by Oliver Sacks, copyright © 1995 by Oliver Sacks. Used by permission of Alfred A. Knopf, a division of Random House, Inc.

named as a combination of naivety and sophistication in autism. Stephen is unable, however, to have esthetic pleasure from the same buildings to which he is attracted and drew incessantly—the thalamic emotion projected on paper was not integrated into the higher level of conscious self and therefore is lost.

When the sensory-motor and thalamic levels that are background become “foreground,” they provide lower degrees of awareness and voluntary control than what is usually considered a conscious state. It would be a special state of consciousness, not fully voluntary but mostly automatic and repetitive activity. It is not about content or meaning, but about *process*, a process of intense concentration on “seeing.” It is some kind of meditation.

Unwittingly, other cases already mentioned in this book come to mind as I reflect on Stephen’s drawings: Grandin in her childhood sitting on the beach and intensely staring at individual grains of sand; the description by the high-functioning autistic adult, Nelson, of his thoughts taking the shape of bricks (while looking at a brick wall of a house near his home); and the savant twins who replied when asked how they remember all events of their life so exactly: “We just see it.”

Sacks also documents about the state of consciousness when Stephen drew, pointing out that, at age five:

> [He] would stare intently at pictures which fascinated him… He would find paper and pencil and scribble, totally absorbed for long periods…his special interest, his fixation, which developed when he was seven, was drawing buildings—buildings in London he had seen on school trips or that he had seen on television or in magazines (Sacks, 1995, p. 198) (see footnote 20).

Sacks himself witnessed Stephen in this state:

> As he took up his pen and started drawing, I held my breath. ‘Don’t worry,’ Chris broke in, ‘you can talk at the top of your voice if you want to. It won’t make any difference—you can’t interrupt him—he could concentrate if the house was fallen down (p. 205) (see footnote 20).

According to Sacks, Stephen was able “to take in the most complex buildings, or cityscapes, in a few seconds, and to hold them in mind, in the minutest detail—indefinitely, it seemed, and without the least apparent effort” (p. 199). When asked to draw Sacks’ house, Stephen “had not studied the house, had made no sketches had not drawn it from life, but had, in a brief glance, taken everything in…and then, in a single, swift line, [drew] it” (p. 205). Sacks describes Stephen’s technique: “Stephen did not make any sketch or outline, but just started at one edge of the paper (I had a feeling that he might have started anywhere at all) and steadily moved across it, as if transcribing some tenacious inner image or visualization” (p. 205).

So, Stephen is an extraverted introvert?! Where is he looking to “see” the subject for his sketch? He is looking “inward,” not “outward”! He “swallowed” the external world, then intensely concentrated on his inner image of it and finally externalized that inner image on paper.

To conclude this discussion, I’ll now compare Stephen’s drawings with the paintings of Franco, the “memory artist” described by Sacks (see Chap. 7).

Both Stephen and Franco have phenomenal memories, and they both are engaged in excessive depiction of buildings. Although both rendered exact representations, neither Stephen's drawings nor Franco's paintings are pure imitation of the physical world. They both convey subjective experience. However, the subjective experiences conveyed by their art are poles apart.

Stephen's is a special state of consciousness due to a disordered LH self whereby RH components of self rise to the surface, leaving him without full awareness or goal-directed activity. RH gestalts as "things in themselves" and saturated with emotion are present in his drawings, but free floating emotion cannot be projected from them.

In Franco's case, "doubling of consciousness" with involuntary remembrance of past experience results from epileptic discharge in the amygdala—temporal cortex network. Emotion from his original experience is re-activated in the recollections. Recall that epileptic discharge intensely activates the temporal lobe, well beyond the range of its physiological activation. Franco's recollections thus are more vivid, and the re-experienced emotions deeper and more intense, than in the original experience. The "free floating" powerful, intensified, and condensed emotion in Franco's paintings *does* get projected onto the viewer.

The buildings in Franco's paintings are his fantasy, an allegory to express his single and most important personal feeling: nostalgia. Franco's nostalgia, magnified, intensified, and deepened by epileptic discharge, helped to unite life experience with visual talent to express the highest aspirations of his personality.[21] Contrastingly, the very absence of a personal center in Stephen is what allows for his remarkable fixation on buildings.[22]

8.4 Autist: Who Is He—Original Imitator or Imitating Original?

> I showed Stephen a portrait by Matisse and asked if he would draw it. He drew it, from the original, swiftly and confidently; it was not wholly, literally accurate, but it was very Matisse-like. When I asked him to draw it again, from memory, an hour later, he drew it differently, and another hour after this, yet differently again; but all his drawings (he did five in all), while different in detail, were strikingly evocative of the original. In some sense, therefore, Stephen had extracted the drawing's 'Matisseness'.... He got the style at once, and his later drawings were improvisations within this style (Sacks, 1995, p. 213) (see footnote 20).

While reading this story in Sacks' book it came to me that the autist's mind is similar to what Jung called the "introverted sensation type."

> [I]ntroverted sensation apprehends the background of the physical world rather than its surface. The decisive thing is not the reality of the object, but the reality of the subjective factor, of the primordial images which, in their totality, constitute a psychic mirror-world (Jung, 1971, vol. 6, p. 395).

---

[21] Recall that when his family returned to a Pontito destroyed by war, 10-year-old Franco promised his mother, "I shall make Pontito again for you, I shall create it again for you."

[22] Imagine what this means for autists—what freedom! No dependency on the object, no ego struggle, no personal emotional bias, no stress of competition, no jealousy. What a vacuum! A piano playing without the pianist. A blind brightness.

What may be the brain correlate of Jung's "background of the physical world" is "seeing" the world through the frame of RH cognitive mechanism. The RH has the ability to see patterns without abstraction. It identifies holistic forms as well as directly and immediately perceives the whole as a continuous totality of unique qualities along with the relations of these qualities. The RH then "sees" patterns without abstracting from the whole, including patterns that make one's individual style. Sacks cites Monet, when this great artist advised an aspiring one to develop his own individual style (which, according to Jung, means to access the "background of the physical world") and to learn to observe the physical world as such (which, according to this work, means to switch "vision" to the sensory-motor level, without the influence of higher abstracting levels):

> Whenever you go out to paint, try to forget what objects you have in front of you—a tree, a house, a field, or whatever….[23] Merely think, here is a little squeeze of blue, here an oblong of pink, here a streak of yellow, and paint it just as it looks to you, the exact color and shape, until it gives you your own naïve impression of the scene before you [cited from Sacks (1995), p. 206].

Note that this advice is valid for artistically gifted individuals. For them, the background patterns of the physical world shimmer through its "real" images, and are conscious from time to time. Another example is what Matisse said about his fascination with cloth, fabric, and textile patterns from Africa and Polynesia: "I never tire of looking at them for long period of time, even the simplest of them, and waiting for something to come to me from the mystery of their instinctive geometry."[24]

In the general population, however, the background of the physical world is buried in the depths of the subconscious. In autists (as we can see in autistic savants, because they can express it), the surface and the background of the physical world (in Jung's terms) are equipollent and indivisible, and both are on the foreground of consciousness.

8.5 Endnotes

Major depressive disorder has been used to illustrate this model of self (Glezerman & Balkoski, 1999). In depression, the continuous line of the LH self is disrupted and its components exposed in a pathological form. Patients with depression not infrequently experience disorder in the sensation of the density or weight of the whole body or of certain body parts. A gradual transition may be traced from the physical feeling of weightlessness to physical–psychical metaphors, reflecting the main symptoms of depression such as emptiness and worthlessness—from a weightless, disappearing body to the worthless, disappearing self. For example, a patient reported, "I often feel that it is not a head any more but an empty shell with nothing

[23] From a neuropsychological point of view, Monet is saying to forget the *whatness* of a thing.

[24] From Matisse: The Fabric of Dreams—His Art and His Textiles. Exhibition of Metropolitan Museum of Art (2005).

inside it and a lot of holes around…my body feels like an empty box with another empty box on the top, instead of my head…I am an empty and hollow nothing" (from Cutting, 1990, p. 351). In nihilistic extremes, a patient reported believing she was dead, and on one occasion described herself as "consisting of mere fresh air or atmosphere" (from Young, Leafhead, & Szulecka, 1994, p. 228).

Another theme in depression is the non-functioning object. This phenomenon can also be traced to a continuum of gradual transitions: from non-working internal organs and body parts to the non-functioning self. For example, a patient reported the experience: "All my internal organs are rotten and decayed…I can feel my left rotting lung, but my right lung is completely dissolved and missing" (from Lukianowicz, 1967, p. 39).

Disorder of functioning self involves self as the *I-concept,* an agent—one that is not thinking, not feeling, not acting, not remembering.

Psychological testing in depressive patients has found that claimed dysfunctions are much milder than experienced (showing, therefore, the subjective nature of the disorder, which is a disorder of self, not a cognitive disorder per se).

Disorder of the concept of self as a member of a social group is expressed by nihilistic ideas on the moral plane. For example, patients have reported: "I have committed the unpardonable sin—the closed mind…. I'm selfish…. I'm certain I'm disliked wherever I go" (from Cutting, 1990).

Subjective experience in depression also illustrates that one's own body is represented as any other object in the external world. Patients with depression who feel that their bodies are not working may feel that other people's bodies are not functioning either. A patient of mine with psychotic depression believed (and blamed himself) that his grandson's internal organs were rotten.

The extreme is what Kraepelin referred to as "ideas of annihilation." A patient reported, "The world has perished; there are no longer railways, towns, money, beds, doctors; the sea runs out. All human beings are dead, poisoned with antitoxic serum, burned, dead of starvation… No one eats or sleeps anymore" [Kraepelin, cited from Cutting (1990), p. 348].

In depression, the body image and concept of self as such are not disordered, but subjective experience, an attitude to oneself, is distorted. This illustration shows that limbic emotion is a fundamental part of LH self, the axis on which the LH self is thread.

Chapter 9
The Left-Hemispheric Self in Autism Revisited

Consciousness is interminable yakking, a frantic effort to keep up appearances, make the game seem always to be your game

(Louis Menand, "True Story", from The New Yorker, Dec., 2003, 108).

We return now to *I-concept* and its disorder in autism. *I-concept* is an idea and an awareness of self as separate from others. *I-concept* is connected with what we usually call consciousness and with language. It is a personal center, and things are happening from its vantage point—*I* see, *I* think, *I* feel, *I* do. The quality of the LH self is nicely expressed in the following metaphor:

> She knows that her mind contains an "invisible center"…. "That was her self," thinks Clarissa, as she stares into the mirror. "Pointed; dartlike; definite. That was her self when some effort, some call on her to be her self, drew the parts together, she alone knew how different, how incompatible and composed so for the world only into one centre, one diamond, one woman who sat in her drawing room"…"[1] [Virginia Woolf's "Mrs. Dolloway" cited from Lehrer (2008), p. 173].[2]

I-concept is designated in language by the first-person pronoun "I." Emergence of "I" in a child's language signifies the development of his consciousness as *one* and apart from others. The autistic child does not say "I," or the appearance of "I" in the language of the autistic child is grossly delayed.

[1] There is no better description showing that experience of personal center is the experience of *oneness*.

[2] Taking a group of modern artists (Marcel Proust, Paul Cezanne, Igor Stravinsky, Gertrude Stein, Virginia Woolf, and a few others) Jonah Lehrer adheres to the idea that the product of art is the result of the artist's attempt to represent his own mental representations, and in doing so, the artist intuitively explores the brain's organization.

T.B. Glezerman, *Autism and the Brain: Neurophenomenological Interpretation*,
DOI 10.1007/978-1-4614-4112-0_9, © Springer Science+Business Media New York 2013

9.1 Phenomenological–Historical Approach to Autism and the Meaning of "I"

We will now analyze, from the neurophenomenological perspective, the phenomenological–historical study of autism by Bosch (1970). Bosch shows that the development of personal identity in childhood can be observed through the development of certain grammatical forms in language, which reflect self-experience. In other words, these grammatical forms are objective markers of an awareness of one's personal identity.[3] Bosch examined the origin of self in history and the development of self in normal and autistic children.[4]

Bosch equates the saying of "I" with the formation of self in children. Why is the pronoun "I" so important? Bosch believes that this linguistic form expresses basic meanings of self-experience. Bosch distinguishes three basic meanings underlying self-experience: *being*, *acting*, and *having*. According to Bosch, these meanings had been experienced long before the emergence of "I" in language:

> If we trace back genetically the path of self-experience, then it gets lost in the darkness of the prereflective, merely lived life.… Along the path to the I, *having* and *acting* have already been lived before they find expression in language, and in 'living them' experience has been gathered (Bosch, 1970, pp. 68, 72).[5]

Along the way to "I," in the process of historical development of language (consciousness), other grammatical forms emerged which expressed basic meanings even before the final stage—the saying of "I." Bosch shows that autistic children not only do not say "I," but also have a disorder of *acting* and *having* revealed in their behavior and in deficiency of other-than-"I" grammatical forms connected with these meanings of self-experience.

I became familiar with Gerhard Bosch study when most of the chapters of this book had already been written. Bosch's approach to autism is very different from the approach taken in this book. Considering autism from the broader, philosophical–phenomenological–historical perspective, Bosch defined it as a specific modification of human existence. Bosch did not make any references to the brain. It is a particularly valuable confirmation of the model presented in this book, that, although our approaches to autism are completely different, we came to similar conclusions, and I believe our views on the nature of autism are in fact complementary.

The meaning of *acting* and *having* that Bosch found to be deficient in autistic children will be interpreted below in the frame of brain systems responsible for the LH self.

---

[3] "Personal identity" would be LH "sense" of self. Note that there are no "objective" signs for the RH self of immediate experience.

[4] Note that what Bosch meant by self is in my definition only part of self, namely the LH self.

[5] From "Infantile Autism—A Clinical and Phenomenological-Anthropological Investigation Taking Language as the Guide" be Gerhard Bosch, Springer-Verlag, NewYork, 1970. With kind permission from Springer.

9.2 Experience of *Acting*, Its Brain Basis, and Autism

Bosch defines the meaning of *acting* as experiencing one's own activity. He indicates that in experiencing one's own dynamic force the subject detaches itself from the material sphere. This leads to development of the purely formal conception of "I" as a center of action. Bosch showed the verbal path to the "I" in the pre-speech stage can be traced to early acting and experimenting of one's own activity, "of being a cause." "After the first experimental step toward independent activity—grasping movements, the child becomes unfatiguable. It grasps, fetches, pushes aside, or throws objects out of its bed with equal joy" (p. 75). Bosch also indicates the pleasure of functioning is experienced in a pure form where "the action does not involve an object but takes the form of movement of the body, for example, crawling, clapping hands, sitting up, falling down… In the second half of the first year of life there are already clear indications that 'own activity,' own desire, are being distinguished from those of others. This initial attempt at delineation is exemplified by the young child's defiance when something it expected is not fulfilled, when the activity of adults breaks in upon its play activity, and it is also exemplified by a clearly expressed satisfaction when the child gets its own way" (p. 75).

Bosch concludes: "We can probably regard the first appearance of the verbally developed 'I' as a clearly distinguished stage in the separation of own action from the action of others" (p. 76).

Bosch notes that in the language of autistic children, not only the source of action—"I"—is not uttered, but other grammatical forms that captured the dynamic force of action, the purposefulness, the in-order-to of action are missing throughout. Linguistic forms that express demand and order are missing too.

Autistic children use predominantly infinitive forms of verbs that describe the process rather than "self-involvement":

> Christa G. [six-year-old] frequently commented on the day's events or certain actions in phrases uttered rapidly and in an undertone to no one in particular. She either named the objects or used infinitive forms. When walking with the author through the ward the girl came to a locked door where the author took a key from his pocket, and she immediately said, 'Key, unlock'.... When she looked at pictures she often designated only the action of the persons by an infinitive. Looking at the picture of a cobbler at work, the child said: 'knock'. A cleaning woman, who was scrubbing the floor on her knees provoked the remark 'clean'. Looking at a man washing his car, the child said, 'spray', and looking at a boy playing with the toy, ships in a washing bowl, she said, 'take bowl'[6] (Bosch, 1970, p. 80) (see footnote 5).

As mentioned before, Bosch distinguishes three basic meanings-experiences underlying saying "I": *being, acting, and having.* He also considers experience of

---

[6] Naming the action—verbs—is connected with the left prefrontal area. Patients with damage to this area present with so-called telegraphic style, where their speech consists of nouns in initial forms. Here is the example of a patient with frontal lobe damage who was unable to use verbs when asked to explain the meaning of words—*gather*—"detail"; *impede*—"the river and rapids, stony" [from Glezerman and Balkoski (1999), p. 229]. This autistic child, Christa, apparently did not have this kind of frontal lobe deficiency.

being to be not disordered in autism. Bosch indicates several language forms that contain "a noting and representing of being" (p. 107). Among them are naming the word, "the individual word that places the object like a picture before other people;... "of the persons it is the objectively observed third person, the he, in which *being* can be most clearly understood, while the you addresses me directly...and the I reveals itself in action" (p. 107). Correspondingly, the autistic child's wish is expressed in language as the representation of fact (intact *being*), but is not directed as an appeal to others (impaired *acting*). Bosch gives an example of this observed in an autistic boy, Hans R:

> Hans R., who was linguistically very productive, but who still had not learnt to say I by the time he was 10... expressed a wish by stating 'he wants to turn' or 'he wants to go home'... When requested several times to come into the room he did not reply with the refusal 'I don't want to come' but with the statement 'there has been enough coming' or 'there should be no coming'.[7] A personal wish or an expression of will is replaced by the statement of a general law, of an impersonal order (pp. 109–110) (see footnote 5).

The meaning of *acting* described by Bosch, corresponds, in my formulation, to LH self as an agent connected with the symbolic level in the brain (left prefrontal cortex).

Remember that the common denominator of frontal lobe function is representation of action. This includes all dynamic schemes: from simple motor programs, to the temporal organization of object action, to more complex behavioral programs, to operations with categorical signs—sorting out, comparing, recombining, etc.—or categorical thinking (thinking is also movement!).

Bosch's account that development of "I" starts with the child becoming aware of his or her own motor activity makes sense in terms of the frontal lobe's evolution in man. In both historical and individual development, there is a stepwise order of the formation and "unfolding" of the frontal cortex: from motor cortex, to premotor cortex, to prefrontal cortex. In correspondence with maturation of the prefrontal cortex, cortical functional networks develop that underlie temporal organization of increasingly complex actions. Hand in hand with cognitive development, LH self as an *agent* is developing: from the sense of "being a cause" of one's own movements and actions to being an actor.

We may think of the evolution of subjective time (temporal organization of action) at higher functional levels as leading to the development of the LH self or LH consciousness. While, according to Bernstein, subjective time is embedded into the very structure of movement at the thalamic level, embodying its rhythmic dynamics, at the sensory-motor level time is speed and tempo, determining the correct moment for the exact movement, and at the gnostic-praxic level time is the temporal order or semantic sequence for the unfolding of object action. Object action is voluntary, with the focus of attention "moving" along the multilinked chain. At this level

[7] The last two sentences have more complicated grammar than the sentence "I don't want to come"—another evidence that the left "cognitive" prefrontal cortex, responsible for grammar and sentence formation, is intact in autism.

(gnostic-praxic) we can begin to speak about LH consciousness. We can refer in this connection to Gazzaniga (2000) who indicates that "serial control of motor output is an important determinant of what we perceive as consciousness" (p. 1362).[8]

Remarkably, Bernstein connected self with the evolution of an effector system and time in the frontal lobe. At the end of his description about the brain's organization of movements and without much detailed discussion, Bernstein outlined the following evolutionary steps of anterior brain function: movements—meaningful multilinked action—behavior—subject.

Bosch states that the experience of *acting* detaches the subject from the material sphere. Indeed, a product of time, "I" as a center of action is detached not only from other objects, but also from the *me-object.* However, at the gnostic-praxic level, a connection with *me-object* is still seen. One can trace the evolution from the gnostic-praxic level's (premotor cortex) awareness of one's own action to the symbolic level's (prefrontal cortex) experiencing "I" as an actor (abstract category of *agentivity*).

What is the experiential quality of the LH self as an agent? It is conscious awareness, not any particular sensory or bodily subjective experience.

Apparently my action and the awareness of my action as mine are separate entities, though both belong to the left prefrontal cortex. Figuratively speaking, disorder in implementing dynamic schemes: disorders of movements, disorders of object actions (praxis), disorders of speech (sentence formation), and even certain disorders of thinking would be considered neurological problems, while disorders of the experience of one's own actions, i.e., disorder of self, must be considered psychiatric disorders.

Realization of dynamic schemes per se involves what Bernstein called cooperation between the anterior (frontal) and posterior (temporal, parietal, and occipital) brain. Being an "actor" for the action of any complexity requires connection between the corresponding "cognitive" area within the left frontal cortex and the OFC system (Fig. 8.1).[9]

Let's return to our hierarchical structure of the LH self (discussed in Chap. 8). At the gnostic-praxic level, the LH self is presented by one's own body as a functional object (mostly visual image) assigned with biological significance from the amygdala (network *BA37-amygdala*). At the symbolic level of the LH self there is

[8] LH sequential processing is explicit. RH experience is not "conscious" in the usual meaning of this word (or at least cannot be directly verbalized) because even complex RH symbolic systems are simultaneous and, thus, implicit. However, we should remember that LH "decoding" is the only way we can understand metaphor, and it is connected with serial unfolding, yet we feel a much broader range of RH content, so, we cannot say that the RH experience is unconscious.

[9] The OFC system is what I designate to be the *amygdala-OFC-MD* network, a "triangle" that fascinates anatomists (Amaral et al., 1992; Jones, 2008) but still awaits a major discovery of its functional significance. I have described this triangle several times in this book, especially in the context of thalamic emotion's contribution to the RH self. It might be useful to repeat this description shortly here. While the MD projects to the vast territory of the prefrontal cortex, the part of the MD that contains fibers from the amygdala projects to the OFC and, in particular, that part of the OFC which receives direct input from the amygdala. Thus, the same area within the OFC receives "double" input from the amygdala: direct and via the MD. Connections in this triangle are reciprocal, except for the amygdala—MD connection, which is unidirectional, from the amygdala to the MD.

no direct input from the amygdala, because the amygdala does not have access to the DLPC, which is the source of the cognitive component of *I-concept*. The subjective, experiential quality of the LH self as an agent comes purely from the OFC. This agency component of LH self must arise from highly processed information from both direct and via the MD amygdala input to the OFC. It is probably also based on additional (or integrated with the MD's amygdala input) information about bodily feelings from the MD.

It seems to me that the model of a specific matrix within parallel-distributed networks proposed by Gloor (1992) is useful for hypothesizing about the difference in localization between cognition and self within the left frontal lobe. According to Gloor, each such matrix consists of widely distributed neurons, with individual neurons participating in many (but not all) matrices. Uniqueness of the matrix for elaboration of a particular function is due not to specificity of the separate neurons, but to specific combinations of neurons from widely dispersed areas.

It might be that the specific matrix for the realization of complex cognitive function consists of neurons located in the prefrontal cortex and neurons in areas of the posterior associative cortex. On the other hand, a specific matrix for "I" as an agent consists of neurons located in the same prefrontal cortical area that is required for cognitive function[10] but also of widely distributed neurons located in the OFC, MD, and amygdala. This unique matrix does not require participation of neurons from the posterior cortex.

In autism, the matrix responsible for self as an agent is disordered while the "cognitive" matrix is spared. However, a missing or weak "sense" of *agentivity* results in the impairment of an active attitude to the surrounding world and thus leads to secondary cognitive impairment.

9.3 Experience of *Having*, Its Brain Basis, and Autism

Bosch analyzed the meaning of *having* as the experience of possession:

> [I]n having an object the subject combines itself with the object in such a way that if the object is lost, something else is lost, too, that has become, to a varying degree, part of the subject… I have money and I am rich, I have skilful hands and I am skilful (p. 70)….' The primary object, the typical object, with which I identify myself and which nevertheless escapes me, is my body. It is as if we were standing here in the most secret and deepest dungeon of having' [from Gabriel Marcel "Phenomenology of Having", cited from Bosch (1970), p. 70] (see footnote 5).

Bosch emphasizes that the prototype of all possessive relationships is one's own body: "The I is able, so to speak, to perceive itself in the image of the object which it has appropriated as its own." (p. 69)

[10] Are they the same neurons that participate in cognitive function? We do not know. They may or may not be.

What can we see here from the point of view of brain networks? Bosch's ideas on *having* clearly fit within the realm of LH cognitive mechanism. One's own body is a separate object as any other object in the external world (the gnostic-praxic level). The "I" is related to one's body (in a possessive relationship), but does not form with it an indivisible whole, as would occur in the RH. The "I" (symbolic level in the LH) is becoming abstract and reflective (more aware of itself): "the saying of I, expresses that the speaker 'has' himself, has come to experience himself" (p. 68). Bosch indicates that, in history, the feeling of self was initially tied up with concrete observation of a person's own body and limbs (here again, observation of one's body as an object indicates LH cognitive mechanism and a "sprout" of the reflective quality of LH self). In language, instead of "I" or "me," it was "my being," "my nature," "my body," "my bosom" or a purely spatial form, a word, the basic meaning of which could be expressed as "central," and words like "heart," "flesh," or "face" stood for the reflexive pronoun "myself." Correspondingly, possessive pronouns emerge before the personal pronoun, "I," both in the history of language and in individual development. In the history of language, there was a state in the course of development of the *I-concept* in which "I" and "mine" as well as "you" and "yours" were not yet separate. The same phenomenon can be observed in individual development. Bosch gives an eloquent example of this:

> Wolfgang [1.5 year old] had one morning in the kitchen found a little bag belonging to his older sister; he dearly wanted to have this little bag, but his sister had always snatched it from him with the words 'No, belongs to me'. When he now met me with it, he held it out to me in one hand and said 'Hilda', as if he were trying to say, 'That really belongs to Hildegard.' But immediately following this he pressed it to him clasping both arms around it and called 'me', which was probably meant to mean 'but now it belongs to me' (p. 73) (see footnote 5).

Bosch notes that the great diversity in conditions of possession (the having of a house, a piece of clothing, of relatives, and of one's own body) and, on the other hand, the "homogeneous" expression of possession by the same possessive pronoun indicates a unifying of the subject of this possession, the creation of a formal entity "I" as a possessor as such. So, in the process of experiencing *having* the "I" becomes a center in relationship to possession.

In autistic children, deficient experience of *having* is expressed in their behavior toward property, as well as in their speech. Possessive pronouns such as "my," "your," etc. and the possessive genitive appear very late, if at all, in autistic children. Below is an example given by Bosch of the autistic child already mentioned earlier, Hans R:

> As late as at the age of 9 the boy stood out in school because he kept taking any objects that interested him from the other children in his class without ever bothering about the owner. In the clinic too, at the age of 10;6 we observed several times that he pounced on a toy and took it away as if he had simply not seen the owner.... If he was spoken to about this and if it was suggested that he should first of all ask the owner for permission, he would leave the object in question where it was and not pay any attention to it.... When his mother went shopping in stores she either had to leave him at home or to keep him close at her side, for otherwise he took down wheels or other objects that interested him from the display stands.... If some of the other children took any of his things away, he became worked up,

> ran angrily up and down the room and said to himself, 'The boy has taken the tin'.... 'The tin should be in front of the bed'.... However, he did not try to prevent anyone taking his toys away, nor did he turn to any adult for help. On such occasions he never used the demanding word 'mine' or 'that is my toy' (p. 77).[11]

While *acting* corresponds to the categorical sign of agentivity in the LH *I-concept*, *having* does not have an equivalent in the categorical component of LH self. As discussed before, categorical signs characterize the object as such and reveal inner, logical connections between objects by their membership in the same category (for example, a cup belongs to the general category of *objects* and to more specific categories of non-animated objects, things, artificial things, tools, dinnerware).

Already mentioned in this book is the fact that besides categorical, inner relations between objects, there are also outer relations, relations by contiguity in space. The meaning of *having* is purely relational, where the "I" (possessor) is a center in relation to possession. In Chap. 6 we discussed that in language certain grammatical constructions express spatial relations between objects. This linguistic function is connected with the left parietal–occipital cortex, responsible for distinguishing signs of spatial relations such as spatial coordinates, direction of movement in space, approaching-withdrawing, expanding-shrinking, part-whole, etc. We also discussed that grammatical forms initially (in the history of language) expressing literal relations between objects in physical space later also expressed possessive relations. The same grammatical construction is used for the following relationships: brother's hand, brother's shirt, brother's neighbor, brother's daughter, which represent part of oneself (part-whole), personal belonging, contiguity in the physical space, and blood relation, respectively, but they are all included into the meaning of *having*. For example, Bosch notes that the transition from *having* as an expression of possession ("I have a house") to *having* as an expression of being part of oneself ("I have a head") is fluid such that having a possession and having a "part" cannot be clearly differentiated. As to blood relation, Bosch emphasizes that one's own body and family members are in an inescapable possessive relationship. Bosch indicates that "in the history of language it is possible to detect stages of the experiencing of possession that vary in proportion to the distance of the possessed object from the person's body" (p. 74). What is very interesting, Bosch illustrates, is that in the preverbal stage the "spatial" concept of *having* develops in action:

> From the verbal expression of wanting-to-have, a desire, we can go back to the first grasping and letting-go again of an object...what infant grasps merges with the grasping hand to a whole...The object...becomes part of the body, bodily property, something 'had'. Just like 'putting-in-the-mouth' of objects, the very early grasping with the hand is also a 'taking-inside-oneself', a 'making-part-of-the-body' of the grasped object. Setting the grasped object apart from the grasping body, however, presupposes that the body can be experienced as an object, and its particular objectliness distinguishes it from the grasped object.

[11] From "Infantile Autism—A Clinical and Phenomenological–Anthropological Investigation Taking Language as the Guide" be Gerhard Bosch, Springer-Verlag, New York, 1970. With kind permission from Springer.

> The objectivization of the body, however, causes it to experience attributes fundamental to that which we call property…. The original 'property' is the own body or what is pressed to the own body (pp. 73, 74) (see footnote 5).

The possessive pronoun then, Bosch indicates, "in conjunction with a grasping, desiring gesture, can express the claim to possession, the 'wishing-to-press-to-the-body'" (p. 75). What does it mean—the concept of *having* develops in action from our neurophenomenological point of view? The parietal–occipital cortex is responsible for the visual–spatial function (static scenes of spatial arrangements of objects) with the left parietal–occipital cortex, in particular, being responsible for distinguishing signs of spatial relations between objects, abstracted from the objects themselves. These "ideas" of spatial relations between objects serve as a meaning for grammatical constructions in language, the latter are stored in the left parietal–occipital region, "classified" by their similarity or opposition, as a so-called paradigmatic series. For example, the paradigmatic series of prefixes and prepositions are organized by the type of simple oppositions—*under* the table–*above* the table; go *in*–go *out;* come *to*–come *away.* Signs of spatial relations are stored in the left parietal–occipital cortex, while the operations with these signs (temporal organization) are a function of the prefrontal region. In history and in individual development, structuring spatial signs into paradigmatic series was developed in the process of operating with them in the frontal lobe. In Bosch's illustration, we see that *having* is experienced through a child's "experimenting," acting out different spatial positions or relations of his own body and a desired object: part-whole ("press-to-the-body," "making-part-of-the-body"), approaching-receding (making the object separate from the body), manipulating with the distance between his own body and the object, etc.

We wondered earlier how it might happen that formal, purely cognitive *signs* of spatial relations between objects in the physical space are the foundation for the grammatical constructions that express any kind of relation—from position of objects relative to each other in physical space to relations between people (see Chap. 6). Now we can understand (thanks to Bosch's observation) that in the historical formation (and in individual development) of grammatical constructions to express interpersonal relationships, not only was it purely a cognitive component of spatial relations, but also an experiential, emotional component (the "desired object") that was participating![12] We can see this process as a cooperation of two areas in the brain, the left parietal–occipital and the left prefrontal regions, simultaneously activated and responsible for "movement" in the physical space toward the "desired object" associated with the experiencing of *having.* In autism, the cognitive component is intact, while the emotional–experiential component is impaired. It is pertinent to recollect here Grandin's introspection (see Chap. 6), where she describes how she "got" an idea, the concept of interpersonal relationships, using the spatial relations between objects.

---

[12] This brings us to the mystery of natural language where, in contrast to other languages such as computer languages, it is not purely a cognitive phenomenon.

Bosch makes an interesting remark that by the experiencing of *acting*—one's own dynamic force—the subject detaches itself more easily from the material sphere than by the experiencing of *having*. This is in accordance with the clarification of *having* as originating from spatial relations between objects in the external physical space and with the corresponding brain mechanisms underlying this process. Although both the experiencing of *acting* and the experiencing of *having* lead to the formation of "I" as the center of reference ("I" as a center of action and "I" (possessor) as a center in relation to possession) the original connection of "possessor" to "possession" is not internal, but external, by contiguity in space.

Bosh emphasizes the exceptional role of *having* in the development of a value system and interpersonal relationships:

> When a child calls imploringly 'me, too', it strives for property. The whole structure of property is only understandable, if we see it as the structure of co-existence. That which I have is distinguished from that which I do not have. Not to 'have' something means, however, that it belongs to and is available to others. It is not until I realize that things are not there 'for me' that those things that I have acquired take on the character of property that must be retained and protected from others' grasp if this property is not to be lost to them. The comprehension of the other person as a 'haver', as someone to whom property refers, pre-supposes that one can put oneself in his place, see things from his standpoint, and achieve a sympathetic understanding of his experience of possession relationship. This understanding enables me to experience not only what property is available to the other person, as opposed to myself, but also to experience what this property is 'worth' to the other person, what it gives to him. With the transition from the simple 'using' of what is available to acquiring it as property, this property is not merely 'acquired', incorporated into a special sphere, but becomes part of a *value structure.* The desire of the other person and my own desire are directed towards the desired object, because this has a *value for us* (pp. 83–84) (see footnote 5).[13]

In autism, there is no movement toward the "desired" object—not assigned with biological significance, the object is not desired.[14]

---

[13] Note the evolution from the desired object to the desired-object-for-us. Bosch goes on to conclude that the external world "becomes not only a world of commonly recognized and noted objects, but a world of values, be they possession values, be they aesthetic values, or ethical and moral values of action" (p. 85).

[14] The following scheme shows the basic steps of active, goal-directed behavior to "get the desired object" indicating its disordered links in autism.

1. Object is recognized (the left temporal-occipital region, BA37)—preserved in autism.
2. Recognized object is assigned with biological significance (the left amygdala—BA37)—impaired in autism.
3. Recognized and biologically significant object becomes "desired" (the left MPC)—impaired in autism because the object is not biologically significant. The left MPC (medial prefrontal cortex), responsible for motivation per se might not be disordered.
4. "Movement" toward desired object to "get" desired object—programs of goal-directed behavior (the left DLPC)—impaired in autism, because the object is not desired. However, the left DLPC, responsible for creating the programs, is preserved in autism. It is what allows the autistic child to learn useful behavioral patterns.

9.4 Comparison of Brain Networks Involved in *Acting* and *Having* and Its Implication for Autism

As our neurophenomenological analysis has shown, *acting* and *having* have slightly different brain bases: *acting*—with pure involvement of the prefrontal cortex connected to the network of the OFC system, and *having—left DLPC-left TPO* network connected to the OFC system.

As the basic radicals of the frontal and parietal areas are different (temporal organization of information versus spatial organization of information, respectively), so we may anticipate the differences in contribution of *acting* and *having* to self. The "temporal" nature of *acting* and the "spatial" nature of *having* is noted by Bosch (although he did not explicitly use these terms) when he indicates that in experiencing one's own dynamic force, the subject detaches itself from the material sphere, and this leads to the development of the purely abstract concept of the "I" as a center of action. On the other hand, Bosch states, the experience of *having* is purely relational. Its framework is spatial organization: "possessor's" central position relative to "possession."

We may think that *acting* is the main contributor to the LH self as an agent, and *having* is the main contributor to the relational component of LH self and to a value system. As *acting* and *having* are provided by partially different brain systems, agentivity (mostly the contribution of *acting*) and the relational component (mostly the contribution of *having*) may be unevenly developed within the LH self of the same person.[15]

Again, *acting* contributes more to the separate "me," individuality, and *having* to the collective "me," or member of the group. The brain network of *acting* must be the most vulnerable in autism, since it is closer to the primary deficit (OFC system). The brain network of *having* includes also the left TPO, which is more peripheral to the primary deficit. The left TPO is relatively intact in autism and can even be used for compensation, as we saw in Grandin's case. Bypassing the amygdala, Grandin (her brain) used the following network—from the right TPO to the left TPO to the left DLPC to the striatum—to gain intellectual understanding of interpersonal relationships and to develop useful behavioral patterns (see Chaps. 6 and 7). Another of Grandin's reports shows us that autistic people may have a strong value system, but that it does not originate from experiencing *having* (i.e., LH self). The compensation here is achieved by using the network: *thalamus-right TPO-right DLPC-left DLPC* (see Chaps. 6 and 7). We understand that this value system is restricted to the global, cosmic, ethical issues, but by no means does it include literal possession values.

In conclusion, while Bosch's phenomenological analysis is in accordance with my theory that disorder of LH self is the cardinal deficit in autism, it also enriches

---

[15] Development of different brain regions varies independently from each other, and so, within the general population, individual variability of the brain is such that people may have prevalence of the prefrontal lobe over the parietal lobe and vice versa.

our understanding of this profound deficit, clarifying the type of experience that underlies the self (in our terminology—LH self). Experiencing *acting* and *having* leads to awareness of self and others. The phenomenological analysis also confirmed a close connection between LH self and language. Meanings of experiencing *acting* and *having* are directly expressed in language. "The world of *I* and *YOU* or *I* and *THINGS* are opening to the child through the parallel development of experiencing *acting* and *having* and their expression in language" (Bosch, 1970, p. 72). Moreover, special grammatical forms are developed to express these basic meanings. It should be emphasized that this experience is not idiosyncratic but connecting *I* and *YOU*, directly translated into language, it becomes conventional.

9.5 Autism—Strange Existence

How must it feel to be a perceiving and thinking subject without awareness of one's own personal identity?

The introspections of Therese Jolliffe (a high functioning autistic individual herself) help us understand the autistic person's experience: "Reality…is a confusing, interacting mass of events, people, places, sounds and sights. There seem to be no clear boundaries, order or meaning to anything" (Jolliffe, Landsdown, & Robinson, 1992, p. 16). Is this bewildering, chaotic experience of reality pictured by Jolliffe a reflection of RH "flow of consciousness"—directly felt reality that is unbearable (this is how it was explained in Chap. 3); or is this chaotic picture of the world a result of a LH deficit, the inability to experience herself as an autonomous individual different from others? Both factors probably contribute to the "picture," but in differing, individually specific proportion that depends on degree of severity, age, degree and type of compensation, special abilities, etc.

How must it feel to not be aware of oneself as a "separate one" and, correspondingly, to be oblivious of "separate others"?

Jolliffe's experience also illustrates an autistic person being oblivious of others:

> I spent a great deal of my time alone in my bedroom and was happiest when the door was closed and I was by myself. I cannot remember ever thinking about where my mother, father, brother and sister were, they did not seem to concern me. I think this is because I did not for a time realize that they were people and that people are supposed to be more important than objects…. I often do not notice a person in a room until they move…it was a long time before I realized that people might actually be speaking to me, so it was a long time before I realized that I too was a person—if somewhat different from others. I never thought about how I might fit in with other people when I was young because I was not able to pick people out as being different from objects (Jolliffe et al., 1992, pp. 12, 14, 16).

As mentioned before, Bosch indicates that reflective quality of the self developed in history by the "objectivization" of one's own body. It was similarly formulated in this book that the abstract *I-concept* (symbolic-categorical level in the LH) is built upon one's own body image as an object, a functional object at the gnostic-praxic level in the LH. It did not happen "at once." In history, the feeling of self was initially tied up with the concrete observation of one's own body and its functional parts. This is demonstrated by the history of language. Take the example, used above, where in early languages such words as my "body," "bosom," "heart," and "flesh" might stand for the reflexive pronoun "myself." This historical stage can be seen in autistic children. Here is an example from Bosch's material: "A 4-year-old girl was noticed repeatedly naming the parts of her body: 'Legs, heel, arms, eyes, nose, mouth, and hair' (always in that sequence). Then she said, 'Anneli is still complete'." (p. 122)[16]

This eloquent example shows us that, lacking the LH *I-concept* to make a separate individual self, this autistic child observes and enumerates her body parts, gathers herself together "piece by piece" to achieve consciousness of self. Even though the girl uses the word "complete," I believe it is *oneness* (not wholeness) that she is searching for. Utilizing the visual sense, she observes herself as some-body in the outside world, as an *object.*[17] Her compensation, however, is not stable. She needs to repeat the enumeration constantly, to "confirm" her existence. Here we are touching upon another facet of LH self—a sense of existence. Autistic children seem unable to experience themselves as existing beings. With age and adequate teaching, autistic children may learn to recognize their selves, but, as Bruno Bettelheim indicates, they live in continuous fear that at any moment they may "disappear." Bettelheim observed autistic children who fight sleep for weeks and months, because they are convinced that by letting go, even if only for a moment, of this tenuous grip on themselves, they would again become nonexistent. As one of them put it, "I feared I just would no longer be" [cited from Bosch (1970)].

To illustrate the specificity of this "strange existence" in autism, we can compare autism with major depression, a disorder in which the existential self is also suffering (see Chap. 8). In depression, the amygdala's connection with both the temporal cortex (BA37) and the OFC is not weakened, and activation of the amygdala is even pathologically increased (which can be connected with a disorder of attitude to the *me-object*). It was hypothesized before (Glezerman & Balkoski, 1999) that in depression there is a disorder of the integration of the LH self's components along the vertical, hierarchical axis in the brain. As the layers of LH self are separated, its internal structure is exposed. The deepest level where one's body is represented as a *weight-object* in the gravitational field gets exposed in an exaggerated and distorted form and expressed through all levels of the LH self. In the norm, this level's contribution to the LH self is a sense of body-mass/density, of fullness, of vitality, of well-being. In depression, patients complain of a feeling of weightlessness or

[16] This same phenomenon where functional, "acting" parts of the body stand for the self can be seen in autistic children's fragmentary perception of the other. Hans R. when 10½ year's old, reacted to his mother's enquiries about his homework in the following way: "He would…suddenly cover his ears or press his hand against his mother's mouth and shout, '*The mouth should be closed*'" (Bosch, 1970, p. 94).

[17] Recall that *wholeness* of the self comes from the sensory-kinesthetic feeling of one's own body-space.

hollowness of the body. This physical feeling, amplified and modified by input from the amygdala, becomes a psychical–emotional experience of emptiness or nothingness, even to disappearance. One patient described herself as "consisting of mere fresh air or atmosphere" (from Cutting, 1990, p. 351). The metaphors of patients' psychical experience preserve the vestibular-kinesthetic flavor—"growing lighter," "growing smaller."

At the symbolic level, a transition is seen from the "small" and "weightless" object (one's body) to a worthless, "bad," sinful self (the moral, relational plane of LH self) and to a disappearing self (existential plane of LH self). The disappearing self in psychotic depression is known as "ideas of annihilation" or nihilistic delusion.

To think about oneself as a bad, guilty, sinful person is a judgment. Judgment was defined (see Chap. 7) as a junction between emotion and concept, related to the *left amygdala—OFC—VLPC* network. It belongs to the LH self as a member of the group, to the "relational" self with its value system. As to nihilistic delusion, it is belief or judgment based on the truly psychotic experience of disappearance.

Now, what can we say? To make judgments, to have beliefs about oneself, one needs to have a separate and different from others self, *I-concept*. To be delusional about oneself, one also needs to have an *I-concept*. We may conclude that patients with depression have a separate and different-from-others self, *I-concept*, and it is strong. It is the creator of their "torture."

The best illustration of the intactness of *I-concept* in depression is Virginia Woolf's exploration of her LH self, reflected in her writing. Woolf herself suffered from severe depression. During remissions, in anticipation of the return of her devastating depression, she was in a constant state of reflection, searching for internal support. She wrote in her diary: "I press to my centre, and there is something there" (from Lehrer, 2008, p. 169). "What she found was the self, 'the essential thing.' It is the fragile source of our identity, the author of our consciousness. If the self does not exist, then we wouldn't exist… The self is a perceiver for our perception…it is the story we tell ourselves about our experiences" (Lehrer, 2008, p. 169).

Making her way through the "wedge of darkness," Woolf "discovered the self's stubborn reality…that inexplicable essence that makes us ourselves, and not someone else—refused to disappear. In fact, the more she investigated experience, the more necessary the self became to her" (Lehrer, 2008, p. 180). "In her modernist novels, she wanted…to show us that we are 'like a butterfly's wing…clamped together with bolts of iron'" (Lehrer, p. 170).

What made Woolf's incredible insight possible? We may consider there to have been at least three factors. It was her depression where, it has been hypothesized, layers of the LH self (integrated in the norm) are separated and can be "looked" at as though through a magnifying glass. Another factor may be a spontaneous attempt of her brain to compensate. Yet, what made the above factors work is her talent. As Lehrer demonstrated in his book, to be an artist is to be able to comprehensively explore one's own brain. Woolf wrote novels where she described the mind, but in doing so she explored the parts of her own brain that contribute to the LH self. Suffering from a distorted LH self, she reached its core and discovered an intact and strong sense of existence and individuality. It was her therapy, her victory over her depression (unfortunately, temporary), and her revolutionary novels (not so temporary).

Thus, the experience underlying a sense of nonexistence in depression and in autism is qualitatively different. Indeed, autism and depression are incompatible. In depression *I-concept* as such is intact, and a sense of oneness is strong. In autism *I-concept* is simply missing or undeveloped. As Bosch indicates, in autism, an important aspect of human development is missing. Missing an *I-concept*, the autistic child must rely on the visual image of his or her body to constantly catch the slipping away "separateness" from others and from the world. Fear of nonexistence in autistic children is caused by instability of LH self. In depression, the root of the experience of nonexistence is disappearance. In psychotic depression, disappearance or ideas of annihilation are beliefs "created" by the LH self in response to a distorted bodily experience (kinesthetic and vestibular), amplified by the increased amygdala activation.

Autistic persons do not have a role, do not have an attitude to their self (in our understanding, LH self); it is undeveloped, but not hated (as in depression). Autists are unable to experience guilt, shame, or embarrassment, but they also do not have any vanity, any desire to show themselves off to others, to impress or please others. They lack what we have automatically: *I* think, *I* feel, *I* do. This awareness of *I* led Descartes to absolutize the LH self: "I think, therefore I am"; while ordinary people do not think about this at all. On the contrary, autistic people are natural philosophers: the question and proof of their own existence is the everyday task for them.

To conclude (1) Interpretation of Bosch's data in terms of brain systems enriches the neurophenomenological analysis of autism conducted in this book, deepening our understanding of the LH self and its disorder in autism. (2) In addition, such interpretation gives us a new insight into the brain systems involved in the LH self and the interaction of cognitive and emotional (experiential) components within the LH (the "bridge" of language grammar and LH self).

Chapter 10
Pieces of Autism's Puzzle Fall into Place (Clinical-Brain Pattern of Autism)

10.1 The Primary (Fundamental) Deficit: Necessary and Sufficient

As argued in previous chapters, the primary disorder in autism is one of consciousness, of the self, and not one of cognition. Autism's fundamental deficit traces back to that part of the LH responsible for our awareness of personal identity. Self as *one* and apart from others, self as agent, is missing or undeveloped. The basis for this LH self can be hypothesized as lying within the LH OFC system: the triangle of left *amygdala/OFC/MD* (Fig. 8.1) as well as the network *LH OFC–LH DLPC–LH TPO*. Dearth of functioning in this network has profound consequences. It touches not only upon LH *I-concept* and complex emotions but also LH object-related emotion at every level, including basic emotion.[1]

The brain basis for the primary deficit in autism, I have proposed, may be an anomalous formation of the specific *matrix* that gives rise to the LH self. The matrix in question would lie within the widely distributed network of the OFC system indicated above. Matrices are characterized by sets of neurons located in different (often remote) regions of the brain that then create a functional system (Gloor, 1992). The neurons participating in a particular matrix can also participate in other matrices. What *is* specific to the matrix is not the individual neurons themselves but a specific combination of neurons coming from widely distributed areas. It was hypothesized

[1] The intrinsic network within the amygdala—lateral nucleus (LN)—basal nucleus (BN)—appears to assign objects with emotional significance (see Fig. 7.1). In Chap. 7, I attempted to show amygdala-related deficiencies in autism: no automatic, near instantaneous response to emotionally significant stimuli (preconscious emotional processing *en route* to consciousness). There are also (1) no attentional preference to emotional versus nonemotional stimuli from presumed lack of amygdala influence on visual processing at its early, preconscious stages (due to its return projections to all levels of the visually related cortex) (Corden, Chivers, & Skuse, 2008; Gaigg and Bowler, 2009); and (2) no superior remembering of emotional versus nonemotional stimuli (amygdala projection to the hippocampus) (Gaigg and Bowler, 2008).

T.B. Glezerman, *Autism and the Brain: Neurophenomenological Interpretation*,
DOI 10.1007/978-1-4614-4112-0_10, © Springer Science+Business Media New York 2013

in Chap. 9 that sets of neurons for the LH self as *one* and *separate* from others are located in the left amygdala, left MD, left OFC, and left DLPC. As this matrix is a highly specific system, its disorder is quite selective and "fine," however it has global consequences.

10.2 The Prevalence of the Right-Hemispheric Situational (Sensory-Motor) Level in Autism as a Direct Consequence of the Primary Deficit

Each cerebral hemisphere has its own unique self. Loss of the LH conscious "I" therefore does not mean full loss of self (or consciousness). In the norm, the LH self dominates and fully occupies the center of consciousness, giving the impression it is the only source of self. In autism, components of the RH self rise to the surface because the deficient LH self is unable to keep them in the background.[2]

What is postulated in this book is new to clinical neuropsychology—it does not suggest autism is a local or diffuse brain disorder, nor even a disorder of connections between brain areas. I am referring to autists as people whose consciousness is dominated by a particular *historical* stratum of the human brain. This work has been like an "archeological" excavation towards understanding deeper levels of the human brain, buried in the norm but surfaced in autism.

The prevalent self (and consciousness) in autism is made of *subjectively felt visual scene-situations* (VSS) of the RH sensory-motor level. In applying to *subjective experience* Bernstein's model of a "space" and "time" framework for each level of movement, it was shown that "consciousness" of the RH sensory-motor level *is* awareness of that level's "space" (VSS) and "time," giving rise to continuity of one's experience of the outside world and one's own life timeline.[3] The degree of awareness at the RH sensory-motor level is assumed to be lower than higher functional levels, again, since awareness and voluntary control decrease as the leading level goes down the brain's vertical axis. Because the sensory-motor level of the *right* hemisphere is leading in the autist, there is no separate understanding of time; time is perceived through movement. Also, objects are not separable from each other or from the visual situation as a whole, and yet autists' picture of the world is exquisitely clear and precise.

Awareness of "space" at this level is an awareness of the visual scene-situation as a moment of experience in the outside world. The VSS, then, is the autist's internal world, his imagination and fantasy. His awareness of this internal content is not complete, and his voluntary control over it is quite limited. As the amygdala in the

[2] See Chap. 2 for a detailed explanation of uniqueness of the RH prevalence in autism compared to other developmental disorders, and also for evidence *against* compensatory RH dominance in response to the local or diffuse LH damage.

[3] Note that continuity of one's identity does *not* arise from this level.

LH assigns biological significance (emotion) to objects, in the RH, emotion is related to the whole situation. The VSS of the RH is saturated with emotions that are "diffusely" related to all objects therein and also are markers of the particular VSS. It is an undifferentiated conglomerate of various emotions that characterize the VSS, because different objects within a situation each evoke particular emotions at the moment the situation was perceived but all are attributed to the entire VSS. As a result, a VSS is characterized not only by perfect fidelity, but also emotional intensity, for different (sometimes opposite) emotions are not able to be differentiated or escape the spatial boundaries of the situation. Although the emotions at this level are potent, they remain locked inside the situation and there is no awareness of them. The vision of the world in the VSS is extremely clear, but there is a break between this experience of the world and the outside world itself, for emotion is not attached to the separate object responsible for evoking it. The VSS can be projected only in an autist's art because the consciousness of autist is such that he himself is not much aware of the VSS—the RH self *cannot* be divided from nonself to be an observer. However, when an autist's sensory-motor level is projected, we can catch a glimpse inside the autistic mind: its subjectivity, imagination, and fantasy. Only when the VSS is projected onto paper as art, can we feel the vibration of its emotional intensity. The autist himself cannot enjoy the beauty, style, and emotional tension in his own work (remember Stephen, the autistic artist described in Chap. 8), for an autist does not express himself in art—his art *is* him.

In addition to having their own style, autistic artists can also reproduce in the likeness of another artist. Recall Stephen's "Matisseness" (Sacks, 1995). Matisse's paintings became Steven's subjectively felt visual scene-situation, part of Steven's self. Here we see the undeniable role of RH holistic processing. In this regard it is interesting to mention a patient with a LH lesion who could hardly recognize any objects in paintings yet realized immediately they were all Van Gogh's [Lhermitte et al. (1973) cited from Cutting (2002)].

In sum, autism's predominance of RH sensory-motor level turns out to be a "subjectivization" of the outside physical world. The following description by Minkowski of the "extroverted" autist eloquently illustrates the above formulation:

> "It cuts itself a piece from the world and merges it, so to speak, intimately with its own person. As soon as it aims at a goal it becomes one with it, as well as with all the external forces that it puts to work to reach the goal. Thus demarcation between self and not self is no longer the surface of the body but has moved outside it". Regardless however, of the fact that it is relocated outside the body, this demarcation can become impenetrable and rigid, an autistic armour in fact [Minkowski (1923) cited from Bosch (1970), p. 49].[4]

[4] We can only guess that before Kanner and Asperger described autism as a separate category, psychiatrists, who of course observed autistic children (and adults), diagnosed them with schizophrenia or schizoid personality disorder. Astute clinicians, as Minkowski was, must have seen a difference and struggled to "differentiate" autism within the framework of schizophrenic nosology. Minkowski tried to differentiate autism in patients with schizophrenia into "introverted" and "extroverted" autism. In doing so, he gave a brilliant metaphor for the different, and separate from schizophrenia, entity—later distinguished by Kanner and Asperger and called *autism*.

An autist's own internal world consists of emotion locked inside the VSS that is inaccessible to him. He does not have complete awareness because he *is* the situation, it is his self. His is a moving picture of the world, its fabric saturated by emotion. The autist possesses directly felt reality: it is mine, made by my emotion, and no one sees the world in the same way. The content of the autist's "me" is the physical world.

The essence of the autist's existence can be understood from the point of view of brain mechanisms: RH sensory-motor situational level functions split from the LH sensory-motor level and from the symbolic level of both hemispheres.

10.3 Stereotypical Behavior in Autism and Its Relation to the Autist's Self

The neurophenomenological analysis conducted in this work closely connects autists' stereotypical behavior with autism's primary deficit. When, for whatever reason, higher hierarchical levels in the brain are not functioning (most commonly due to prefrontal cortex's damage) goal-directed behavior becomes impossible and is replaced by stereotypical behavior.[5] In autism, lack or underdevelopment of agentivity (primary deficit) causes severe restrictions of spontaneous activity, and goal-directed behavior is necessarily replaced by stereotypical behavior.[6]

I have also argued that the autist's separate, un-integrated "layers" of RH self are released in autism as a consequence of the primary deficit. As this RH self resides in lower hierarchical levels, its actions would be more automatic—stereotypical rather than voluntary, with a lower degree of awareness.

Also, recall that RH representations are implicit or simultaneous. The RH is not realized in actions based upon, but separate from, the content of consciousness. The RH instead reveals itself via externalization or "acting out" its holistic content.[7] RH automatisms must then be the repeated playing out of RH holistic content. Below we will explore types of stereotypical behavior connected with different levels of the exposed RH self.[8]

10.3.1 The Thalamic Level: Subjectively Felt Body (I-Space)

"Acting out" this part of the RH self can be seen mostly in LFA with their well-known *rhythmical repetitive body movements* (see examples in Chap. 5). Such actions

[5] See Chap. 5.

[6] Although, as noted several times already, the cognitive part of the LH DLPC (inner programming for behavior of various complexity) might be intact.

[7] See Chap. 6 for examples of aborigine rituals (Levy-Bruhl, 1930) as an illustration of externalized RH representations.

[8] Recall that the content of consciousness for each level of self is determined by the "space" and "time" framework of experience specific for each level.

"play out" their own bodies' inner rhythmical pattern. Stereotypical movements involving release of automatisms from the thalamic level have an even lesser degree of awareness and voluntary control than those of the sensory-motor level (striatum). The content of the autist's consciousness during this type of stereotypical behavior is the body itself, or the kinesthetic feeling that create his body's spatial boundaries.

10.3.2 The Sensory-Motor Level: Subjectively Felt Reality (Non-I-Space)

What I have argued to be specific to autism—*sameness behavior* and *preoccupation with parts of objects*—are in fact an externalization and playing out of the content of RH consciousness of the sensory-motor level: VSS as a moment of experience in the outside world. Making automatisms from the VSS and the objects within it becomes a compensation of the autistic brain for the RH flow of consciousness, the constantly moving RH world—to "stop the moment." In the case of sameness behavior, it is preservation of the VSS: *RH IT (BA37)—RH DLPC—RH Striatum.*

Preoccupation with parts, with its intense sustained attention to the object[9] *within* the VSS (*RH DLPC*), is the autistic brain's tool for distinguishing a discrete moment of experience out of RH continuous flow of time (*LH DLPC*) and repeating it (*LH Striatum*).

10.3.3 Subjectively Felt Non-I-Space Combined with the integrated I-Space/Non-I-Space

"Acting out" the thalamo-parietal version of RH self in autism is usually found alongside of the above-discussed levels of RH self (thalamic and sensory motor). The RH self connected with the sensory-motor level, combined with the thalamo-parietal version of the RH self, makes possible the astounding achievements of autism—the savant's talent for numbers and calendar calculation. Before we discuss externalization of the RH self "combined," we have to dwell on the visual spatial situation connected with the right TPO.

10.3.3.1 "Geometric" Brain and Visual Spatial Situation

The VSS presents its different aspects depending on the functional specialization of the right inferotemporal region—IT (BA37) and the right inferoparietal region—TPO

[9] Or part of the object connected with movement.

(BA39,40). Here we will discuss the visual gestalt connected with the right TPO, which we called the visual spatial situation (VSpS). Objects in the VSpS are "shapes," holistic forms that are stored as perceived within the VSpS. Location and position of object shapes within the VSpS are also unique for each VSpS and preserved as perceived. "Space" of the VSpS itself is not divisible from the spatial configurations it contains. VSpS as a *whole* can be understood as the continuous totality of unique qualities (*parts*). It (*whole*) is directly and immediately perceived as the relations of its *parts.* The difference of the VSpS from the "generic" VSS is also that VSpS is thought to combine the functioning of the sensory-motor level and some transition to the symbolic level (probably equivalent to what is called the gnostic–praxic level in the LH) when it performs immediate and simultaneous recognition—"matching" of visual images—of "shapes" from different VSpS by their appearance (holistic forms).

Remember Kanner so astutely indicated that "sameness" is understood in the context of *whole-part.* VSpS is an ideal "case" for the RH *whole-part* perception, and indeed "sameness" behavior in autism is manifested here in full power. The autist is preoccupied with pattern recognition. He directly and instantaneously "sees" relations between *part* and *whole* without distinguishing *part* out of *whole*. An illustration of this is the superior performance of autistic people on Raven's Matrices test (see Chap. 6). In the Raven test, subjects are presented with a series of visual patterns from which a piece is missing. The task is to find the missing piece from the several offered choices, in other words, to find the *part* to *complete-the-whole.* The high speed of the autist's performance rules out LH analytical (logical-sequential) processing in autism.

10.3.3.2 Prime Numbers as "Parts" for Autistic Savant Calculator

It was hypothesized in Chap. 6 that, for autistic savant calculators, prime numbers are a kind of visual–spatial configuration. Primes would then be holistic forms within a VSpS. Savants' intense and sustained attention to prime numbers is thus a preoccupation with *parts* of the *whole* of a VSpS. Preoccupation with parts is one aspect of their experience with prime numbers. There is another aspect of primes for autistic savants. These "parts" are particularly entrancing because primes bear the inner rhythm of autists' *I-space* (by projection into the *non-I-space* of VSpS). Thus, prime numbers are a holding area for autists' subjective experience of a moment in the outside world (*time* of sensory-motor level) and inner rhythm (*time* of thalamic level). In "calculating" with primes, autists "act out" their highest form of self—the RH thalamo-parietal one—and in doing so, they experience an ecstatic belonging. When the twins volleyed primes back and forth to one another, they were in mystical communion with each other (as the aborigines exchanged sacred objects) as well as with the physical world [see Sacks (1995), in Chap. 6].

10.3.3.3 Autistic Calendar Calculator Savants, the Clinical-Brain Pattern of Autism in Its Extreme[10]

The autistic savant calendar calculator "acts out" the chronological sequences of VSSs[11] and their corresponding VSpSs.[12] Sequence of time is incorporated in these RH spatial images, and *pattern* of sequence is implicitly contained in the file of the repetitious VSpSs. "Seeing" instantaneously the pattern in his periodically repetitious experience, the autistic savant discovers "sameness" in the flow of VSS. Preoccupied with "sameness," the autist-savant re-plays patterns of repeated sequences of month–year configurations, making thus a "file" of automatisms (stored in the striatum).[13] Although infinitely complicated, autistic savant calendar "calculation" is automatic, repetitive activity, of the same mechanism as other patterned sequences stored in the striatum. However, there is a difference, I believe. In calendar "calculation," repetitive patterns of external space (*non-I-spaces*) are integrated with internal space (*I-space*) into a continuous whole by the inner rhythm of one's own body kinesthetic space. "Seeing" and playing with what I have called "archetypal" calendar (see Chap. 6) is then an externalization of the thalamo-parietal version of RH self.[14] It must be accompanied by the mystical experience of belonging, of being one with the physical world, i.e., empathic identification with the physical world (autistic-communication!).

What happens if the RH TPO version of self is exposed but a strong involvement of the LH DLPC also is taking place? In this regard, again, reports of Temple Grandin are curious and instructive. In her case, the network *LH DLPC–LH Striatum* prevails, but a disjoint RH thalamo-parietal version of RH self is also strong. Her experience of "belonging," of unity with the world is "covered" by LH "rationalization" of RH metaphor—identification: *good IS right and IS physical order* (world order/harmony). But it is *autistic* rationalization—a literal "reading" (autistic understanding) of her own primary experience—it is not a replacement of an unacceptable feeling by an acceptable one, but the absence (removing) of emotional evaluation altogether from LH interpretation (no participation of LH limbic emotion). Grandin still immediately, directly perceives the harmony of the physical world, presumably projecting her RH thalamo-parietal self. It is part of her talent as a scientist. We see here two disjoint components at work—LH interpretations and RH experience of inner rhythm. She is a "passionate creature" (as Sacks characterized her) due to her spiritual experience; however, we absolutely cannot imagine her doing calendar "calculation," too strong is her LH DLPC.

---

[10] Calendar savants were discussed in detail in Chap. 6. Here I will just enumerate the involved networks and their functional meaning.

[11] The network is *RH IT-RH DLPC.*

[12] The network is *RH TPO-RH DLPC.*

[13] The network is *RH DLPC—RH Striatum.* The *LH DLPC—LH Striatum* contribution varies in degree in different calendar savants, but this network does not contribute to the power or scale of the talent.

[14] The network is *thalamus (LP)-RH TPO-RH DLPC.*

10.4 Pattern Recognition in Autism and the Myth of Weak Central Coherence

Now we finally can return to the 6-year-old boy presented in Chap. 1, and explain the dissociation in his performance on the two tasks, both measuring higher cortical visual function. His flawless performance on the overlapping figure test (Fig. 1.1a) can be explained by his immediate recognition of the *parts* without distinguishing them out of the *whole* (RH holistic processing). At the same time, he was not able to "reach" LH hierarchical meaning—to separate the figure out of the background, Fig. 1.1b[15] (LH analytic processing).

Several studies have shown superiority of autistic people in the Embedded Figures Task (EFT) (Jolliffe & Baron-Cohen, 1997; Ring et al., 1999; Shah & Frith, 1983). In EFT, subjects are presented with the drawing of a simple target shape and a complex figure with the simple shape hidden within it. The subject is then asked to find the simple shape in the larger design. I interpret autists' superb performance on the EFT as their "seeing" the *part* in the *whole,* and the mechanism is pattern recognition and identification of holistic forms, a manifestation of *sameness* behavior.

Autists' superior performance on the EFT was the basis for the weak central coherence theory (WCC). According to WCC theory, local, detail-oriented information processing is characteristic for autism at the cost of global, gestalt-like processing (Happe, 1999; Shah & Frith, 1983). WCC had been one of the leading theories of autism until it was challenged by later studies that did not show impaired global processing in autism (Burnette et al., 2005; Mottron, Burack, Iarocci, Belleville, & Enns, 2003; Mottron, Dawson, Soulieres, Hubert, & Burack, 2006). However, the crucial point here is that WCC, yet another cognitive theory, a psychological model, does not correlate directly with established brain functions or known brain networks. For example, fMRI during performance on EFT revealed greater activation of the RH visual occipital areas (BA18,19) and inferior temporal cortex (BA37) in the autistic group as compared to controls (Ring et al., 1999).[16] Despite the clear prevalence of RH activation, authors argued that autistic performance was based on "feature" analysis, "disembodying" the target shape out of the complex figure. It was shown in previous chapters how the autist immediately recognizes *parts* within the *whole* or manipulates with parts of objects. *Parts* are not "features" or "details" for the autist but RH *gestalts* by themselves. Furthermore, all studies showed that, in addition to their superior accuracy, autistic subjects spot the embedded figures more quickly than other groups. Autistic children tend to use immediate strategy, very rarely visually searching. In contrast, visual search was the predominant strategy in the control group. For example, in Shah & Frith's (1983) study, 19 out of 20 normal children used visual search.

---

[15] His answer was: "iron and water."

[16] Interestingly, in this study *shapes* were treated by the autistic brain as objects, i.e. processed by the RH IT (BA37), within the VSS saturated with affect, while in the control group, activation was bilateral and mostly in the TPO (BA39,40). This may give us a clue to the intense emotional experience of shapes and numbers observed in autism (recall Elly's example in Chap. 6).

Having said all this, I am quick to acknowledge that WCC does touch upon the cardinal issue in autism, namely part-and-whole. Part-and-whole is a protruding thread to take hold of and unravel the psychopathology and neuroanatomy of autism.

10.5 Pathological Disinhibition of the Striatum and Stereotypical Behavior in Autism

As discussed in Chap. 5, it seems that dopamine regulation in the striatum plays some role in the frontal cortex's inhibitory influence on striatal activity. Interestingly, Saunders, Kolachana, Bachevalier, and Weinberger (1998) found that neonatal (developmental) lesions in *limbic* structures of monkeys' temporal lobes disrupt prefrontal lobe regulation of striatal dopamine. They found that, whereas infusion of amphetamine into the DLPC led to a reduction of caudate nucleus dopamine overflow in normal animals and those with *adult* limbic lesions, monkeys with neonatal lesions showed an *increase* of dopamine release from the caudate. According to another study (Malkova, Mishkin, Suomi, & Bachevalier, 1997), some monkeys with early limbic lesions show increased locomotor stereotypies not seen in animals with similar adult lesions. What can we extrapolate from this data for our purposes? Namely, is the primary deficit in autism—disorder in the LH *amygdala-OFC* network—also a cause of striatal disinhibition?

Lipska and Weinberger (1995) found that developmental lesions in the medial temporal lobe in rats contribute to increased locomotor stereotypies in response to pharmacological (administration of amphetamine) or environmental (novelty) stress, and degree of response depended on the size of the lesion. However, rats of one particular strain did not have stereotypies in any experimental circumstances. Neither novelty nor amphetamine evoked locomotor stereotypies in these animals, even those with large developmental lesions. It seems an additional factor, genetic vulnerability of the striatum, was needed for increased stereotypies in response to developmental lesions in the medial temporal lobe. Is genetic vulnerability of the striatum to pathological disinhibition an additional, but independent factor that may play a role in stereotypical behavior in autism?

10.6 Autism Clinical-Brain Pattern

I'll now reiterate the chain of events in the brain that give rise to the pattern of autism as hypothesized in this book (see Table 10.1).

The primary deficit is the disordered development of the specific matrix responsible for the LH self. The matrix lies within the following widely distributed network: *LH amygdala-OFC* along with the triangle of connections between *LH amygdala–MD–OFC* (see Fig. 8.1) *and LH OFC–DLPC–LH TPO.* This system's failure causes an underdevelopment of the LH construct of self, with profound

Table 10.1 Brain pattern of autism

Chain of events	Network
1. The primary deficit: deficiency of the left hemisphere self	Left hemisphere: amygdala-OFC, amygdala-MD-OFC, OFC-DLPC-TPO
2. Direct consequence of the primary deficit: "release" of the right hemisphere self	Right hemisphere: IT-DLPC-striatum, thalamus (LP)-TPO-DLPC-striatum
3. Secondary brain functional reorganization as an attempt to spontaneous compensation for right hemisphere excess	Right hemisphere-left hemisphere: RH DLPC-LH DLPC-LH striatum

deficits in LH object-related limbic emotion. Herein lie the "negative" symptoms of autism. On the superficial, behavioral level, an underdeveloped LH self is manifested by an inability to socialize and communicate in a conventional manner (this is what the behavioral-statistical DSM-IV refers to in its social and communication criteria for the diagnosis of autism).

The direct consequence of the primary deficit is surfacing of the RH self and a resultant "acting out" of RH consciousness. Exposure of RH self accounts for the "positive" symptoms of autism, the underlying networks of which are *RH IT–RH DLPC–RH Striatum* and *Thalamus (LP)–RH TPO–RH DLPC–RH Striatum.* Table 10.1 shows that the *LH DLPC–LH Striatum's* role in the autistic pattern is a secondary one. It represents a functional reorganization to compensate for the dominating RH "flow" of consciousness in the autist. The role of this particular network is to transform RH continuous flow of time into a sequence of discrete moments of experience—*to stop and master* (repeat) the moment (see Chaps. 3 and 5).

Both brain networks involved in RH self "acting out" and those involved in functional reorganization as an attempt at spontaneous compensation are the brain basis of autistic functioning, to be found independent of age, degree of severity, and IQ. Autistic functioning can then be categorized as sameness behavior and preoccupation with parts of objects. The complexity of this autistic behavior lies on a continuum: from repetitive manipulation with objects (see Chaps. 3 and 5 for examples) and insistence on sameness expressed as preservation of the spatial–temporal context of VSS (see Chaps. 3 and 5 for examples), to obsessive collecting, autistic classification,[17] and preoccupation with pattern recognition (superiority on Raven's Matrices and EFT), to autistic savant-calculators' preoccupation with prime numbers and autistic savant calendar calculators finding "sameness" in periodically repetitive sequences of month–year configurations.

To my knowledge, the model proposed in this book is the only theory that can explain all DSM-IV symptoms of autism with a description of their brain mechanisms.

The symptoms of the social and communication domains result from the primary deficit (lack of social identity)—"negative" symptoms. However, the communication

[17] DSM-IV defines these as "restricted (or circumscribed) interests."

domain is more complex and heterogeneous. While lack of spontaneous speech and "impairment in the ability to initiate or sustain a conversation with others" are "negative" symptoms, "stereotyped and repetitive use of language or idiosyncratic language" must be considered "positive" symptoms. Furthermore, they have different sources. Stereotypical speech is a manifestation of patterned motor (articulatory) sequences and should be included into the domain of repetitive behavior. Idiosyncratic language results from RH word meaning. Pronominal reversal (or history of it), such a crucial symptom of autism and a direct indication of underdevelopment of LH self (as *one and apart from others*), is not listed in the DSM-IV.

The symptoms of the restricted and repetitive behavior (RRB) domain are "positive" symptoms that are heterogeneous in their brain mechanisms and specificity to autism. Release of automatisms from the striatum is the source of nonrhythmical stereotypical movements, patterned motor sequences in autism. Release of automatisms from the thalamic-pallidum system is the source of rhythmical stereotypical body movements in autism. Both types of stereotypical movements are diagnostic, but not specific for autism.

Sameness behavior and "persistent preoccupation with parts of objects" are specific for autism. Their complex brain pattern and significance for autistic functioning were discussed earlier. Obsessive collecting, or autistic classification (restricted interests in DSM-IV) were not analyzed in this book, so I will now discuss briefly this particular manifestation of RRB. In Chap. 2, brain basis for "meaning" and classification of the external world was discussed in detail. It requires LH analysis and synthesis at the gnostic–praxic and symbolic (categorical) levels. These levels are not functioning properly in autism.

Autistic classification does not use an internal (abstracted from the objects that are "classified") principle, i.e., a functional or categorical signs. Thus, what is called autistic classification is *pseudo-classification*. It is *collecting* singular, physical things in all their (physical) variations to preserve *sameness*. For example, Elly (discussed in Chap. 6) made a picture book of "four words in three colors: SING, SANG, SUNG, and SONG…each word had its own page, the three-inch letters formed of the snips, SING and SANG in coral, a different shade for each letter, SUNG and SONG in green to correspond…" (Park, 2001, pp. 72–73). Elly used and perceives words (together with color) as variations in physical-sensory (visual and auditory) properties not as semantic-grammatical ones. This example is not different from the autist collecting meteorites, or names and types of lizard (Baron-Cohen & Wheelright, 1999), or "learning [collecting] the Latin names of every plant and their optimal growing conditions" (Baron-Cohen, Ashwin, Ashwin, Tavassoli, & Chakrabarti, 2010, p. 44). "Collecting" as a particular manifestation of the "generic" sameness and preoccupation with objects (parts) emphasizes a peculiar, chimeric aspect of brain reorganization in autism (Table 10.1, point 3). Instead of each hemisphere operating with its own units (as in the norm), in autism the RH forces its units (gestalts) upon LH operations—to enumerate (make inventory) of the same object in all its variations. Such peculiarity of interaction between the cerebral hemispheres could be the basis for a nonconformist approach to surrounding world and may become in some cases a basis for original discoveries.

10.7 Beyond Autism Clinical-Brain Pattern: Role of the Left Dorso-Lateral Prefrontal Cortex—Striatum Network in Autism

Two independent factors may play a role in a nonspecific disinhibition of the striatum in autism: the primary deficit[18] and a genetic vulnerability of the striatum to pathological disinhibition (see Sect. 4).

The hypothesized primary deficit being in the LH, disinhibition secondary to dopamine dysregulation would likely be expressed mostly in the *left* striatum. It can be supposed that *left* striatal disinhibition is responsible for release of the particular automatisms—patterned motor sequences[19]—that are highly diagnostic (although not specific) for autism. This type of motor stereotypy (*patterned motor sequences*) found in autists depends on age, degree of severity, and IQ. Simple motor stereotypies are usually characteristic of young autistic children, and with age they change into more complex *patterned behavioral sequences*. For example, compensating for a deficiency of emotional experience, Grandin *voluntarily* manipulated her brain record, collecting visual pictures of different types of social interactions. It was her "library of videotapes." She fixated her attention on VSSs, in particular *behavioral* responses to social–emotional situations,[20] re-playing them in her mind, and thus making automatisms. These stereotypical patterned behavioral responses require the striatum; however, the entire network involved is more complicated: *RH IT—RH DLPC—left DLPC—left Striatum.* The example of Temple Grandin illustrates what can be the highest achievement of HFA. In Chap. 7 we saw how other HFA used this same network while performing emotion recognition tests, although their activity was less complex and they were unable to introspect.

However, in autists with low IQ, simple motor stereotypies appear to be more persistent with age (Esbensen, Seltzer, Lam, & Bodfish, 2009). Progression to more complex behavioral patterned sequences (connected with the LH DLPC) can be explained by cortical maturation, which proceeds from motor, to premotor, and, finally, to prefrontal cortex. It was discussed in Chap. 7 that the degree of involvement of LH DLPC in autists' activity is correlated with verbal IQ and also dictates which LFA child will become HFA in later life. Here again arises the *LH DLPC–LH Striatum* network's significant role in the autist's adaptation and compensation. The purely cognitive part of the LH DLPC provides compensation for "negative" symptoms, deficit of the LH self. However, the *LH DLPC–LH Striatum* network does not belong to autism's brain pattern—oppositely, the compensation connected with the LH DLPC "strives" to overcome the autistic pattern, to break the boundaries of autistic functioning, and to make up for the autistic deficiency of agentivity by building a

[18] If we extrapolate experimental data of dopamine system dysregulation by prefrontal cortex due to early limbic lesions.

[19] They are also called in this book *nonrhythmical repetitive movements* (see Chap. 5).

[20] This would be the temporal (action) aspect of the VSS (RH DLPC) connected with action programming (LH DLPC).

repertoire of behavioral responses to social–emotional situations. Furthermore, lacking genuine emotional experience (LH limbic emotion), the highest achievement of LH DLPC's compensation is intellectual insight and introspection (reflective ability). That autists can achieve a level of self-reflection and make a storehouse of responses to social interactions is more evidence that the LH DLPC is not damaged in autism. Alas, the LH DLPC in autism functions only in tandem with the left striatum. Here we have to recall the not specific, but highly diagnostic for autism, disinhibition of the left striatum. In autism, the striatum always predominates in the *LH DLPC–LH Striatum* network (there is no "top-down" regulation). Thus, even the use of the sophisticated cognitive part of the brain, the LH DLPC, ends up being a stereotypical activity. The involvement of this compensatory cognitive network (*LH DLPC–LH Striatum*) is highly variable in different autistic subjects, and it does not seem to be at work in those with severe autism.

Note that this compensatory network is separate and independent (different localization within the LH DLPC and the LH Striatum) from the previously considered *LH DLPC-LH Striatum* network that counteracts the RH DLPC's chaos-inducing flow of consciousness. The latter network is part of the autistic clinical-brain pattern, giving rise to "positive" symptoms of autism.

To summarize (1) the compensatory network *LH DLPC–LH Striatum* plays a crucial role in the compensation and social adaptation of an autistic child, determining which autistic child will become a HFA. (2) This network is neither necessary nor sufficient for the formation of the autistic brain pattern as such. It is the most remote from the primary deficit. In this regard, use of verbal IQ for distinguishing more homogenous groups for genetic studies is not justified.[21]

Finishing our discussion on the autism clinical-brain pattern, it should be noted that recent factor analytic studies of autism fully correspond to this book's neurophenomenological analysis, confirming my division between specific and not-specific to autism types of stereotypical behavior. According to these studies, the RRBI (restricted, repetitive behaviors and interests) domain in DSM-IV is composed of two underlying factors: Insistence on Sameness (IS) and Repetitive Sensory and Motor Behaviors (RSMB) (Cuccaro et al., 2003; Shao et al., 2003; Szatmari et al., 2006). RSMB roughly correspond to both simple stereotypical movements (patterned motor sequences) as well as "thalamic" stereotypical bodily movements of this book's definition. Only IS is independent of age, IQ, and autism severity (Hus, Pickles, Cook, Risi, & Lord, 2007). There is clear evidence for familial aggregation of IS but not the RSMB construct across studies (Szatmari et al., 2006). Furthermore, Lam, Bodfish, and Piven (2008) distinguished a third factor, circumscribed interests (CI), in addition to repetitive motor behaviors (RMB) and insistence on sameness (IS). CI, as discussed earlier, is a particular manifestation of preoccupation with objects, in my formulation.[22] Both IS and CI showed significant familial associations in this study. Finally, factor analysis has failed to demonstrate

[21] On the other hand, autistic savants can be very important for genetic studies of autism because they represent the autistic brain pattern in its extreme.

[22] Lam et al.'s CI factor includes object attachment, so it is closest to my formulation.

the existence of separable categories paralleling the three DSM-IV criterion domains (social deficits, language deficits, and stereotypic behaviors). Instead, deficits across all three domains appear attributable to a singular underlying factor (Constantino & Todd, 2005).

It needs to be emphasized, however, that only a neurophenomenological analysis allowed for establishing connections between stereotypical behavior and the sensory, emotional, and social-communicative symptoms of autism at the level of their brain mechanisms and, thus, combining them into a cohesive, pathogenetically meaningful pattern.

10.8 Individual Variability Within Autism Clinical-Brain Pattern

The individual variability of cortical regions and their development in the general population is tremendous. Furthermore, in each individual, the degree of development of separate cortical regions is relatively independent of one another (Blinkov & Glezer, 1968; Schleicher, Morosan, Amunts, & Zilles, 2009). Anatomist Blinkov noted frontal lobe prevalence in some people's brains, while the parietal and occipital lobes dominated in others. He also found that cortical fields within frontal and parietal lobes developed unevenly in each individual.

Blinkov concluded that such brain structure variability between and within individuals must be reflected in functional individual differences (Blinkov, 1955).[23]

[23] Individual variability of human brain can be best illustrated by its extreme manifestation in extraordinary individuals. For example, there is a sufficient proof of correlation between Albert Einstein's mathematical genius and unique morphological features of his brain.

Einstein described his thinking as mostly nonverbal "associative play," with the ability to voluntarily combine and spatially organize visual and motor-kinesthetic images. He used mathematical terms only to write down results (Hadamard, 1949).

A detailed report of Einstein's brain was later published (Witelson, Kigar, & Harvey, 1999). It showed the parietal region to be 15% larger than average. In both hemispheres of Einstein's brain, the posterior ascending branch of the Sylvian fissure is confluent with the postcentral sulcus. Consequently, there is no parietal operculum. These unique features suggest extensive development of the posterior parietal lobes early in brain morphogenesis, thereby constraining the posterior expansion of the Sylvian fissure and resulting in a larger expanse of the inferior parietal lobule (BA40,39). A further consequence of this morphology is that the full supramarginal gyrus lies behind the Sylvian fissure, undivided by a major sulcus (Witelson et al.). Furthermore, it was hypothesized that a gyrus, in general, develops within a region of functionally related cortex to allow for efficient axonal connectivity between opposite walls of the gyrus; by contrast, sulci separate cortical areas having less functional relatedness. Authors concluded that "in this context, the compactness of Einstein's supramarginal gyrus [mostly BA40]...may reflect an extraordinary large expanse of highly integrated cortex within a functional network" (p. 2152).

The described above region corresponds to the temporal–parietal–occipital region—TPO, the term used in this book. We remember that this area is responsible for visual–spatial ability, mathematical ideation, and spatial schemes for movement.

The autistic brain pattern, then, is always superimposed on the uniqueness of each autistic individual's "premorbid" brain.

I will illustrate individual variability within the autistic brain pattern analyzing drawings of several autistic savants. But first, one autist's[24] exceptional drawing ability as a variant of preoccupation with parts will be analyzed.

E.C. is a 36-year-old autistic male with IQ 88 (discrepancy of verbal IQ 75 and nonverbal IQ 105) examined and described by Mottron and Belleville (1995). E.C. has outstanding abilities for perspective and three-dimensional drawings of inanimate objects (Fig. 10.1).

When constructing perspective in paintings or drawings, artists normally follow conventional rules for graphic representation—so-called perspective systems. Not using the rules governing perspective creation often results in distorted depictions. Mottron and Belleville show that E.C. did not use conventional rules in creating perspective, but his perspective construction was nevertheless extremely accurate. Similar to other autistic savants, E.C. could not explain his strategy, but merely repeated his drawings stereotypically, even failing art school because he was not able to modify his drawing routines (Mottron & Belleville, 1995). Mottron and Belleville note that systems of rules for graphic representation of three-dimensional space appeared at a particular time in human history and that the *perception* of perspective might have different underlying mechanisms. I agree with these authors that direct visual perception of perspective is different from the perspective systems. Moreover, Nikolaenko and Deglin (1984) show that the RH is responsible for forming perceptual space, an iconic image of real physical space.[25] Their study delineates several characteristics of RH perceptual perspective, including its limitation to the locus of space near the observer and its objects being volumetric and oriented vertically.[26] Both characteristics of RH perspective can be seen in E.C.'s drawing (Fig. 10.1). His chair is on the foreground, falling towards the viewer, similar to Nadia's *horse* that charged from the page (see Fig. 3.2–3.5). Additionally, the exceptional preciseness, "metricity" and "geometricity," of the sensory-motor level

---

Ivanov (1978) writes of individual variability in the intriguing example of Sergei Eisenstein's brain. A worldwide cinema producer, Eisenstein described his thinking as a continuous tape of visual images and events which he sometimes mentally "outlined with his hand as drawings." He confessed that words were not in the focus of his sharp attention. In Eisenstein's brain, size of the right hemisphere exceeded significantly the size of the left hemisphere (according to V.V. Ivanov, information about Eisenstein's brain was given to him by the noted neuropsychologist, Alexander Luria, in personal communication).

[24] This autistic savant has not yet been discussed in this book.

[25] Nikolaenko and Deglin's work was mentioned in Chap. 3. These authors examine drawings by patients with one cerebral hemisphere functioning in isolation (the other hemisphere was temporarily inhibited) and thereby "reconstruct" how the LH and the RH are different from each other in the way they "see" external space.

[26] In the LH's perspective, there is a loss of object volumeness, coupled with an exaggerated depiction of depth of space due to excessive use of a linear perspective.

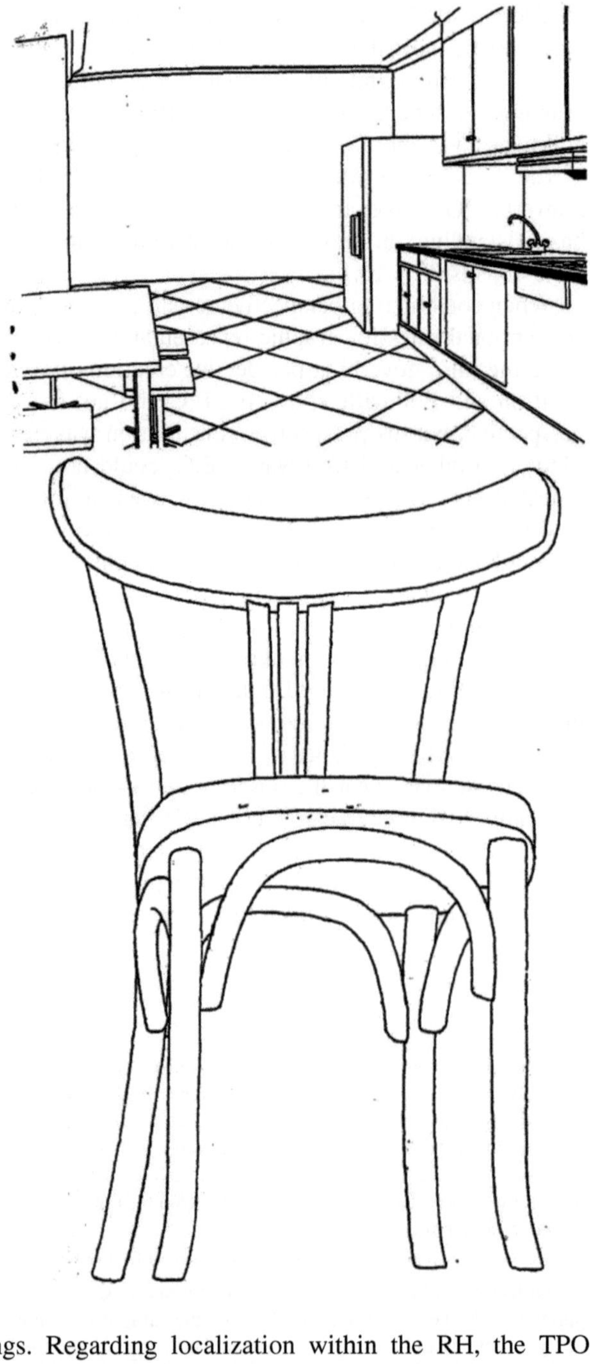

Fig. 10.1 E.C.'s drawings. From L. Mottron and S. Belleville, Perspective production in a savant autistic draughtsman, *Psychological Medicine*, 1995, 25, 639–648. 2009 © Cambridge Journals, reproduced with permission

characterizes E.C.'s drawings. Regarding localization within the RH, the TPO ("geometric" part of the posterior cortex) must play a leading role in E.C.'s skills.

Now we will speak about the intriguing rotational aspect of his drawings, important for our purposes of understanding autists' preoccupation with moving parts.

E.C. was extremely accurate in the most difficult perspective task available, namely to draw objects rotated in three-dimensional space. E.C. could reproduce the geometric modifications objects sustain when rotated in space, and he did so without using perspective systems. In one of the experiments he was asked how he completed the ellipses (this task required mental rotation) to which E.C. answered he "just saw them in front of him" (Mottron & Belleville, 1995, p. 644). Mental rotation of objects must play a significant role in E.C.'s talent. Moreover, I believe E.C. represents a particular and extreme case of what all autists are doing, namely acting out their preoccupation (fear and fascination) with movement and moving objects.[27]

The autist's preoccupation with moving objects or parts is often expressed in *rotating* movement, i.e., any round objects.

The following examples illustrate preoccupation with rotation in LFA. H., a 10-year-old autistic boy described by Bosch (1970), preferred to occupy himself with anything that turned.

> [He] showed a particular interest in round and rotatable objects.... At home he had filled the box with a very varied collection of wheels, and when visitors were present he would always say, 'He wants a wheel.' Whenever he entered a room he immediately spied any round or rotatable objects in it and pounced on them.... In his drawings, too, he preferred to depict wheels or trace circles with his pencil or brush. He mostly drew wheels or cars or sewing-machines, the latter because of the large flywheel. He introduced circular shapes into other pictures, even if they did not properly belong there.... Faced with an unpleasant situation...he would suddenly cry out, 'Spin' or 'Sewing-machine'. We gained the impression that the intended effect of this cry was to calm him down. Words like bicycle, escalator, or scooter seemed to have a similar significance [from Bosch (1970), pp. 7–8].[28]

According to functional MRI studies (Cohen et al., 1996), the most important area participating in mental rotation of complex objects is BA7 (sometimes spreading to the bordering BA40), but activation is also observed in BA39 and the circumscribed area within BA19 known as V5/human MT. BA7 in the superior parietal region, as discussed in Chaps. 3 and 5, belongs to the RH attentional system, providing sustained attention to visual stimuli. It is also involved in visuospatial processing and has been suggested to play a role in determining the position of an object with respect to the observer, consistent with the theory that figures are rotated as *solid* three-dimensional objects (Cohen et al. 1996). BA39 and BA40 are the major players in shape and spatial relation processing, and area V5/human MT is responsible

[27] Superior performance of autistic subjects on recognition of reversed (upside-down) faces has been claimed as evidence of "local" processing in autism. Hobson, Ouston, and Lee (1988) were the first to doubt this conclusion, indicating other strategies might be available. They give the example that when normal adults were presented with upside-down faces, some expressed the view that they were "mentally rotating" the faces. It seems to me very likely that the autistic brain "chose" this strategy during the task as an acting out of autistic fascination with movement and with rotation in particular.

[28] From "Infantile Autism—A Clinical and Phenomenological-Anthropological Investigation Taking Language as the Guide" be Gerhard Bosch, Springer-Verlag, New York, 1970. With kind permission from Springer.

for the perception of movement (even illusory movement).[29] We can thus see that the regions activated during mental rotation are also the "acting" (possibly hyperactivated) networks demonstrated in this book to be at play in autism.

Thus, in this particular autistic savant, individual-specific features *within* the autistic brain pattern can be distinguished. Neurophenomenological analysis of his drawings reveals the prevalence of the right-posterior cortex belonging to the *where*-system: visual–spatial BA39,40, visual attention BA7, and dynamic form—"moving target"—V5 within BA19.

Below, several other autistic savants' drawings will be analyzed. While all these drawings similarly reflect the RH sensory-motor level, they differ based on individually specific variations of underlying regions and networks (within the right posterior cortex) involved in each autist's talent.

In Nadia's drawings (see Nadia's drawings and their detailed analysis in Chap. 3), an experienced moment of the external world (time of the sensory-motor level in Bernstein's terms) is vividly reproduced. Her "horse," "rooster," and "leg" are representations of her VSS, suggesting involvement of RH BA37. Nadia's drawings also evoke powerful kinesthetic experience, a readiness for movement. This suggests projection of her thalamic level and, in particular, her "muscle" body space (the posterior part of the thalamic level or the "space" of this level in Bernstein terms).

In Stephen's drawings (see Chap. 8 for his drawings and their detailed analysis), feeling of harmony induced by a periodic repetition of lines and forms was interpreted as the projection of his inner rhythm from the thalamic level. The prevalent networks for his drawings would be *RH thalamo-pallidum* system-*RH TPO-RH DLPC.* Instead of the posterior part of thalamic level being projected (as was hypothesized in Nadia's drawings), in Stephen's drawings it is then the anterior part, the rhythmical pattern of one's own body or the "time" of the thalamic level.

Elly's[30] painting (Fig. 10.2) displays triumphant celebration of form and color, saturated with intense emotion. Her painting thus suggests RH BA39,40 (preoccupation with holistic geometric forms) and, bordering with BA37, the associative occipital area BA19 (namely, V4 part of BA19 responsible for static form and color) are the prevailing cortical regions. Note, Elly was also distinguished by her superior

[29] There is confusion in the literature regarding the functions of STG and V5/human MT areas. They both are characterized as "movement perception" areas. Confusion came from not paying attention to the core function and origin of these areas. The core function of STG is attention, detection of change, which is perceived as movement. The core function of V5/human MT is visual perception of *dynamic form* ("moving target"), and it belongs to the phylogenetically older Magno system and projects to the *where*-system, BA39. In contrast, the other circumscribed area within BA19, known as V4, is responsible for *static form* and color. It belongs to the newer Parvo system and projects to the *what*-system, BA37, where the visual form is assigned with meaning (object recognition) (Livingstone and Hubel, 1988; Mishkin, Ungerleider., & Macko, 1983).

[30] Elly was discussed in Chap. 6 for her outstanding visual–spatial, calculating, and pattern recognition abilities.

Fig. 10.2 Elly's drawing of St. Paul's and St. Andrew's Methodist Church. From *Exiting Nirvana* by Clara C Park. Copyright © 2001 by Clara Claiborne Park. By permission of Little, Brown and Company. All rights reserved

performance on the Raven's Matrices and her outstanding calculating ability, both of which relate to BA39,40. The *intensity* of Elly's paintings reminds us of Franco's paintings (Chap. 7). However, Franco's work symbolized his nostalgia while in Elly's paintings no such emotion or idea is evoked. Because of this, although Elly's paintings are vivid and splendid, they seem to come to us from another world.[31]

We see splendor of color in Jeroen Pomp's painting (Fig. 10.3), yet this painting is very different from Elly's drawing. His is "populated" not by forms, but by objects. In Elly's painting, the viewer is charmed by its simultaneous vividness and otherworldliness, but Pomp's painting induces a very different feeling. The viewer is struck by the intrusively close—bright and uproaring—physical reality. However, the energy of Pomp's painting does not come from the tension of physical movement, as we observed in Nadia's drawings. It comes from the tension of emotion imprisoned within the visual scene-situation. Particularly striking in his painting are the poor boundaries of its depicted animals and plants. It reminds us of Therese Jolliffe's introspection on her autistic experience: "Reality is a confusing interacting mass of events, people, places, sounds and sights. There seem to be no clear boundaries, order or meaning to anything." In this regard, one of Nadia's drawings (Fig. 3.15) which was interpreted as the projection of a chaotic RH world—no boundaries between images—is similar to Pomp's painting. We can come to the conclusion that the prevalent brain network underlying Pomp's talent is *RH BA37-RH BA19* (V4 part of BA19).

Two more examples of autistic art (Figs. 10.4 and 10.5) show yet another variant of localization within the right posterior cortex. In James Castle's drawing (Fig. 10.4), we see the visual–spatial situation in a pure form. Here, a static aspect, "sameness," is more involved than in another "geometric" savant, E.C. (see above in this chapter) in whom preoccupation with objects (mental rotation) predominates. In Castle's drawing, a more narrow area within the right posterior cortex is projected than in E.C. BA39,40 would be the most appropriate "place" as the brain basis to produce this drawing.

Finally, in Gregory Blackstock's drawing, we see a graphical depiction of "collecting"—autistic classification. The network for it would be *RH IT-RH DLPC-LH DLPC-LH Striatum.*

To summarize, although all savants are "acting out" their experiences within the autistic brain pattern, as hypothesized in this book, in each savant a particular combination of cortical and subcortical regions prevail and thereby determine each savant's not replicable, individual style.

---

[31] Franco's later paintings (see Chap. 7) are also more distant. However, this phenomenon derived from his increasing preoccupation with a "mission" at the cost of the intensity of his feeling. The likely shift in networks would be from *LH amygdala-OFC* to *LH OFC-LH DLPC*. The preaching nature of these paintings induces the emotion of irritability in the viewer, especially if one has known the richness of his earlier works.

Fig. 10.3 Autistic artist, Jeroen Pomp, untitled painting (zoo). From *Outsider Art and the autistic creator*, by R. Cardinal, in: Autism and Talent, U. Frith and F. Happe, eds., 2010, Oxford University Press, New York. Courtesy of Galerie Hamer, Amsterdam and by permission of Oxford University Press

Fig. 10.4 Autistic artist, James Castle, untitled drawing in soot and saliva (artist's workplace). From *Outsider Art and the autistic creator*, by R. Cardinal, in: Autism and Talent, U. Frith and F. Happe, eds., 2010, Oxford University Press, New York. Reproduced by permission of Oxford University Press

Fig. 10.5 Autistic artist, Gregory Blackstock, cover of his book *Blackstock's Collections* (2006). From *Outsider Art and the autistic creator*, by R. Cardinal, in: Autism and Talent, U. Frith and F. Happe, eds., 2010, Oxford University Press, New York. Reprinted by permission of Princeton Architectural Press and Oxford University Press

10.9 Autism Boundaries from the Point of View of Brain Networks

Anatomist Blinkov (1938, 1955) noted that structure of the cortex is changing along its surface continuously and gradually, but in its changing, which can be caught if you glance over the cortex on its significant extent and compare its distant parts, there are some places, "critical moments" where structure has suddenly and abruptly changed along the lines penetrating all layers. Similarly, modern cytoarchitectural studies emphasized that the transition in the laminar pattern is a key feature of border localization between neighboring areas (Schleicher et al., 2009). Borders of cortical cytoarchitectural fields are defined based on *quantitative* changes in cortical lamination. For example, transitions in the laminar pattern are detected in several steps using a computer controlled scanning connected to a microscope. Moving a sliding window across the cortical ribbon, dissimilarity of laminar pattern is calculated in each position. Maximal dissimilarity of corresponding laminar patterns represents two distinct clusters of cytoarchitectural profile and thus separate neighboring cortical sectors as belonging to the different cyroarchitectural fields. An example of the border between BA17 and BA18 has been already given (see Chap. 1) as a drastic *qualitative* change in laminar pattern—BA17's lamina IV, broad and subdivided into sublaminae, abruptly changes to a thinner and less differentiated lamina IV in BA18 (Schleicher et al., 2009).

Thus, dialectics of both quantitative and qualitative changes are clearly seen in the cytoarchitectonic differentiation of the human cortex. There must be a genetic and evolutionary basis to this dual nature, and it must also be reflected in functional differences.

Blinkov (1955) distinguished within each cortical area unique architectural features and transitional ones, the latter being shared in common by neighboring areas. He noted that specific features are usually prevalent in the center of a field, while transitional formations occupy a field's peripheral part. The ratio between specific and transitional formations determines, according to Blinkov, a cytoarchitectural field's level of development. The extent of transitional formations varies between individuals and might be a source for inter-individual variability of cortical fields in the general population. Is this variability based on genetic polymorphism? Are there polygenes of small effect?

Recently, compelling experimental evidence has been obtained for genes intrinsic to the cortex that control the gradual emergence of area-specific properties during development. However, none of these genes' expression is limited to a single area. A cortical area is not defined by the expression of specific genes restricted to this one area. Instead, a cortical area is defined by the expression of a unique combination of genes, each single gene also being expressed in other areas. Furthermore, each layer within a brain area has its own specific profile of gene expression. Each gene differentially expressed in the cortex has different expression patterns in each layer (O'Leary & Nakagawa, 2002). It is intriguing how this new knowledge fits to the clear architectural profile of cortical fields: area identity (a unique subset of genes) versus architectural features a given area shares with other areas (each cortical-differentiating gene is also expressed in other areas). Especially tempting for speculation is the similarity between these genetic findings and the construct of brain matrices used in this book to hypothesize about the primary deficit in autism. A matrix represents sets of widely distributed neurons that might be located in areas remote from each other and belonging to different functional levels in the brain; however, these sets are united into a functional system. Specificity of matrix is not defined by the specificity of individual neurons (they may participate in other matrices), but *specific combinations of neurons*. Analogously, genetic determination of area identity is provided by a *unique subset* of genes, but not by genes specific to a given area. What is interesting is that a matrix's constellations of neurons are situated in different areas and, therefore, their realization is controlled by the differing subset of genes. What is the factor that unites them into one functional system? It must be the pleiotropic effect of a gene (or few major genes). And this gene or genes must be involved in disruption of the system. This deduction flows well with the disruption of the system in autism being of one system at different levels that nevertheless produces a cohesive syndrome.

10.9.1 Categorical Versus Dimensional Approach to Autism

There is a trend toward the dimensional approach to autism. Autistic traits are considered to be continuously distributed in the general population, with autism being an extreme end in this continuum (Constantino & Todd, 2000, 2005; Spiker, Lotspeich, Dimiceli, Myers, & Risch, 2002). The Social and Communication Disorders Checklist (SCDC) (Skuse, Mandy, & Scourfield, 2005) and Social Responsiveness Scale (SRS) (Constantino & Todd, 2005) were proposed to measure heritable, dimensional autism-related traits in the general population. Deficits in reciprocal social behavior were found in nonautistic relatives of autistic probands. These deficits were interpreted as traits that are milder, but qualitatively similar to those in autism. The expression of mild, nonpathological autistic characteristics among relatives of autistic people is referred to as the broader autism phenotype (BAP) (Pickles et al., 2000; Piven, Palmer, Jacobi, Childress, & Arndt, 1997). The broader definition of autism's heritable phenotype may lead to autism's boundaries being increased (Reisinger, Cornish, Fombonne, 2011), but in fact the meaning of BAP is unclear. For example, in one study, BAP features were found in 22% of Down syndrome families who were the control group in this study and supposed to have no predisposition to autism (Losh, Childress, Lam, & Piven, 2008). "Social" BAP is controversial. This behavioral continuum can have a heterogeneous nature. For example, emotional detachment in schizoid personality disorder, disorder of object-related emotion in autism, and isolation of affect in OCD all look, on the behavioral surface, like poor sociability. Studies of sibling correlation have consistently shown that repetitive behaviors are correlated among affected siblings while impairments in social reciprocity are not (Szatmari et al., 2006). In linkage studies, it is also usually repetitive behaviors for which significant linkage results are reported (Shao et al., 2003). Furthermore, in factor analytic studies, strong familial association was found for IS and CI constructs (Lam et al., 2008; Szatmari et al., 2006),[32] which is not surprising considering sameness behavior and preoccupation with parts of objects are (as postulated in this book) the "acting out" of the RH self, a direct result of autism's primary deficit. On the contrary, deficit in social reciprocity is less specific and, therefore, more remote from the heritable phenotype. But most important is the question: What does BAP have to do with autism—this 'strange existence', a disorder of *one-ness*, lack of agentivity, and fundamental change of an attitude to the surrounding world—a qualitatively different content of consciousness? Autism cannot be envisioned as a result of the accumulation of small, quantitative changes. Instead, autism is the result of a break in continuity of behavioral traits observed in the general population, a new, qualitatively different category. Rutter (2011) justly indicates: "The existence of the broader phenotype raises the query of how it becomes transformed into 'autism proper'. Is this simply a measure of the severity of the genetic liability or there some kind of two-hit mechanism and, if there is, what is the other influence? We do not as yet know" (p. 398).

We cannot fully disregard the dimensional approach to autism. From the very first descriptions by Kanner and Asperger, certain peculiarities of personality and cognitive profile in parents of autistic children were noted. It might be reasonable to explore within the "social" BAP, phenomenological features which are closer, more specific to autism's fundamental deficit (as proposed in this book).

We do have an appropriate example in the literature where authors examined parents of autistic children and found manifestations that were, although mild, somewhat closer qualitatively to what is formulated in this book as the disorder of the LH self and object-related limbic emotion. In search of a more specific phenotypic trait for autism, Szatmari et al. (2008) used the construct of alexithymia in examining parents of autistic children. Parents of autistic probands demonstrated greater alexithymia on the Toronto Alexithymia Scale (TAS-20). Alexithymia—difficulty identifying and describing feelings—can be considered a difficulty in naming and categorizing one's own emotion, i.e., problems in the network *LH amygdala—LH OFC* (see Chap. 7). Interestingly, parental alexithymia scores were not correlated with their child's score of autistic symptoms in the social domain, but on the contrary, correlation was found among parental alexithymia scores (especially fathers) and severity of repetitive behavior in autistic probands.

It can be accepted that multiple "quantitative" genes, with the effect of each individual gene being very weak, play a role in autism. These are polygenes responsible for the continuous distribution of autistic traits in the general population (BAP). However, there needs to be an additional factor (one gene with pleiotropic effect or the interaction of a few major genes) to give such a brutal blow to the brain, to change its whole development. It must be a much rarer event than what gives rise to BAP.

[32] *IS* insistence on sameness. *CI* circumscribed interests; *preoccupation with objects and their parts* is used in this book in a broader sense that includes CI.

Fig. 10.6 Comorbidity in autism and overlapping brain networks

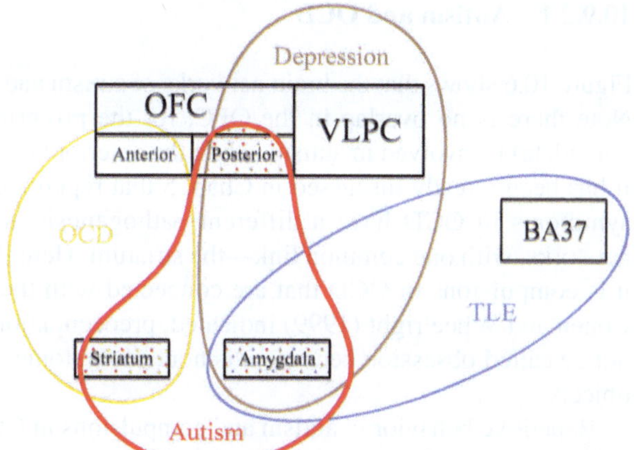

10.9.2 Comorbidity in Autism and Overlapping Brain Networks

It is a relatively small territory of the brain that can become an arena for many human tragedies: several distinct networks are situated in this territory, and a disorder of each of them gives rise to a particular psychiatric disorder. Figure 10.6 shows some of them: TLE (network *amygdala-BA37*), Depression (network *amygdala-OFC/VLPC*), OCD (network *OFC-Striatum*), autism (*amygdala-OFC-DLPC-Striatum*).

TLE is characterized by a deepening and intensification of emotion and a hypersocial personality. Pathogenesis of such intensification is epileptic discharge stimulating the temporal cortex and amygdala beyond the physiological range (Gloor, 1990; Penfield & Perot, 1963). In Depression, there is a deepening of human-specific emotion and a distortion of the LH self connected with hyperactivation of the left amygdala and left VLPC (Drevets et al., 1992). The pathogenesis of depression was hypothesized as a disorder of the hierarchical structure of the LH self (Glezerman & Balkoski, 1999). OCD is characterized by exaggerated harm concern (hyperestimation of risk) connected with pathological hyperactivation of the OFC and Striatum (Zald & Kim, 1996b).

As we can see, these are different disorders, separate diagnostic categories, and none of them is characterized by underdevelopment of the LH self—the primary deficit of autism proposed in this book.

When contrasting these disorders with autism in previous chapters, I made a point that different imbalances within the same system (for example, *amygdala-OFC*) result in profoundly different conditions.

How then can we explain co-occurrence of autism with OCD, affective disorders, and epilepsy? I will discuss this problem in connection with overlapping brain networks (Fig. 10.6).

10.9.2.1 Autism and OCD

Figure 10.6 shows that the brain networks of autism and OCD overlap in the striatum. Note there is no overlap in the OFC, for the posterior OFC (connected with the amygdala) is involved in autism, while the anterior OFC is part of the OCD network. It has been already discussed in Chap. 5 that repetitive behavior in autism and OC symptoms in OCD have a different pathogenesis and distinct underlying brain networks, with one common link—the striatum. Here, I want to make two points (1) it is compulsions in OCD that are connected with the striatum; and (2) as Baron-Cohen and Wheelright (1999) indicated, preoccupation with objects in autism cannot be called obsession because it is not ego-dystonic, indeed, it is *fascination* with objects.

Repetitive behavior in autism and compulsions in OCD are qualitatively different phenomena that, in fact, might involve different areas within the striatum as projections from different cortical areas. Nevertheless, there are studies showing increased occurrence of OC traits or OCD in parents (especially, fathers) of children with autism. For example, Bolton, Pickles, Murphy, & Rutter (1998) found that OCD was more common in first-degree relatives of autistic probands (3%) compared to relatives of Down syndrome probands (0%). Furthermore, highly significant comorbidity between OCD and BAP was observed in this study. Hollander, King, Delaney, Smith, & Silverman (2003) reported higher severity of repetitive behavior in autistic children who had one or both parents with OC traits or OCD. Curiously, in this study, authors did not find more severe expression of repetitive behavior such as preoccupation with parts of objects, which is specific to autism, in autistic probands.

Do autism and OCD share the common genes? Not necessarily. There is a possibility of marital selection, for OCPD and subthreshold autistic traits, although different by their phenomenological origin, have a similar behavioral manifestation.

It is compulsions in OCD that are connected with the striatum. However, it is obsessions that are specific for OCD, not compulsions. Nestadt et al. (2000) conducted a family study of OCD and concluded that obsessions are more specific to the phenotype than are compulsions. The authors emphasized that compulsions alone should not be included in the phenotype for studies of OCD. If autists and patients with OCD do share common genes, these genes are likely not specific for either condition. What might be common in OCD and autism is a genetic vulnerability of the striatum to pathological disinhibition.

10.9.2.2 Autism and Epilepsy

Approximately a quarter of autistic individuals develop epilepsy (Hara, 2007; Volkmar & Nelson, 1990). Interestingly, in many cases seizures did not begin until adolescence (Rutter, 1970). This is a much later age of onset than that for children with idiopathic mental retardation developing seizures for the first time [Bolton et al. (in press) cited from Rutter (2011); Goulden, Shinnar, Koller, Katz and Richardson (1991)]. The neuropathological meaning of these findings might be that organic factors are not likely to be involved in epilepsy comorbidity in autism. Next, even more interesting is the fact that epilepsy in autistic individuals is *not* associated with a family history of epilepsy. On the other hand, epilepsy in an autistic proband was associated with the likelihood of a nonautistic relative having the BAP [Bolton et al. (in press) cited from Rutter (2011)]. It looks as though epilepsy in autistic probands is intrinsically connected with their autism. It is not that epileptic genes are involved in autism. It is rather autistic genes "products"—disruption of particular networks in the LH and/or pathological disinhibition of the RH sensory-motor level—creating the conditions for epileptic activity. This assumption is supported by the following facts. Abnormal EEG epileptiform activity without a history of seizures is even more frequent than clinical seizures in people with autism. In one study of 889 ASD patients without a history of clinical seizures, 540 (60.7%) subjects had epileptiform EEG activity. The most frequent sites of epileptiform abnormalities were localized over the right temporal region (Chez et al., 2006).

10.9.2.3 Autism and Depression

Studies of comorbidity in autism have shown that autistic individuals may develop new onset of affective symptoms and, also, the risk to affective disorders is increased in relatives of autistic probands. However, liability to depression in families with an autistic proband does not increase liability to autism. It has been concluded that new onset of affective disorder in autistic individuals is not an intrinsic feature of an autistic liability (Bolton et al., 1998; Hutton, Goode, Murphy, Couteur, & Rutter, 2008).

How can we explain the accumulation of depression in families of autistic probands? One possibility is, again, marital selection.

In this book, depression as a nosological category and autism are considered incompatible conditions. In depression, the LH self—the *oneness*, is exposed, but in autism the LH self is missing or underdeveloped (see Chap. 9 for a comparison of autism and depression's neurophenomenology). One may suppose that even though autism and depression share impairments in the left amygdala and left prefrontal cortex (Fig. 10.6) as their brain mechanisms, their actual networks within these structures are separate.

One of the purposes of this work was to delineate the boundaries of autism as a diagnostic category. These boundaries are the "autistic clinical-brain pattern." It remains distinct even when overlapped with the brain networks of other psychiatric disorders.

10.9.3 RH Prevalence in Nonautistic People

It has recently become popular to "re-name" some outstandingly talented people as having Asperger syndrome (one such example is the writer Vladimir Nabokov). What is ignored here is the reality that in the general population and, especially, in talented people, prevalence of the RH can be observed as a variant of brain individual variability.[33] In this regard, the formulation of a primary deficit as necessary and sufficient is crucial. Autism can be diagnosed only in the presence of a LH self deficit.

I will illustrate RH prevalence and autistic-like features in Vladimir Nabokov's "case," using his writing.

1. Increased role of the RH attentional system[34]—*concentration on the moment.* Nabokov described how he sometimes remembers the past in his foreword to *Speak, Memory*: "I discovered that sometimes, by means of intense concentration, the neutral smudge might be forced to come into beautiful focus so that the sudden view could be identified, and the anonymous servant named" (Nabokov, 1989, p. 12).[35]
2. Exposed visual scene-situation.
   The following example of visual-situational associations is Nabokov's impression of the name of a character in his review of Gogol's *Government Inspector*:

   > Khlestakov's very name is a stroke of genius, for it conveys to the Russian reader an effect of lightness and rashness, a prattling tongue, the swish of a slim walking cane, the slapping sound of playing cards, the braggadocio of a nincompoop and the dashing ways of a lady-killer (minus the capacity for completing this or any other action) (Nabokov, 1971, p. 55).

   We see here experiential polymodal (synesthetic?) visual-auditory-action-situations, however, the images are highly symbolic-polysemantic, requiring supramodal levels of both hemispheres. They also convey personal attitudes, values, feelings—it is not purely RH VSS with impersonal emotion confined within the visual scene-situation.
3. Preoccupation with objects.
   Well known is Nabokov's fascination, even preoccupation, with butterflies. Nabokov spoke of it as some kind of meditation (intense focus on the object's sensory-physical, not utilitarian, features) and described a change in state of consciousness:

   > And highest enjoyment of timelessness—in a landscape selected at random—is when I stand among rare butterflies and their food plants. This is ecstasy, and behind the ecstasy is something else, which is hard to explain. It is like a momentary vacuum into which rushes

[33] Reports of Einstein and Eisenstein's brains as extreme variants within the human brain's individual variability have already been discussed (see footnote 32), and correlation has been shown between the brain's unique morphological features and both men's creative achievements: mathematical genius in the case of Einstein, a unique "visual-motor" talent in the case of Eisenstein.

[34] Discussed in detail in Chaps. 3 and 5.

[35] A striking similarity with the above example can be found in one split-brain patient's reaction to a stimulus presented to his RH. This patient who had a posterior commissurotomy was discussed in a previous work (Glezerman and Balkoski, 1999), the example was taken from Sidtis et al. (1981). When presented with the written word "knight" to his left visual field, i.e., to his RH, the patient could not immediately name the stimulus but described a visual scene-situation: "'I have a picture in mind but can't say it.... Two fighters in a ring.... Ancient...wearing uniforms and helmets... on horses trying to knock each other off...Knights?' The patient described his thought processes in the following way: 'It's like things are moving around constantly, and I'm trying to narrow it down to something that will just stop. I'm seeing a whole general picture but one thing is almost in the middle'" (p. 156).

all that I love. A sense of oneness with sun and stone. A thrill of gratitude to whom it may concern—to the contrapuntal genius of human fate or to tender ghost humoring a lucky mortal (Nabokov, 1971, p. 139).

This experience must involve not only the RH sensory-motor level, but also the symbolic level of both hemispheres. It reminds us more of the experiential seizures described by Dostoevski (see Chap. 7) than the trance of autistic people manipulating objects—the latter is characterized by a decrease of awareness.

4. Resemblance of holistic forms (RH identification of objects according to their form even if their content is different).
   Gogol's *Dead Souls* is traditionally considered a social satire with grotesque depictions of provincial landowners. Nabokov takes a completely different and unique approach, asserting that Gogol's writing has a "huge, seething, prodigiously poetic background..." (Nabokov, 1971, p. 55). Nabokov reveals the internal fabric of Gogol's writing—a transformation of the images of Gogol's own dream-like world. Nabokov shows that throughout the text of *Dead Souls*, one *form* (object), that is, *circle* (wheel) takes a different appearance, each time acquiring a new symbolic meaning. "Dead Souls ...[is] a closed circle...with the theme of wheel cropping up at each new revolution on round Chichikov's part" (p. 76). According to a neurophenomenological interpretation, we deal here with RH associations based on the resemblance of holistic form.

Here are few examples:

In the very beginning of Gogol's novel, the main character, Chichikov is arriving in the town N., with two peasants standing nearby making "certain remarks which however referred more to the carriage than to the person seated therein. 'Look at that wheel there,' said one. 'Now what do you think—would that wheel hold out as far as Moscow if need be, or would it not?' 'It would,' answered the other. 'And what about Kazan—I think it would not last that far? 'It would not,'—answered the other. Upon this the conversation came to a close" [Gogol in Nabokov's translation, cited from Nabokov (1971), p. 75]. Nabokov interprets this conversation as a kind of to-be-or-not-to-be meditation in a primitive form, and then, remarks: "...rotundity of Chichikov is bound to come to grief, being symbolized by the rotundity of that doubtful wheel" (p. 76).

Nabokov (1971) points out the transformation of Sobakevich's head ("...a man's face, and round and broad it was") into a pumpkin, then, transformation of pumpkin into a special kind of balalaika [made from pumpkin], "and finally the placing of that *balalaika* in the hands of a young villager" (p. 81) to symbolize a poetic facet of Sobakevich's nature.

The inner arrangement of Chichikov's traveling chest "is not a box at all but a circle in hell and the exact counterpart of Chichikov's horribly rotund soul..." (p. 90).

Gogol described Korobochka's carriage as "being in sooth more like a fat-cheeked very round watermelon set upon wheels" and Nabokov noted that it has "a certain subtle correspondence to the description of round Chichikov's box" (p. 93).

Nabokov emphasizes a symbolic meaning of Gogol's images of Chichikov leaving town N. and disappearing: "...the blurred wheelspokes form a circular void...And now all you can see afar is a whirl of dust boring a hole in the air" (p. 112).

We may think that Nabokov achieves an empathic understanding of Gogol's writing through symbolic identification (projecting his own RH symbolic thinking). Indeed, the meaning of circle (wheel) is intriguing for Nabokov. He continues to contemplate on this topic in his autobiographical book *Speak, Memory*, now in philosophical form: "The miraculous paradox of smooth round objects conquering space by simply tumbling over and over, instead of laboriously lifting heavy limbs in order to progress, must have given young mankind a most salutary shock" (Nabokov, 1989, p. 301), and more—"The spiral is a spiritualized circle. In the spiral form, the circle, uncoiled, unwound, has ceased to be vicious; it has been set free" (p. 276).

Returning to Nabokov's interpretation of *Dead Souls*, it presents not just preoccupation with round objects (which autists have) but a very complex polysemantic symbol where each transformation ("turn") of the wheel brings a new meaning.

What can we conclude? All that was described earlier looks like autistic features,[36] except that Vladimir Nabokov was not autistic. His prevalent RH functioning included both the sensory-motor level and, to a significant degree, symbolic level. His LH self was intact,[37] and he had superb categorical and logical thinking.

10.10 Schizoid Personality and Autism

> Many schizoid folk are like Roman houses and villas, which have closed their shutters before the rays of the burning sun; perhaps in the subdued interior light there are festivities (Kretschmer, 1925/1999, p. 150).
>
> Asked about his dreams he said…that he dreamt about nice things, e.g., sewing-machines, lamps, grinding machines, wheels, flashlights (Bosch, 1970, p. 29).

Analyzing trends in Autism Spectrum Disorder diagnoses for the past 15 years, Rosenberg, Daniels, Law, Law, and Kaufman (2009) found a significant increase in Asperger syndrome (AS) diagnosis and suggest that a rise in AS diagnoses accounts for the increased prevalence of autism. They also hypothesize that AS is overdiagnosed, with AS diagnosis serving as a broad diagnosis for high functioning children with behavioral problems. I would argue that reliance on behavioral methods for diagnosis might present difficulties in differentiating mild forms of autism, in particular Asperger syndrome, from Schizoid Personality Disorder. Dramatic examples of such misdiagnoses exist. Sugihara, Tsuchiya, and Takei (2008) asked three individuals with Schizophrenia Spectrum Disorder to fill out the Broad Autism Phenotype Questionnaire (BAPQ) designed to capture qualitatively similar but

---

[36] Vladimir Nabokov had also synesthesia and a history of a prodigious calculating gift. The latter suddenly and mysteriously disappeared after a high fever with delirium at the age of 7 (Nabokov, 1989).

[37] Importantly, Nabokov describes his vivid memory of his "awakening of consciousness," "The inner knowledge that I was I," which felt like revelation and coincided with the emerging understanding of temporal order (all are the LH frontal lobe functions).

milder traits of autism. All three individuals scored as BAP positive. Reaven et al. (2008) presented three children with Childhood-Onset Schizophrenia with ADOS (Autism Diagnostic Observation Schedule) and ADI-R (Autism Diagnostic Interview-Revised). All three children met both ADOS and ADI-R criteria for an autism spectrum diagnosis.

Asperger syndrome and Schizoid Personality Disorder overlap in their unsociability and emotional deficit (schizoid's emotional detachment, autist's lack of emotional contact).

However, despite similarity in "end results" on the behavioral surface, autism and schizoid personality are fundamentally different phenomena (experiential conditions) with different brain bases. Schizoid personality roughly corresponds to Karl Jung's introverted personality type. It has already been suggested that the brain mechanisms of extraversion–introversion personality trends are the relative predominance of either the "introverted" thalamic level (*I-space*) or the "extraverted" sensory-motor level (*non-I-space*) within individuals (Glezerman, 1986; Glezerman and Balkoski, 1999; see Chap. 8 for more details). Individual differences in these two domains constitute a population's continuum of personality type—from extraverted to introverted, with schizophrenic autism as its extreme pathological end.[38]

Ernst Kretschmer (1922/1925/1999) was the first to describe schizoid personality. He understood it as a continuum that includes schizothymic personality (variant of norm), schizoid personality disorder, and schizophrenia. The DSM-IV represents Kretschmer's schizoid personality disorder in an oversimplified and incomplete form, so that what Kretschmer defined as the core phenomenological features of schizoid personality, independent of the degree of severity, needs to be clarified.[39]

First and foremost is the introverted nature of the schizoid personality. Kretschmer used the term "autism," defining it as the living inside oneself. He borrowed this term from Bleuler, as did Kanner and Asperger.

> Schizoid men have a surface and a depth…. One cannot know what he feels…[b]ut what their feeling is, whether it be a banality, a whim, an indecency, or a pearl of fairy lore, that is for no one—that is for them alone…. He smiles in an embarrassed way, rather ironically. He says nothing. One day there appears a volume of poetry that he has written, full of an exquisite feeling for nature, with every blow that some fat lout has given him as he passed by moulded into an inner tragedy, and the polished rhythms flowing in full of quiet (Kretschmer, 1999, p. 151)

In Chap. 8, I struggled and failed to reconcile the autist of Kanner and Asperger with either the introverted or extraverted personality. Indeed, although the thalamic level is what is introverted in us, with the proportion of extraversion depending on the amount of sensory-motor level contribution, both these are background levels,

[38] Use of term "pathological" indicates not only quantitative differences, but also a new *quality*, which requires its own pathological network, as hypothesized in a previous work (Glezerman and Balkoski, 1999).

[39] Kretschmer did not distinguish schizoid and schizotypal personality disorders as does the DSM-IV. Instead, he demarcated different subtypes *within* schizoid personality disorder.

integrated at the symbolic level. It is the symbolic level, not the lower levels, that is connected with the formation of personality. In autism, the thalamic and sensory-motor levels are not integrated into the symbolic level. Split from higher levels, each level is exposed and represents the autist's "consciousness." The autist therefore has no personality, no character in the usual sense of these terms, owing to the core problem in autism—lack or underdevelopment of agentivity. Thus, a neurophenomenological analysis belies any similarity between autism (pervasive developmental disorder) and the "autism" of schizophrenia spectrum disorder. Another cardinal feature of the schizoid personality Kretschmer emphasized is the predominance of a so-called psychesthetic proportion. As mentioned in Chap. 8, Kretschmer distinguished in human temperament two dimensions: mood, or "diathetic proportion," and sensitivity, or "psychesthetic proportion." It was suggested before (Glezerman & Balkoski, 1999) that sensitivity dimension comes from the thalamic level's contribution to human-specific complex emotion, namely *thalamic* emotion.[40] Kretschmer indicated sensitivity scale qualities are the most important foundations upon which a schizoid personality is built. It is in the sensitivity dimension (and *thalamic* emotion) where partial overlap between autism and schizoid personality at the brain level occurs, creating much confusion in distinguishing these conditions.

Thalamic emotion is the sensational feeling of one's own body (or attitude to one's own bodily sensations). It is not conscious[41] and has nothing to do with external stimuli. In the norm, *thalamic* emotion is incorporated into the RH symbolic system and plays an essential role in the formation of the RH self (Glezerman & Balkoski, 1999).

Kretschmer distinguished three groups of superficial schizoid peculiarities of character (1) unsociable, quiet, reserved, serious (humourless), eccentric; (2) timid, shy, with fine feelings, sensitive, nervous, excitable, fond of nature and books; and (3) pliable, kindly, honest, indifferent, dull-witted, silent. He emphasized that the characteristics in the first group are "absolutely the most common, in that they run like scarlet thread through the whole schizoid characterology" (Kretschmer, 1999, pp. 155–156), including groups two and three. One may conclude that the first group reflects the general *introverted* nature of schizoid character. Groups two and three reflect its inner essence—*sensitivity* scale: "the phenomena of psychic over-sensitivity, from mimosa-like, timid fineness of feeling to a continual state of passionate excitation" (group two), and "psychic insensitivity, dullness, and lack of spontaneity" (group three). Kretschmer emphasized the majority of schizoids are not either over-sensitive or cold, but that they are over-sensitive *and* cold *at the same time*, differing markedly in relative proportions. Even in the cold schizoid with paucity of affective response, there is "behind the unresponsive, numbed exterior, in the innermost sanctuary, a tender personality-nucleus with the most vulnerable nervous sensitivity, which has withdrawn into itself, and lies there contorted"

[40] Thalamic emotion is contradistinct to *limbic* emotion, the latter being related to Kretschmer's "diathetic proportion."

[41] *Thalamic emotion* can emerge to the surface of consciousness in pathology.

(p. 157). Kretschmer's "psychesthetic proportion" refers to the mixture in which, in any given schizoid, these hyperesthetic and anesthetic elements of the sensitivity scale are combined with one another.

I will now attempt to analyze the difference between *thalamic* emotion (Kretschmer's sensitivity scale) in autism and schizoid personality.

Schizoid hypersensitivity is *psychological.* It arises from the symbolic level of a fully developed RH self. In autism, *thalamic* emotion is not incorporated into the symbolic system, rather the thalamic level is exposed, and its emotion remains *physical*—a feeling tone of bodily sensations (see Chap. 8). In Chap. 8, I cited Asperger's remarks about autistic children's natural appreciation for art. What should be clarified here is that autists could in fact "distinguish between art and kitsch" (see Chap. 8) *physically*, but did so without much awareness of their choice, or to be more exact, with the low awareness of the thalamic level. Hypersensitivity/hyporeactivity of autistic children to external sensory stimuli is connected with exposure of the sensory-motor level. In this case, hypersensitivity/hyporeactivity is a manifestation of and compensation for the exposed RH "flow of consciousness." Thus, while in schizoid personality hypersensitivity is *psychological* and an essential part of the personality, in autism hypersensitivity is *physical.*

The damaging effect of the behavioral approach, which ignores fundamental differences between HFA and schizoid personality, comes in the grotesque form of diagnosing extraordinary, talented individuals (among them such giants of Western civilization as philosophers Spinoza and Kant and composers Mozart and Beethoven) with Asperger's Syndrome (Fitzgerald, 2004, 2005). Without considering the fundamental problems in autism (underdevelopment of the self as one and apart from others, underdevelopment of agentivity, lack of character, and social identity) the application of an inventory of autistic criteria to people who not only have a great talent, but also a great personality, becomes absurd. I will give a few examples.

After judging Kant's every day schedule to be "sameness" behavior, Michael Fitzgerald concludes that it is a description of a classically autistic person. He then goes on, in attempting to prove this assertion, to cite Kant's own words, which actually demonstrate the philosopher's full self-awareness of his traits along with strength of personal resolve. "All change makes me anxious, even if it seems to contribute greatly to the improvement of situation. I believe I must pay attention to this instinct of my nature, if I still want to lengthen somewhat the thread, which fate spins very thinly and delicately for me" (from Fitzgerald, 2005, p. 123). In arguing that Kant displayed another criterion for Asperger's Syndrome—narrow interests and obsessiveness—Fitzgerald again cites him: "I have already prescribed the route I want to take. I will begin my course and nothing shall prevent me from continuing it" (p. 122). At the very least this statement indicates strong goal-directedness (agentivity). Moreover, Fitzgerald cites Kant's biographer, Kuehn, who described him as autonomous and self-reliant—a self-made character.[42] This example shows

[42] According to Kretschmer, "Kant…in his private personality displays the schizothymic type…in its highest and purest form, with Spartan freedom from desires, childlike simplicity, and the most genuine idealistic morality" (1999, p. 247).

that criteria for stereotypical behavior in autism have meaning only in connection with the primary deficit and in the context of the autistic clinical-brain pattern. Even though any repetitive activity or habitual behavior becomes automatic and involves the striatum, it is not specific. Keeping strict schedules in Kant's case may have been a compensation for schizoid oversensitivity. Also, how else could he have done his colossal work if he did not implement routines with iron discipline which then became automatic activity that could unload the field of consciousness for the major work. In fact, a recent phenomenological study of creativity shows that "routine is often experienced as providing a 'holding' structure that allows the artist to be more spontaneous and exploratory in his approach" (Nelson & Rawlings, 2007, p. 234).

In the very beginning of his chapter about Lewis Carroll, Fitzgerald declares that Carroll had Asperger's syndrome. He gives examples from Carroll's biography and recollections of his contemporaries to justify his assertion. For example, to fulfill the criterion that Carroll showed impairment in reciprocal social interaction Fitzgerald writes "...despite his shy and sensitive nature, [Carroll] did form friendships and during the 1860s led an active social life, with a particular fondness to the theatre and art exhibition. Though his interests filled every minute of his day and kept him busy, Carroll's life was lonely and isolated."[43]

More quotes Fitzgerald uses about Carroll actually demonstrate well the *hypersensitivity* of a schizoid man.

> The dramatist A.W. Dubourg described him as a 'quiet, retiring, scholar-like person, full of interesting and pleasant conversation, oftentimes with an undercurrent of humour, and certainly with a sense of great sensitiveness with regard to serious side of life'.... An Oxford colleague, Frederick York Powell, recalled 'the quiet humour of his voice, the occasional laugh.... He was not a man who laughed, though there was often a smile playing about his sensitive mouth' (Fitzgerald, 2005, pp. 57, 58, 61).

Fitzgerald goes on then to redefine Asperger's syndrome: "Lewis Carroll shows that a person with Asperger's syndrome has a capacity for enormous imagination, even if of an immature kind" (p. 65).

Lastly, we'll turn to Fitzgerald's treatment of composers. He cites Mozart saying "I am completely absorbed in my music, so to speak, that it occupies my thoughts the whole day long, that I like speculating, studying, reflecting" (p. 155); and also Beethoven: "I live entirely in my music" (p. 162). He then sentences these men to having had "narrow interests and obsessiveness" characteristic for Asperger's syndrome. It is probably not relevant to start a "scientific" discussion with the above argumentation, it is rather a topic for Jonathan Swift (alas, he also did not escape the fate of being diagnosed by Fitzgerald as having Asperger's syndrome). Maybe it is more adequate to finish by citing Goethe who attached great importance to the artist's personality: "All that matters is the artist, that in his life he should feel no other felicity that that of his art, live in it, bury himself in the tools of his creation, with all of his sensibility and all of his force" (Goethe [1772], 1980, in: *Goethe on Art*, ed. and trans. by John Gage. Berkeley, University of California Press).

---

[43] According to Kretschmer, social attitude of the schizoid persons may be expressed in one of three ways (1) simply unsociable; (2) sociable within a small closed circle or else; (3) superficially social, without deeper psychic rapport with their environment.

10.11 Creativity and Autism

10.11.1 Definition of Creativity

Psychometric psychology defines creativity as the product of intelligence and personality (Eysenck, 1995). Two aspects of creativity are indicated: *trait*-creativity and *achievement*-creativity. In general, trait-creativity is an associative process, a thinking style termed *divergent thinking*. Trait-creativity points to potential—a necessary, but not sufficient cause for creative productivity. *Achievement*-creativity, on the other hand, stands behind exceptional real-world creative works, achieved only by a minority of creative individuals (McCrae, 1987; Stavridou & Furnham, 1996). Achievement creativity derives from trait-creativity, intelligence, and multiple other components acting synergistically (Eysenck, 1995). The ability to associate remote domains into new useful combinations accounts for creative thinking. The more mutually remote the elements of the new combination, the more creative the process or the conception will be (Mednick, 1962). Correspondingly, divergent thinking tests (in contrast to a single correct answer on intelligence tests) ask for as many appropriate answers as possible, and score for unusual and original responses (McCrae, 1987).

An enormous amount of literature links creativity and genius with psychopathology. However, actual psychosis in fact negatively correlates with creativity, while certain desirable personality traits (for example, ego-strength) are positively associated with creativity and genius. Responding to this paradox of genius and psychopathology, Eysenck introduced the concept of *psychoticism.* According to Eysenck and Eysenck (1976), psychosis lies at the extreme end of the distribution of *psychoticism*, a hypothetical personality trait conceived as a continuum, ranging from normal to psychotic. Numerous studies have found a link between psychoticism and trait-creativity[44]: high scores on unusual answers in divergent thinking tests had a significant positive correlation with the *psychoticism* scale in the Eysenck Personality Questionnaire (EPQ).

Eysenck believes that the overinclusive cognitive style[45] constitutes the basis for creativity and is also characteristic both of people who scored positively on the

[44] Eysenck believes psychiatric abnormality to be dimensional rather than categorical. His continuum, which includes both schizophrenia and manic-depressive psychosis, parallels Kretschmer's "categorical" continuum (described earlier) that ranges from schizoid personality (variant of norm) to schizoid personality disorder to schizophrenic psychosis. Kretschmer based his continuum on his sensitivity scale. Thus, if creativity indeed connects in some way with psychopathology, it would be mostly with schizophrenia spectrum disorders, not autistic spectrum disorders. Furthermore, creativity itself manifests not as psychiatric illness but as a latent trait in normal people.

[45] Eysenck seems to interchange *overinclusive* thinking with *divergent* thinking, which creates confusion. In psychiatry, overinclusive thinking implies pathologically excessive and irrelevant detailization, often referring to circumstantial or even to concrete thinking. In divergent thinking, associations are not irrelevant, they are *not known* before the creative person "discovers" them.

psychoticism scale, but not actually psychotic, and of schizophrenics.[46] It was suggested that the physiological mechanism behind overinclusive thinking is the decreased capacity to screen irrelevant stimuli[47] out of attentional focus (called *latent inhibition*—LI). Indeed, LI, known to be decreased in schizophrenia, was attenuated in normal people with high *psychoticism* scores and in creative people (Carson, Peterson, & Higgins, 2003; Stavridou & Furnham, 1996). Reduced LI was then concluded to be a predisposing factor for both psychosis as well as creativity. "The highly creative individual may be privileged to access a greater inventory of unfiltered stimuli during early processing, thereby increasing odds of original recombinant ideation" (Carson et al., 2003, p. 505). Eysenck (1995) even suggests that originality is conceptually similar to loosening of associations (LOA) in schizophrenia. Carson et al. add that "such looseness is presumably a byproduct of the failure of an inhibitory filtering mechanism, functioning to limit associations to those relevant to current task performance" (p. 500). All such assumptions seem reductionistic. Making associative process (thinking) dependent on early selective attentional processing ignores the hierarchical organization of brain functions. The associative process (be it normal or pathological) has its own "internal rules" connected with the *symbolic* level in the brain—analytical cognitive mechanism in the LH, holistic in the RH. Indeed, we deal here not with *thinking* but with the *RH attentional system.*[48] Anatomically distinct, the RH activating system sustains attention to the constantly changing sensory stimuli from the external world, which is necessary for maintaining an alert state. Persons are usually not aware of all these stimuli. Attenuation of LI in both psychosis and creativity, only tells us there is some kind of *disinhibition* of RH flow of consciousness in both conditions. Interestingly, and in line with what has been said earlier, there is evidence that the creative process is connected with decreased arousal. Arousal is controlled by the midline activating system, more connected with LH focus of consciousness and voluntary selective attention.

In Barron (1969) study, creative persons received high scores in Minnesota Multiphasic Personality Inventory (MMPI) dimension on which psychotic persons likewise score high. Creative persons, however, also scored high on MMPI's ego-strength dimension, which negatively correlates with the "psychotic" dimension in normal population; a truly psychotic person always scores low on the ego-strength dimension. Similar results were obtained by Fodor (1994) who presented creativity tests to four categories of persons (1) psychosis-prone, high ego-strength; (2) psychosis-prone, low ego-strength; (3) not psychosis-prone, high ego-strength; (4) not psychosis-prone, low ego-strength. Only psychosis-prone, high ego-strength persons scored high in creativity tests. Barron (1993) refers to creativity as "controllable oddness" where oddness of thought or feeling combines with an ability to reconsider

[46] For example, Andreason and Powers (1975) found that highly creative writers were overinclusive on a task designed to measure overinclusion in schizophrenia.

[47] Stimuli, with no apparent value for current motivations and goals.

[48] Lateralization of attentional system in the human brain was discussed in Chaps. 3 and 5.

and reformulate—to produce a socially communicable original meaning. Interestingly, a low level of arousal was found during creative inspiration, but not during the elaboration stage (Marindale & Hasenfus, 1978). Thus, we see here a strong interaction exists between RH and LH, with a significant role played by LH self (agentivity), in *achievement*-creativity. Creativity poses a peculiar, paradoxical situation where opposing hemispheres potently cooperate while both remain dominant, in contrast to the general population where the presumably lateralized psychoticism (RH) and ego-strength (LH) are in reciprocal relation.

What did we learn from the above studies about creativity that can contribute to our neurophenomenological point of view?

Divergent thinking (DT) underlies *trait*-creativity, and it also correlates with the *psychoticism* personality trait in psychometric studies.

Both DT and *psychoticism* (ranging from normal to psychotic) share decreased latent inhibition (LI), the latter I interpreted as readiness of RH "flow" of consciousness to surface. What distinguishes creative people is the paradoxical *combination* of *psychoticism* personality trait and high *ego-strength*; these two traits usually have strikingly *negative* correlation in the general population. While the psychoticism personality trait might be connected with RH prevalence, *ego-strength* parallels the concept of LH agentivity proposed in this book. A peculiar situation is found in creative people: the cerebral hemispheres, strongly specialized, have strong complementary interaction where neither are dominant, but reign powerfully together. Lastly, DT is measured by the ability to unite *remote* associations into new and useful combinations. This view of DT calls to mind the concept of metaphor as uniting remote domains to create a polysemantic symbol. As hypothesized in this book, it is the RH "law" of identification (not logical rules) that unites remote domains into a new meaning. Is this also the mechanism underlying DT?

10.11.2 Creativity and RH/LH Interaction

Maslov (1983) interpreted the mathematical results of the theory of deductive systems in terms of the functional asymmetry (FA) of the human brain and the psychology of creativity. He indicated that two cognitive mechanisms, Left and Right, reflect two different aspects of cognition (1) operating according to a fixed program within a given deductive system ("study" of one story), and (2) "organization" of ascent to the next level (change of system). Maslov suggested that each ascent to the next story requires a "right hemispheric creative act." The sequential shifting of variables (LH mechanism of a deductive system unfolding) is interrupted by a "brake," which Maslov called "intuitive puncture." His "intuitive puncture" is characterized by an "iterative deepening into the meaning of the whole object" (p. 19).[49] When LH

[49] Maslov's deepening into the meaning of the whole object can be understood from a neurophenomological point of view as projection of RH self and formation of RH polysemantic symbol.

functioning resumes, there is a priority for the analysis of new gestalts that emerged during the "brake" (RH functioning). This leads to re-organization of the deductive system and formation of the next story, a deductive system with a new fixed program (LH functioning). This is how Maslov envisioned the creative process.

We see here the intense working of both hemispheres, but how does it differ from the reciprocal and complementary interactions of LH and RH in the norm? The cerebral organization of sentence formation is one example of complex interhemispheric interaction, as described in detail in a previous work (Glezerman & Balkoski, 1999, pp. 203–214). The "brake" Maslov described is a "brake" in the LH "focus" of consciousness (or what we usually call consciousness). "Intuitive puncture" on the other hand may represent intrusion of RH "flow" of consciousness. This temporary change in consciousness is, I believe, how creative process differs from even sophisticated (requiring complex LH/RH interactions) intellectual processes. Showing the significance of each hemisphere for creative process, Maslov warned that each hemisphere has its limitation for creativity. The limitation of LH cognitive mechanism is dogmatization of the given deductive system and its fixed program. The limitation of RH cognitive mechanism is the inclination to the formation of more and more refined versions of the singular and closed-in-itself text. And with all the importance of RH primacy in the creative process, it is eventually the LH who builds the next "story," i.e., new ideas.

What do we learn from Maslov's model of creativity for our purposes? The RH is responsible for the recognition of a "new gestalt," which coincides with change in consciousness: the intrusion of RH "flow" of consciousness to the surface and direct access to RH symbolic self.

10.11.3 How Creative Process is Experienced by the Artist

Recent phenomenologic studies of creativity are in correspondence with Maslov's "objective" reasoning about deductive system development. I will analyze some of these works from a *neurophenomenological* point of view. Nelson and Rawlings (2007) interviewed several professional artists (musicians, writers, visual artists). In each case, the artists' descriptions of certain artistic experiences are analyzed for the "logic" behind them. Artists reported an increasing "connection" with their work and a unity to their sense of self which was felt via the sense of a new synthesis of elements. For example, one artist experienced "losing the sense of distinction between herself and the object she was depicting" (p. 244). The recognition of a new gestalt, or in Maslov's words, "intuitive puncture" is experienced as a new synthesis "providing coherence to previously loose elements of the art work—they are now felt to be where they 'belong', where they 'should' be" (p. 229). The artist is experiencing that the *piece* already "'exists' in some sense and…he is discovering its character and structure…a process referred to by one writer as discovering and following the 'rhythm' of the piece" (p. 233). Thus, we can see here RH identification of the artist with his artistic work. As such, the art object becomes part of the symbolic system of the artist's RH self, a metaphor of the self. Recognition of a new gestalt was accompanied by the sense of self as undivided whole, and, curiously, in some artists was also accompanied by sense of *purity*. The latter seems to be related to the sensitivity dimension (*thalamic emotion*).[50]

Together with sense of wholeness of self, there was also a sense of dissolution of personal boundaries and sense of "connection" with something beyond the self. Thus, we clearly recognize the exposed RH self: experience of wholeness, continuity of the RH symbolic system, identification with the object of art. Simultaneously, the participants reported suspension of self-awareness, as one participant referred to as a "bracketing of the self." Also, the "decision makings," "worrying," and "deliberations" inherent to conscious experience seemed to break down during such artistic moments. Thus, we can see here "diminishing" and even temporary disappearing LH self.

Finally, when the artist was in "intuitive mode," he or she was experiencing a sense of ease in the artistic activity where ideas for the piece emerged fluently and constantly, seemingly of their own accord. There was also effortless use and a high level of technical skills without much awareness of doing technical aspects of work. Here we recognize the RH and lower level automatisms, freed from the control of higher levels. "The artistic activity is felt to be propelling itself forward as having a self-generating momentum" (p. 234). Isn't this reminiscent of the "flow" of RH symbolic system—RH DLPC unfolding during autistic savant's calendar calculation (hypothesized in Chap. 6)?

The study also shows that during the creative process, there is a movement between intuitive and analytical modes. During a more critical, analytical approach to work, the artist is an *agent* (creator), and LH complex personal emotions play a significant role. Such emotions would include a feeling of satisfaction or success and a "sense of fulfillment…based on an awareness that this particular creative experience fulfilled the individual's ongoing commitment to his art form, his reasons for being an artist" (p. 238).

For a neurophenomenological perspective, a study by Reinders [(1992) cited from Nelson & Rawlings (2007)] is also of importance. Reinders investigated the phenomenology of artistic creativity in three internationally recognized artists from the fields of choreography, painting, and musical composition. This study indicates that in the beginning of the creative process, the artist has a "sense of lack" and a vague, intuitive sense of the artistic object. There is a desire to discover the intuited object to fulfill the sense of lack. The artist actively manipulates his artistic material, and in this exploratory process, the artist recognizes the "demands of the artistic object," artistic constellations that bring the work closer to realizing the intuited intentional object. Still, the discovery of a new gestalt (Maslov's "intuitive puncture")

[50] "The sense of purity can manifest itself in a number of ways, including the sense of growing a 'higher' awareness, a 'raising of consciousness', either on an individual or on a collective level, and also in the sense of being 'in touch with oneself'… or the sense that a 'purer' form of being is being revealed…[to the artist]" (Nelson and Rawlings, 2007, p. 233), the latter is provided by "the sense of self as undivided whole"(p. 233).

is spontaneous and felt to be beyond the artist's control. Sense of recognition is strong and characterized by a "global bodily felt sense of meaning" [cited from Nelson & Rawlings (2007), p. 67].

These two studies beautifully illustrate what was discussed earlier. Specifically they reveal (1) a strong "agent" (LH self) exploring and searching for access to the RH gestalt, as well as (2) an involvement of the RH *symbolic* level along with *thalamic* emotion in the creative process.

The artist's experience during intuitive mode can be interpreted as a disappearing of the "agent" and a dominating RH self with its continuous "flow" of RH symbolic system and automatic technical performance.

10.11.4 Neurophenomenology of Creativity: Role of the Thalamic Level and RH Symbolic System

A brain basis for the inner source of creativity was hypothesized in a previous work (Glezerman & Balkoski, 1999). This hypothesis feels like a "missing piece" that could be easily integrated with what has been said already about creativity in this chapter. A short description of this hypothesis follows. *Originality*, a necessary aspect of creativity, derives from the brain level of internal space—the kinesthetic space of one's body with its inner rhythm, unique for each individual.[51]

Kinesthetic bodily feelings provide subjective experience of *I-space*: the boundaries and continuity of one's own body which is the foundation for the experience of the self as *whole*. A subjective attitude (feeling) to one's own bodily sensations was defined as *thalamic* emotion. *Thalamic* emotion, in turn, corresponds to the sensitivity dimension in Kretschmer's two-dimensional approach to human emotion. *Thalamic* emotion is not conscious. It is directly incorporated into the RH symbolic system, playing a basic role in the formation of the RH self. Much of the hypotheses about thalamic emotion are based on psychopathology where *thalamic* emotion comes to the surface of consciousness. In schizophrenia, pathological bodily sensations (*coenesthesia*) are observed. *Coenesthesia* was interpreted as a "marker" of breakdown in the continuity of *I-space* boundaries, and, therefore, a disorder of the experience of self as whole. Further, it was suggested that, as an attempt at spontaneous compensation, pathological bodily sensational feelings (coenesthesia) were re-integrated into the symbolic system. This new and pathological symbolic system was hypothesized as the brain mechanism underlying Bleuler's autistic thinking (Glezerman & Balkoski, 1999). In setting forth these ideas about schizophrenia, observations of *coenesthesia* in otherwise intact individuals were also considered

[51] Although there is consensus that a tendency toward introversion is one of several factors that describe the creative person, the brain basis for this personality trait has not yet been explored in the literature.

[Djupre, cited from Anufriyev (1979)].[52] Unable to describe their feelings directly, these individuals use various metaphoric expressions. The ability to convey one's own experience by means of figurative comparisons corresponds to a person's intellectual level and wealth of imagination. Djupre stated that such patients often understand the incompleteness, the imperfection of their descriptive tales. He suggested that language cannot express these absolutely new, unusual sensations, which are unique and idiosyncratic for each patient and not related to past experience.

A proposed explanation for the uniqueness of these sensations to the individual is that they belong to the *introverted* thalamic level, and, as such, they are not subject themselves to external influence. Further, Bernstein's thalamic level has no input from the visual modality (see Chap. 1). The development of language in phylogenesis, on the other hand, is closely associated with the visual modality (Glezerman & Balkoski, 1999; also see Chap. 2). Thus, the sensations of the thalamic level, without direct access to language for expression, are somehow translated into visual images which can then be expressed in language, though generally imperfectly. Metaphors seem to capture thalamic emotion more accurately than direct efforts at expression. Why metaphor? First of all, thalamic emotion implies both feeling (content) and physical sensation (form). Thalamic emotion itself is metaphor! Second, *thalamic* emotion is directly incorporated into the RH symbolic system, "saturating" the RH symbolic self. From all this, it was speculated that the translation of thalamic emotion into visual images, and then into language, is a universal phenomenon, and a major source of creativity in art forms as well as all kinds of creative thinking. Considering bodily feelings of the thalamic level to be the boundary of *I-space*, one can see how their inability to be directly accessed creates a constant drive for their translation or decoding into knowable forms and for their understanding or expression through projection onto external space where they can be seen, touched, or heard and in doing so perhaps some semblance of transcendence of *I-space* boundary is achieved. It must be emphasized again that this is not a literal translation of bodily sensations but an expression reliant on metaphor, behind which lies RH visual symbolic thinking. Indeed, artists and great thinkers must have a highly developed thalamic level and concomitant substantial drive to decode this space of the self, for they create infinite numbers of metaphors (symbols of *I-space*) with their particular media, including not only art forms such as the visual arts, music, and literature but also scientific "media"—the language of physics, mathematics, etc.[53]

Applying this hypothesis to the material discussed earlier on studies in creativity, we can attempt to trace creative process from the depth of the thalamic level to the RH "creative act" or new gestalt, which is just another metaphor of *I-space* integrated into continuity of the RH symbolic system. Recognition of a new gestalt

---

[52] These rare cases, described in literature, were considered to be due to inborn anomaly of the thalamus.

[53] Recall Einstein's thinking as nonverbal "associative play" with visual and motor kinesthetic image (see footnote 23 in section eight of this chapter).

(RH individual symbol) is marked by a change in consciousness, which is followed by the "restoration" of LH consciousness during the elaboration stage, where "new ideas" are formed.

What can be said now about a purported link between madness and genius? In schizophrenia, the boundaries of *I-space* are broken, and pathological bodily feelings are "re-integrated" into the symbolic system (resulting in a distorted view of the world) which dominates the patient's consciousness (autistic thinking). Autistic thinking is expressed literally as disorganized thinking, i.e., loose associations (LOA),[54] or through LH interpretations (delusions).

In creativity, there must be a significantly increased sensitivity (right-shift on Kretschmer's sensitivity scale). There is a tension: increased bodily feelings demand expression, they are about to emerge, to burst into consciousness, but they do not. Instead, bodily sensations find a media for their expression. Boundaries are thin, but not broken. LOA and divergent thinking are qualitatively different processes. LOA presents a pathological production of the RH symbolic level in the brain, split off from the LH. If divergent thinking reflects a creative process, there must be strong interaction between the RH symbolic level and the LH symbolic (categorical) level. We can now see that confusion surrounding madness and creative genius has a real basis. Both creative process and schizophrenia involve structures in the brain responsible for the cerebral organization of the RH symbolic self. Can the person in the premorbid stage of schizophrenia be creative? It is conceivable. Similarly, modern literature relates schizotypal personality with creativity, and a strong overlap in schizotypality and creative experience was suggested (Nelson & Rawlings, 2010). However, clinical stage and psychosis itself (fundamental disorder of RH self) is not compatible with creativity. Creativity's "trade mark" is a heightened functioning of the RH self.

Eminent creativity and its extreme—genius—are very rare events because they require many components, not only the highest development of the thalamic level, but a very high development of "performers," multiple cortical regions in both RH and LH pertinent to the particular media. As discussed before, the different brain areas vary independently from each other.

10.11.5 *Miracle of Autistic Savant Talent and Myth of Autistic Creativity*

Are autistic savants creative? In Table 10.2 the possible brain mechanisms underlying the phenomena of human creativity and autistic savant talent are compared.

As indicated in the table, creativity occupies an intermediate position between two distinct domains: cognition and personality (Eysenck, 1995; McCrae, 1987).

---

[54] Recent fMRI studies of schizophrenic patients show reversed lateralization in activation of posterior temporal region during spontaneous thought-disordered speech, i.e., LOA (Kircher et al. 2002).

Table 10.2 Creativity and autistic savant talent

Creativity	Autistic talent
Product of intelligence and personality	Extreme manifestation of the autistic pattern
Right hemisphere "flow" of consciousness is "ready" to surface at times	Right hemisphere "flow" of consciousness is dominant
Strong interaction of the right hemisphere symbolic and left hemisphere symbolic (categorical) levels	Right hemisphere sensory-motor level
Change in consciousness with right hemisphere symbolic self emerging to the surface is only temporary	Right hemisphere situational self is a prominent part of the autist's self
High IQ	High IQ inhibits autistic savant's talent
Right hemisphere flow of symbolic systems	Right hemisphere flow of VSS

In terms of brain mechanisms, psychometric definition of these domains roughly corresponds to cortical functions, mostly left hemispheric, including the concept of LH self as discussed in this book. Highly creative individuals generally score very high on measured intelligence (Barron & Harrington, 1981). Autistic savant talent is considered in this book to be on a "different plane" than creativity, for it is an extreme manifestation of the autistic clinical-brain pattern (described earlier in this chapter). Autistic savant activity is fully automatic and stereotypical. Higher IQ (especially verbal IQ) indirectly indicates involvement of higher levels in the brain, which are characterized by conscious, effortful, sequential activity that is antagonistic to, or may even suppress, the magnificent autistic savant performance, precisely because this performance is connected with the "unleashing" of lower levels in the RH. Indeed, both Nadia (see Chap. 3) and the twins (see Chap. 6) lost their talent after they underwent treatment to become "more socially acceptable." Nadia "started talking—and stopped drawing." The twins "deprived of their numerical 'communion' with each other…lost their strange numerical power" (Sacks, 1987, pp. 209–210). In Chap. 6, the highest achievements of calendar calculation was illustrated by the examples of twins with severe intellectual impairment (Sacks, 1987), a savant who was not able to speak or even communicate through signs (Hermelin & O'Connor, 1990), and a savant with IQ 50 (Howe & Smith, 1988). On the other hand, a neurophenomenological analysis of savants showed that, while the LH may play a role in savants having higher IQs, it determines neither the quality, nor the degree of the savant's talent (see Chap. 6).

A creative person has a temporary access to the RH "flow" of consciousness. The autistic savant "lives" in that flow (Table 10.2). Change in consciousness during creative processes is characterized by exposure of the RH symbolic level, whereas, in the autistic savant, consciousness of the RH situational (sensory-motor) level predominates. In creativity, the RH "creative act," or Maslov's "intuitive puncture" transpierces all layers—from the thalamic level to the symbolic. The autistic savant mostly has the situational level, split from the thalamic and with no evidence of symbolic level's participation, to work with (Table 10.2). Despite all these differences, there is an overlap between the RH "creative act" and autistic savant activity.

Clearly, "sacred" objects of aborigines, preoccupation of autistic savants with prime numbers (see Chap. 6), and the above-mentioned writer who discovers the "rhythm" of a piece (artistic object) all hold in common the subjective experience of identification of *I-space* and *non-I-space* with projection of the inner rhythm onto *non-I-space*. Additionally, the effortless automatic performance during intuitive mode of creative process is similar to the autistic savant's low-awareness and automatic activity.

In creativity, there is an intense work by both cerebral hemispheres along with a strong LH/RH interaction, resulting in some peculiar situation where each hemisphere seems to be strongly specialized, but neither is dominant. LH/RH interaction is illustrated by the alternation of intuitive with analytical mode during creative process. Returning to Nelson and Rawlings' findings:

> The artist is being both active and receptive: active in the sense that his actions are informed by his artistic intuition of the intentional object and receptive in the sense of remaining sensitive to the emerging artistic configuration. Another paradox at work is 'distant-engagement'—that is, an alternation between immersion in the manipulation of material and distancing oneself…in order to clarify the overall structure or gestalt of the object (Nelson and Rawlings 2007, pp. 221–222).

Arthur Koestler (1978) even speaks about simultaneous use of intuitive and rational processes in creative activity: "By living in both planes at once, the creative artist or scientist is able to catch an occasional glimpse of eternity looking through the window of time" (p. 146). As hypothesized in this book, autistic savant activity is very different from all this. The savant's talent comes from mostly RH functioning: externalization of his RH self, and it is his particular mode of existence and empathic communication with the physical world.

The autistic savant's activity cannot meet *achievement*-creativity criterion. He does not produce a socially communicable *new* meaning. The main cause of this failure: there is no *creator*. Again, autistic savant activity reflects the very core of autism—deficit of the LH self. There is no aspiration of personality to "ascend to the next story" (using Maslov's expression), no awareness of personal meaning, no personal feelings of satisfaction, success, fulfillment. Also, "new meaning" cannot be achieved because the LH categorical level in autism, in general, is not functioning sufficiently.

The RH "creative act" or "new gestalt" (in Maslov's terms) provides the possibility for the LH to build the "next story." In particular, a new gestalt, the RH symbol, allows for more specific categories to be distinguished within the categorical classification of the world. This amounts to further differentiation of existing concepts ("new ideas").[55]

The limitations of autistic savants for creativity are something akin to what Maslov indicates to be a limitation of the RH cognitive mechanism—"inclination to the formation of more and more refined versions of the singular and closed in itself text." However, the autistic savant's RH activity is not what Maslov calls a "RH creative act": "iterative deepening into meaning of the whole object" (polysemantic

[55] Similar interdependence was suggested for the phylogenetic and ontogenetic development of the RH symbolic and LH symbolic (categorical) levels in the brain (Glezerman and Balkoski, 1999).

RH symbol in my formulation). The autistic savant is functioning mostly at the RH situational level, and so there are no "more refined versions." Furthermore, the autistic savant activity is, in principle, opposite to being creative. Whereas in creativity there is a strong motivation to "ascend" to the "next story," in the autistic savant, there is a strong desire to preserve *sameness.*

We can see that autistic savant talent remains a mystery. However, neurophenomenological analysis revealed that autistic savant talent is a qualitatively different phenomenon from what we understand to be human creativity.

10.11.5.1 Are Achievements of HFA/AS Creative?

Can HFA/AS be creative? The achievements of HFA/AS and those of autistic savants have different (even opposite) sources. Autistic savant talent is a product of the autistic pattern, what it "can do" in its extreme. HFA/AS's success is determined by the degree to which they are able to overcome the autistic pattern, to be "not autistic." In HFA/AS's case we are speaking about compensatory re-wiring of brain networks. It depends on the "premorbid development" of the LH DLPC in each individual and how much this region can be employed, so that, even in tandem with the LH striatum, its functioning will allow a greater degree of voluntary control, intellectual insight, and some ability to reflect, coexisting with the autist's RH visual situational world. This is what we see in Temple Grandin. If such autists still manage "to be in touch" with their inner rhythm projected onto *non-I-space*, they will be able to "collect" a tremendous amount of visual–spatial patterns and manipulate them. This can lead to significant technical achievements. However, we should not forget that this activity is mostly stereotypical. Most importantly, the deficit of LH self (agentivity) and of both RH symbolic and LH symbolic (categorical) thinking are not compatible with the phenomenon of what is understood to be human creativity.

Thus, when we speak about autistic talent, there is only one: autistic savant talent, which amounts in reality to a talent at being autistic.

References

Adolphs, R., Tranel, D., Damasio, H., & Damasio, A. (1994). Impaired recognition of emotion in facial expressions following bilateral damage to the human amygdala. *Nature, 372*, 669–672.

Adolphs, R., Tranel, D., Damasio, H., & Damasio, A. R. (1995). Fear and the human amygdala. *Journal of Neuroscience, 15*, 5879–5891.

Akshoomoff, N. (2006). Autism spectrum disorders: introduction. *Child Neuropsychology, 12*, 245–246.

Alajouanine, T. (1963). Dostoevski's epilepsy. *Brain, 86*, 209–221.

Alderson-Day, B., & McGonigle-Chalmers, M. (2011). Is it a bird? Is it a plane? Category use in problem-solving in children with autism spectrum disorder. *Journal of Autism and Developmental Disorders, 41*, 555–565.

Alexander, G. E., & Crutcher, M. D. (1990). Functional architecture of basal ganglia circuits: neural substrates of parallel processing. *Trends in Neurosciences, 13*, 266–271.

Allison, T., McCarthy, G., Belger, A., Puce, A., Luby, M., Spencer, D. D., & Bentin, S. (1994). What is a face? Electrophysiological responsiveness of human extrastriate visual cortex to human faces, face components, and animals faces. *Society for Neuroscience Abstracts, 20*, 316.

Allison, T., Puce, A., & McCarthy, G. (2000). Social perception from visual clues: role of the STS region. *Trends in Cognitive Science, 4*, 267–278.

Amaral, D. G., Price, J. L., Pitkanen, A., & Carmichael, S. T. (1992). Anatomical organization of the primate amygdaloid complex. In J. P. Aggleton (Ed.), *The amygdala: Neurobiological aspects of emotion, memory, and mental dysfunction* (pp. 1–66). New York: Wiley-Liss.

Andreason, N. J. C., & Powers, P. C. (1975). Creativity and psychosis: an examination of conceptual style. *Archives of General Psychiatry, 32*, 70–73.

Angyal, A. (1936). The experience of the body-self in schizophrenia. *Archives of Neurology and Psychiatry, 35*, 1029–1053.

Anufriyev, A. K. (1979). *Pathology of coenesthesia and affective disturbances with its equivalents: Psychosomatic disorders in cyclothymic and cyclothymic-like states* (Vol. 87, pp. 8–24). Moscow: R.S.F.S.R Ministry of Health, Moscow Institute of Psychiatry Works.

Asperger, H. (1991). 'Autistic' psychopathy in childhood. In U. Frith (Ed.), *Autism and Asperger syndrome* (pp. 37–92). Cambridge, UK: Cambridge University Press.

Aylward, E., Bernier, R., Field, A., Grimme, A., Dawson, G. (2004). Autism during the viewing of familiar faces. *Poster presented at the International meeting for Autism Research.* Sacramento, CA.

Bachevalier, J., & Loveland, K. A. (2006). The orbitofrontal-amygdala circuit and self-regulation of social-emotional behavior in autism. *Neuroscience and Biobehavioral Reviews, 30*, 97–117.

Bailey, A., Le Coueur, A., Gottesman, L., Bolton, P., Simonoff, E., Yuzda, F. Y., et al. (1995). Autism as a strongly genetic disorder: evidence from a British twin study. *Psychological Medicine, 25*, 63–77.

T.B. Glezerman, *Autism and the Brain: Neurophenomenological Interpretation*,
DOI 10.1007/978-1-4614-4112-0, © Springer Science+Business Media New York 2013

Balonov, A., Barkan, D., & Deglin, V. (1979). *Unilateral ECT*. Leningrad: Meditsina.

Bar-Haim, Y., Shulman, C., Lamy, D., & Reuveni, A. (2006). Attention to eyes and mouth in high-functioning children with autism. *Journal of Autism and Developmental Disorders, 36*, 131–137.

Baron-Cohen, S., Ashwin, E., Ashwin, C., Tavassoli, T., & Chakrabarti, B. (2010). Talent in autism: Hyper-systemizing, hyper-attention to detail and sensory hyper-sensitivity. In F. Happe & U. Frith (Eds.), *Autism and talent* (pp. 41–51). Oxford: Oxford University Press.

Baron-Cohen, S., & Belmonte, M. (2005). Autism: a window onto the development of the social and the analytical brain. *Annual Review of Neuroscience, 28*, 109–126.

Baron-Cohen, S., Ring, H. A., Bullmore, E. T., Wheelwright, S., Ashwin, C., & Williams, S. C. R. (2000). The amygdala theory of autism. *Neuroscience and Behavioral Reviews, 24*, 355–364.

Baron-Cohen, S., & Wheelright, S. (1999). 'Obsessions' in children with autism and Asperger syndrome. *British Journal of Psychiatry, 175*, 484–490.

Baron-Cohen, S., Wheelwright, S., Griffin, R., Lawson, J. & Hill, J. (2002). The exact mind: empathizing and systemizing in autism spectrum conditions. In *Hand Book of cognitive development* (ed. Goswani). Oxford: Blackwell.

Barron, F. (1969). *Creative person and creative process*. New York: Holt, Rinehart and Winston.

Barron, F. (1993). Controllable oddness as a resource in creativity. *Psychological Inquiry, 4*, 182–184.

Barron, F., & Harrington, D. M. (1981). Creativity, intelligence, and personality. *Annual Review of Psychology, 32*, 439–476.

Barton, J., Cherkasova, M., Hefter, R., Cox, T., O'Connor, M., & Dara Manoach, D. (2004). Are patients with social developmental disorders prosopagnostic? Perceptual heterogeneity in the Asperger and socio-emotional processing disorders. *Brain, 127*, 1706–1716.

Barton, J., Cherkasova, M., Press, D., Intriligator, J., & O'Connor, M. (2003). Developmental prosopagnosia: a study of three patients. *Brain and Cognition, 51*, 12–30.

Barton, J., Press, D., Keenan, J., & O'Connor, M. (2002). Lesions of fusiform face area impair perception of facial configuration in prosopagnosia. *Neurology, 58*, 71–78.

Bechara, A., Damasio, H., Damasio, A. R., & Lee, G. R. (1999). Different contributions of the human amygdala and ventromedial prefrontal cortex to decision-making. *Journal of Neuroscience, 19*, 5473.

Beggeer, S., Rieffe, C., Terwogt, M. M., & Stockmann, L. (2006). Attention to facial emotion expression in children with autism. *Autism, 10*, 37–51.

Bein, E.S. (1961). Paraphasias in different forms of aphasia. In: *Clinic and pathophysiology of aphasia* (eds. E.B. Schmidt and P.A. Tkachjov). Moscow: Medgiz.

Belmonte, M. K., Cock, E. H., Anderson, G. M., Rubenstein, J. L. R., Greenough, W. T., Beckel-Mitchener, A., Courchesne, E., Boulanger, L. M., Powell, S. B., Levitt, P. R., Perry, E. K., Jiang, Y. H., DeLorey, T. M., & Tierney, E. (2004). Autism as a disorder of neural information processing: directions for research and targets for therapy. *Molecular Psychiatry, 9*, 646–663.

Ben Shalom, D., Mostofsky, S. H., Hazlett, R. L., Goldberg, M. C., Landa, R. J., Faran, Y., McLeod, D. R., & Hoehn-Saric, R. (2006). Normal physiological emotions but differences in expression of conscious feelings in children with high-functioning autism. *Journal of Autism and Developmental Disorders, 36*, 395–400.

Bentin, S., Allison, T., Puce, A., Perez, E., & McCarthy, G. (1996). Electrophysiological studies of face perception in humans. *Journal of Cognitive Neuroscience, 8*, 551–565.

Benton, A., & Tranel, D. (1993). Impairment in facial recognition. In K. M. Heilman & E. Valenstein (Eds.), *Clinical neuropsychology* (pp. 176–182). New York: Oxford University Press.

Bernier, R., Dawson, G., Panagiotides, H., & Webb, S. (2005). Individuals with autism spectrum disorder show normal responses to a fear potential startle paradigm. *Journal of Autism and Developmental Disorders, 35*, 575–583.

Berns, G. S., Cohen, J. D., & Mintun, M. A. (1997). Brain regions responsive to novelty in the absence of awareness. *Science, 276*, 1272–1275.

Bernstein, N. (1947). *On the construction of movements*. Moscow: Medgiz.

Bernstein, N. A. (1990). *Physiology of movements and activity*. Moscow: Nauka.

Bernstein, N. A. (1997). *Biomechanics and physiology of movements*. Moscow—Voronezh: The Institute of Practical Psychology.

Blair, R. J. (1999). Psychophysiological responsiveness to the distress of others in children with autism. *Personality and Individual Differences, 26*, 477–485.

Blair, R. J. R., Frith, U., Smith, N., Abell, F., & Cipolotti, L. (2002). Fractionation of visual memory: agency detection and its impairment in autism. *Neuropsychologia, 40*, 108–118.

Blinkov, S. M. (1938). Structural variability of the cerebral cortex: middle temporal region of adult man. *Moscow, Brain Institute, 314*, 313–362.

Blinkov, S. M. (1955). *Temporal region of the brain in man and monkey*. Moscow: Meditsina.

Blinkov, S. M., & Glezer, I. I. (1968). *The human brain in figures and tables*. New York: Plenum Press.

Blonsky, P. D. (1935). *Memory and thinking*. Moscow-Leningrad: Ogiz.

Boddaert, N., Belin, P., Poline, J.B.. et al. (2001). Temporal lobe dysfunction in childhood autism: a PET auditory activation study. *Pediatric Radiology*, 31 [suppl.I]: S3.

Boddaert, N., & Zilbovicius, M. (2002). Functional neuroimaging and childhood autism. *Pediatric Radiology, 32*, 1–7.

Boddaert, N., Zilbovicius, M., Belin, P., et al. (2000). Temporal lobe dysfunction in autism: a PET auditory activation study. *Society for Neuroscience Abstracts, 26*, 2007.

Bodfish, J. W., Symons, F. J., Parker, D. E., & Lewis, M. H. (2000). Varieties of repetitive behavior in autism: comparisons to mental retardation. *Journal of Autism and Developmental Disorders, 30*, 237–243.

Bogyo, L., & Ellis, R. (1988). Elly: A study of contrasts. In L. K. Obler & D. Fein (Eds.), *Exceptional brain* (pp. 265–276). New York: Guilford Press.

Bolte, S., Feineis-Matthews, S., & Poustka, F. (2008). Brief report: emotional processing in high-functioning autism—physiological reactivity and affective report. *Journal of Autism and Developmental Disorders, 38*, 776–781.

Bolton, P., Pickles, A., Murphy, M., & Rutter, M. (1998). Autism, affective and other psychiatric disorders: patterns of familial aggregation. *Psychological Medicine, 28*, 385–395.

Bosch, G. (1970). *Infantile autism: A clinical and phenomenological-anthropological investigation taking language as the guide*. New York: Springer.

Boucher, J., Lewis, V., & Collis, G. (1998). Familiar face and voice matching and recognition in children with autism. *Journal of Child Psychology and Psychiatry, 39*, 171–181.

Bowler, D. M., Gaigg, S. B., & Gardiner, J. M. (2008). Subjective organization in the free recall learning of adults with Asperger's syndrome. *Journal of Autism and Developmental Disorders, 38*, 104–113.

Brodmann K. (1909). *Vergleichende Localisationslehre der Grosshirnrinde in ihren Prinzipien dargestellt auf Grund des Zellenbaues*. Leipzig: J.A.Barth.

Bruneau, N., Roux, S., Adrien, J. L., & Barthelemy, C. (1999). Auditory associative cortex dysfunction in children with autism: evidence from late auditory evoked potentials (N1 wave—T-complex). *Clinical Neurophysiology, 110*, 1927–1934.

Buck, R., & Duffy, R. (1980). Nonverbal communication of affect in brain-damaged patients. *Cortex, 16*, 351–362.

Burnette, C. P., Mundy, P. C., Meyer, J. A., Sutton, S. K., Vaughan, A. E., & Charak, D. (2005). Weak central coherence and its relation to theory of mind and anxiety in autism. *Journal of Autism and Developmental Disorders, 35*, 63–73.

Buxbaum, J. D., Silverman, J. M., Smith, C. J., Kilifarski, M., Reichert, J., & Hollander, E. (2001). Evidence for a susceptibility gene for autism on chromosome 2 and for genetic heterogeneity. *American Journal of Human Genetics, 68*, 1514–1520.

Campbell, R., Landis, T., & Regard, M. (1986). Face recognition and lip-reading: a neurological dissociation. *Brain, 109*, 509–521.

Caramazza, A., & Shelton, J. R. (1998). Domain-specific knowledge systems in the brain: the animate-inanimate distinction. *The Journal of Cognitive Neuroscience, 10*, 1–34.

Carson, S. H., Peterson, J. B., & Higgins, D. M. (2003). Decreased latent inhibition is associated with increased creative achievement in high-functioning individuals. *Journal of Personality and Social Psychology, 85*, 499–506.

Celani, G., Battacchi, M. W., & Arcidiacono, L. (1999). The understanding of the emotional meaning of facial expressions in people with autism. *Journal of Autism and Developmental Disorders, 29*, 57–66.

Cesaro, P. (1987). Thalamic pain. In J.-M. Besson, G. Guilbound, & M. Perchanski (Eds.), *Thalamus and pain* (Proceedings of the international symposium on thalamus and pain, pp. 259–270). Amsterdam, New York, Oxford: Excerpta Medica.

Chez, M. G., Chang, M., Krasne, V., Coughlan, C., Kominsky, M., & Schwartz, A. (2006). Frequency of epileptiform abnormalities in a sequential screening of autistic patients with no known clinical epilepsy from 1996 to 2005. *Epilepsy & Behavior, 8*, 267–271.

Cohen, M. S., Kosslyn, S. M., Breiter, H. C., DiGirolamo, G. J., Thompson, W. L., Anderson, A. K., Bookheimer, S. Y., Rosen, B. R., & Belliveau, J. W. (1996). Changes in cortical activity during mental rotation: a mapping study using functional MRI. *Brain, 119*, 89–100.

Constantino, J. N., & Todd, R. D. (2000). Genetic structure of reciprocal social behavior. *The American Journal of Psychiatry, 157*, 2043–2045.

Constantino, J. N., & Todd, R. D. (2005). Intergenerational transmission of subthreshold autistic traits in general population. *Biological Psychiatry, 57*, 655–660.

Corden, B., Chivers, R., & Skuse, D. (2008). Emotional modulation of perception in Asperger's syndrome. *Journal of Autism and Developmental Disorders, 38*, 1072–1080.

Costall, B., Naylor, R. J., & Olley, J. E. (1972). Stereotypic and anticataleptic activities of amphetamine after intracerebral injections. *European Journal of Pharmacology, 18*, 83–94.

Courchesne, E., Chisum, H., & Townsend, J. (1994). Neural activity—dependent brain changes in development: implications for psychopathology. *Development and Psychopathology, 6*, 697–722.

Craig, A. D. (2002). How do you feel? Interoception: the sense of the physiological condition of the body. *Nature Reviews Neuroscience, 3*, 635–666.

Craig, A. D. (2003). Interoception: the sense of the physiological condition of the body. *Current Opinion in Neurobiology, 13*, 500–505.

Critchley, H. D., Daly, E. M., Bullmore, E. T., Williams, S. C. R., Amelsvoort, T., Robertson, D. M., Rowe, A., Phillips, M., McAlonan, G., Howlin, P., & Murphy, D. G. M. (2000). The functional anatomy of social behavior (Changes in cerebral blood flow when people with autistic disorder process facial expressions). *Brain, 123*, 2203–2212.

Critchley, H., Daly, E., Phillips, M., Brammer, M., Bullmore, E., Williams, S., Van Amelsvoort, T., Robertson, D., David, A., & Declan, M. (2000). Explicit and implicit neural mechanisms for processing of social information from facial expressions: a functional magnetic resonance imaging study. *Human Brain Mapping, 9*, 93–105.

Cuccaro, M. L., Shao, Y., Grubber, J., Slifer, M., Wolpert, C. M., Donnelly, S. L., Abramson, R. K., Ravan, S. A., Wright, H. H., DeLong, G. R., & Pericak-Vance, M. A. (2003). Factor analysis of restricted and repetitive behaviours in autism using the Autism Diagnostic Interview-R. *Child Psychiatry and Human Development, 34*, 3–17.

Cutting, J. (1990). *The right cerebral hemisphere and psychiatric disorders*. Oxford: Oxford University Press.

Cutting, J. (1997). *Principles of psychopathology*. Oxford: Oxford University Press.

Cutting, J. (2002). *The living, the dead, and the never-alive: Schizophrenia and depression as fundamental variants of these*. Mill Wood, The Forest Publishing Company.

Dalton, K. M., Nacewicz, B. M., Johnstone, T., Schaefer, H. S., Gernsbacher, M. A., Golsmith, H. H., Alexander, A. L., & Davidson, R. J. (2005). Gaze fixation and the neural circuitry of face processing in autism. *Nature Neuroscience, 8*, 519–526.

Damasio, A. (2003). Mental self: the person within. *Nature, 423*, 227.

Damasio, A. R., & Anderson, S. W. (1993). The frontal lobes. In K. M. Heilman & E. Valenstein (Eds.), *Clinical neuropsychology*. New York: Oxford University Press.

Damasio, H., Grabowski, T. J., Tranel, D., Hichwa, R. D., & Damasio, A. (1996). A neural basis for lexical retrieval. *Nature, 380*, 499–505.

Damasio, A. R., & Maurer, R. G. (1978). A neurological model for childhood autism. *Archives of Neurology, 35*, 777–786.

Damasio, A. R., Tranel, D., & Damasio, H. (1990). Face agnosia and the neural substrates of memory. *Annual Review of Neurocsience, 13*, 89–109.

Darwin, C. (1872). *The expression of emotions in man and animals*. London: Murray.

Dawson, G. (1988). Cerebral lateralization in autism: Clues to its role in language and affective development. In D. L. Molfese & S. J. Segalowitz (Eds.), *Brain lateralization in children: Developmental implications* (pp. 437–457). New York: Guilford Press.

Dawson, G., Webb, S., & McPartland, J. (2005). Understanding the nature of face processing impairment in autism: insights from behavioral and neurophysiological studies. *Developmental Neuropsychology, 27*, 403–424.

De Fosse, L., Hodge, S. M., Makris, N., Kennedy, D. N., Caviness, V. S., McGrath, L., Steele, S., Ziegler, D. S., Herbert, M. R., Frazier, J. A., Tager-Flusberg, H., & Harris, G. J. (2004). Language-association cortex asymmetry in autism and specific language impairment. *Annals of Neurology, 2004*(56), 757–766.

Devlin, J. T., Russell, R. P., Davis, M. H., Price, C. J., Moss, H. E., Fadili, M. J., & Tyler, L. K. (2002). Is there an anatomical basis for category-specificity? Semantic memory studies in PET and fMRI. *Neuropshychologia, 40*, 54–75.

Dobrochotova, T. A. (1974). *Emotional pathology in focal brain disorders*. Moscow: Meditsina.

Dobrochotova, T.A. & Bragina, N.N. (1977). *Functional asymmetry and psychopathology in focal brain disorders*. Moscow: Meditsina.

Drevets, W.C., & Raichle, M.E. (1995). Positron emission tomographic imaging studies of human emotional disorders. In: *The Cognitive Neuroscience* (ed. M.S. Gazzaniga). MIT Press, Cambridge, MA, 1153–1164.

Drevets, W., Videen, T., Price, J., Preskorn, S., Carmichael, S., & Raichle, M. (1992). A functional anatomical study of unipolar depression. *Journal of Neuroscience, 12*, 3628–3641.

Dworzynski, K., Happe, F., Bolton, P., & Ronald, A. (2009). Relationship between symptom domains in autism spectrum disorders: a population based twin study. *Journal of Autism and Developmental Disorders, 39*, 1197–1210.

Ekman, P. (1982). *Emotion in the human face*. Cambridge: Cambridge University Press.

Ekman, P., & Friesen, W. V. (1975). *Unmasking the face*. Englewood Cliffs, NJ: Prentice-Hall.

Ellis, H. D., & Young, A. W. (1988). Training in face-processing skills for a child with acquired prosopagnosia. *Developmental Neuropsychology, 4*, 283–294.

Esbensen, A. J., Seltzer, M. M., Lam, K. S. L., & Bodfish, J. W. (2009). Age-related differences in restricted repetitive behaviors in autism spectrum disorders. *Journal of Autism and Developmental Disorders, 39*, 57–66.

Eysenck, H. J. (1995). Creativity as a product of intelligence and personality. In D. H. Saklofske & M. Zeidner (Eds.), *International handbook of personality and Intelligence* (pp. 231–247). New York: Plenum Press.

Eysenck, H. J., & Eysenck, S. B. (1976). *Psychoticism as a dimension of personality*. London: Hodder & Stoughton.

Fitzgerald, M. (2004). *Autism and creativity: Is there a link between autism in men and exceptional ability*. Brunner-Routledge: Hove and New York.

Fitzgerald, M. (2005). *The genesis of artistic creativity (Asperger's syndrome and the arts)*. London and Philadelphia: Jessica Kingsley Publishers.

Fodor, E. M. (1994). Subclinical manifestations of psychosis-proneness, ego-strength, and creativity. *Personality and Individual Differences, 18*, 635–642.

Fog, R., & Pakkenberg, H. (1971). Behavioral effects of dopamine and p-hydroxyamphetamine injected into corpus striatum in rats. *Experimental Neurology, 31*, 517–529.

Frith, U. (1996). Cognitive explanation of autism. *Acta Paediatrica Supplement, 416*, 63–68.

Frith, U. (2008). *Autism. A very short introduction*. New York: Oxford University Press.

Fuchs, T. (2002). The challenge of neuroscience: psychiatry and phenomenology today. *Psychopathology, 35*, 319–326.

Fuster, J. M. (1985). The prefrontal cortex and cortical integration. In A. Peters (Ed.), *Cerebral cortex* (Vol. 4, pp. 151–177). New York: Plenum Press.

Fuster, J. M. (2002). Frontal lobe and cognitive development. *Journal of Neurocytology, 31*, 373–384.

Gaigg, S. B., & Bowler, D. M. (2008). Free recall and forgetting of emotionally arousing words in autism spectrum disorder. *Neuropsychologia, 46*, 2336–2343.

Gaigg, S. B., & Bowler, D. M. (2009). Brief report: attenuated emotional suppression of the attentional blink in autism spectrum disorder: another non-social abnormality? *Journal of Autism and Developmental Disorders, 39*, 1211–1217.

Gaigg, S., Gardiner, J., & Bowler, D. (2008). Free recall in autism spectrum disorder: The role of relational and item-specific encoding. *Neuropsychologia*, 46, 983–992.

Garreau, B., Zilbovicius, M., Guerin, P., Samson, Y., Syrota, A., & Lelord, G. (1994). Effects of auditory stimulation on regional cerebral blood flow in autistic children. *Developmental Brain Dysfunction, 7*, 119–128.

Gazzaniga, M. S. (1995). In M. S. Gazzaniga (Ed.), *The cognitive neurosciences* (pp. 1049–1061). Cambridge, MA: MIT Press.

Gazzaniga, M. S. (2000). Consciousness and the cerebral hemispheres. In M. S. Gazzaniga (Ed.), *The new cognitive neurosciences*. Cambridge, MA: MIT Press.

Gerrard, S., & Rugg, G. (2009). Sensory impairments and autism: a re-examination of causal modeling. *Journal of Autism and Developmental Disorders, 39*, 1449–1463.

Geschwind, N. (1986). Interictal behavioral changes in epilepsy. *Epilepsia, 24*(Suppl. 1), S23–S30.

Glezerman, T. B. (1983). *Brain dysfunctions in children*. USSR Academy of Sciences, Department of Physiology, Nauka: Moscow.

Glezerman, T. B. (1986). *Psychophysiological grounds for intellect deterioration in aphasia: aphasia and intellect*. Moscow: USSR Academy of Sciences, Department of Physiology.

Glezerman, T. B., & Balkoski, V. I. (1999). *Language, thought, and the brain*. New York: Kluwer Academic/Plenum Publishers.

Glezerman, T., Balkoski, V., Miller, A. (1996). A neuropsychiatric model of self. In *X World congress of psychiatry abstracts*. Madrid, 2, 400–401

Gloor, P. (1990). Experiential phenomena of temporal lobe epilepsy: facts and hypotheses. *Brain, 113*, 1673–1694.

Gloor, P. (1992). Role of the amygdala in temporal lobe epilepsy. In J. P. Aggleton (Ed.), *The amygdala: Neurobiololgical aspects of emotion, memory and mental dysfunction* (pp. 505–538). New York: Wiley-Liss.

Gloor, P., Oliver, A., Quesney, L. F., Andermann, F., & Horowitz, S. (1982). The role of the limbic system in experiential phenomena of temporal lobe epilepsy. *Annals of Neurology, 12*, 129–144.

Goldenberg, G., Podreka, I., Steiner, M., & Willness, K. (1987). Patterns of regional cerebral blood flow related to memorizing of high and low imagery words. *Neuropsychologia, 25*(3), 473–485.

Gomot, M., Giard, M-H., Adrien, J-L., Barthelemy, C., & Bruneau, N. (2002). Hypersensitivity to acoustic changes in children with autism. *Psychophysiology*, 39, 577–584.

Goulden, K. J., Shinnar, S., Koller, H., Katz, M., & Richardson, S. A. (1991). Epilepsy in children with mental retardation: a cohort study. *Epilepsia, 32*, 690–697.

Grandin, T. (1995). *Thinking in pictures*. New York: Vintage Books. A division of Random House.

Grandin, T. (2010). How does visual thinking work in the mind of a person with autism? A personal account. In F. Happe & U. Frith (Eds.), *Autism and talent* (pp. 141–149). New York: Oxford University Press.

Graybiel, A. M. (1998). The basal ganglia and chunking of action repertoires. *Neurobiology of Learning and Memory, 70*, 119–136.

Hadamard, J. (1949). *The psychology of invention in the mathematical field*. New Jersey: Princeton University Press.

Hadjikhani, N., Joseph, R. N., Snyder, J., Chabris, C. F., Clark, J., Steele, S., et al. (2004). Activation of the fusiform gyrus when individuals with autism spectrum disorder view faces. *Neuroimage, 22*, 1141–1150.

Halgren, E. (1992). Emotional neurophysiology of the amygdala within the context of human cognition. In J. P. Aggleton (Ed.), *The amygdala: Neurobiological aspects of emotion, memory and mental dysfunction* (pp. 191–228). New York: Wiley.

Happe, F. G. (1999). Autism: cognitive deficit or cognitive style? *Trends in Cognitive Sciences, 3*, 216–222.

Hara, H. (2007). Autism and epilepsy: a retrospective follow-up study. *Brain & Development, 29*, 486–490.

Harris, J. C. (1995). *Developmental neuropsychiatry*. New York: Oxford University Press.

Harris, G. J., Chabris, C. F., Clark, J., Urban, T., Aharon, I., Steele, S., McGrath, L., Condouris, K., & Tager-Flusberg, H. (2006). Brain activation during semantic processing in autism spectrum disorders via functional magnetic resonance imaging. *Brain and Cognition, 61*, 54–68.

Hasselmo, M. E., Rolls, E. T., Baylis, G. C., & Nalwa, V. (1989). Object-centered encoding by face-selective neurons in the cortex of the superior temporal sulcus of the monkey. *Experimental Brain Research, 75*, 417–429.

Heaton, P., Hermelin, B., & Pring, L. (1999). Can children with autism spectrum disorders perceive affect in music? *Psychological Medicine, 29*, 1405–1410.

Heilman, K. M., Bowers, D., Speedie, L., & Coslett, H. B. (1984). Comprehension of affective and nonaffective prosody. *Neurology, 34*, 917–921.

Heilman, K. M., Watson, R. T., & Valenstein, E. (1985). Neglect and related disorders. In K. M. Heilman & E. Valenstein (Eds.), *Clinical neuropsychology* (pp. 243–293). New York: Oxford University Press.

Herbert, M. R., Harris, G. H., Adrien, K. T., Ziegler, D. A., Makris, N., Kennedy, D. N., Lange, N. T., Chabris, C. F., Bakardjiev, A., Hodgson, J., Takeoka, M., Tager-Flusberg, H., & Caviness, V. S. (2002). Abnormal asymmetry in language association cortex in autism. *Annals of Neurology, 52*, 588–596.

Herbert, M. R., Ziegler, D. A., Deutsch, C. K., O'Brien, L. M., Kennedy, D. N., Filipek, P. A., Bakardjiev, A. I., Hodgson, J., Takeoka, M., Makris, N., & Caviness, V. S. (2005). Brain asymmetries in autism and developmental language disorder: a nested whole-brain analysis. *Brain, 128*, 213–226.

Herbert, M. R., Ziegler, D. A., Deutsch, C. K., O'Brien, L. M., Lange, N., Bakardjiev, A. I., Hodgson, J., Adrien, K. T., Steele, S., Makris, N., Kennedy, D., Harris, G. H., & Caviness, V. S. (2003). Dissociations of cerebral cortex, subcortical and cerebral white matter volume in autistic boys. *Brain, 126*, 1182–1192.

Hermelin, B. & O'Connor, N. (1967). Remembering words by psychotic and subnormal children. *British Journal of Psychology*, 58, 213–218.

Hermelin, B., & O'Connor, N. (1990). Factors and primes: a specific numerical ability. *Psychological Medicine, 20*, 163–169.

Hills, A. E., Rapp, B., & Caramazza, A. (1995). Constraining claims about theories of semantic memory: more on unitary versus multiple semantics. *Cognitive Neuropsychology, 12*, 175–186.

Hirstein, W., Iversen, P., & Ramachandran, V. S. (2001). Autonomic responses of autistic children to people and objects. *Proceedings of Royal Society, London, 268*, 1883–1888.

Hobson, R. P., Ouston, J., & Lee, A. (1988). What's in a face? The case of autism. *British Journal of Psychology, 79*, 441–453.

Hollander, E., King, A., Delaney, K., Smith, C. J., & Silverman, J. M. (2003). Obsessive-compulsive behaviors in parents of multiplex autism families. *Psychiatry Research, 117*, 11–16.

Horwitz, W. A., Kestenbaum, C., Person, E., & Jarvik, L. (1965). Identical twin 'idiot savants'—calendar calculators. *American Journal of Psychiatry, 121*, 1075–1079.

Howe, M. J. A., & Smith, J. (1988). Calendar calculating in 'idiots savants': how do they do it? *British Journal of Psychology, 79*, 371–386.

Huber, G. (1957). Die coenasthetische Schizophrenie. *Fortschr. Neurol. Psychiatrie*, 25, 491–520.

Huber, G. (1992). 'Cenesthetic schizophrenia'—a subtype of schizophrenic disease. *Neurology Psychiatry and Brain Research, 1*, 54–60.

Hudenko, W. J., Stone, W., & Bachrowski, J. A. (2009). Laughter differs in children with autism: an acoustic analysis of laughs produced by children with and without autism. *Journal of Autism and Developmental Disorders, 39*, 1392–1400.

Hurlburt, R. T., Happe, F., & Frith, U. (1994). Sampling the form of inner experience in three adults with Asperger syndrome. *Psychological Medicine, 24*, 385–395.

Hus, V., Pickles, A., Cook, E. H., Jr., Risi, S., & Lord, C. (2007). Using the autism diagnostic interview—revised to increase phenotypic homogeneity in genetic studies of autism. *Biological Psychiatry, 59*, 438–448.

Hutt, C., & Hutt, J. (1965). Effects of environmental complexity on stereotyped behaviors of children. *Animal Behavior, 13*, 1–14.

Hutton, J., Goode, S., Murphy, M., Couteur, A. Le, & Rutter, M. (2008). New-onset psychiatric disorders in individuals with autism. *Autism, 12*, 373–390.

Ivanov, V. V. (1978). *Asymmetry of the brain and semiotic systems*. Moscow: Sovetskoje Radio.

Iversen, S. D. (1971). The effect of surgical lesions to frontal cortex and substantia nigra on amphetamine response in rats. *Brain Research, 31*, 295–311.

Izard, K. (1991). *The psychology of emotions*. New York: Plenum Press.

Izard, C. E., & Buechler, S. (1979). Aspects of consciousness and personality in terms of differential emotions theory. In R. Plutchik & H. Kellerman (Eds.), *Emotion: Theory, research and experience*. New York: Academic Press.

Jackson, J. H. (1880). On right and left-sided spasm at the onset of epileptic paroxysm and on crude sensation warning and elaborate mental states. *Brain, 3*, 192–206.

Jackson, J. H. (1958). *Selected writings of John Hughlings Jackson*. New York: Basic Books.

Jacobson, R. (1986). Disorders of facial recognition, social behavior and affect after combined bilateral amygdalotomy and subcaudate tractotomy—a clinical and experimental study. *Psychological Medicine, 16*, 439–450.

Jakobson, R. (1971a). Linguistics in its relation to other sciences. In UNESCO (Ed.), *Main trends of research in the social and human sciences* (pp. 419–463). Paris—The Hague: Mouton.

Jakobson, R. (1971b). Two aspects of language and two types of aphasia disturbances. In R. Jakobson (Ed.), *Selected writings* (Vol. 2, pp. 239–259). Paris—The Hague: Mouton.

Jarvinen-Pasley, A., Pasley, J., & Heaton, P. (2008). Is the linguistic content of speech less salient than its perceptual features in autism? *Journal of Autism and Developmental Disorders, 38*, 239–248.

Jaspers, K. (1968). The phenomenological approach in psychopathology. *British Journal of Psychiatry, 114*, 1313–1323.

Jaspers, K. (1997). *General psychopathology*. Baltimore and London: The John Hopkins University Press.

Jennings, W. B. (1973). *A study of the preference for affective cues in autistic children* (Unpublished PhD thesis). Memphis State University, Memphis, TN.

Jolliffe, T., & Baron-Cohen, S. (1997). Are people with autism and Asperger syndrome faster than normal on the embedded figures test? *Journal of Child Psychology and Psychiatry, 38*, 527–534.

Jolliffe, T., Landsdown, R., & Robinson, C. (1992). Autism: a personal account. *Communication, 26*, 12–19.

Jones, E. G. (2008). Triangular connectivity of the mediodorsal nucleus and amygdala. In E. G. Jones (Ed.), *The thalamus, II*. Cambridge: Cambridge University Press.

Joseph, R. M., & Tanaka, J. (2003). Holistic and part-based face recognition in children with autism. *Journal of Child Psychology and Psychiatry, 44*, 529–542.

Jung, C. G. (1971). *Psychological types* (The collected works of C. G. Jung, Vol. 6). Princeton, NJ: Princeton University Press.

Just, M. A., Cherkassky, V. L., Keller, T. A., & Minshew, N. J. (2004). Cortical activation and synchronization during sentence comprehension in high-functioning autism: evidence of underconnectivity. *Brain, 127*, 1811–1821.

Kandel, E. R., & Schwartz, J. H. (1985). *Principles of neural science*. New York: Elsevier Science.

Kanner, L. (1943). Autistic disturbances of affective contact. *Nervous Child, 2*, 217–250.

Kanner, L. (1946). Irrelevant and metaphorical language in early infantile autism. *American Journal of Psychiatry, 103*, 242–246.

Kanner, L. (1951). The conception of wholes and parts in early infantile autism. *American Journal of Psychiatry, 108*, 23–29.

Kanner, L. (1958). The specificity of early infantile autism. *Acta Paedopsychiatrica, 25*, 108–113.

Katznelson, S. D. (1972). *Typology of language and verbal thinking*. Leningrad: Nauka.

Katznelson, S. D. (1986). *General and typological linguistics*. Leningrad: Nauka.

Keller, F., & Perisco, A. M. (2003). The neurobiological context of autism. *Molecular Neurobiology, 28*, 1–22.

Kinsbourne, M. (1980). Do repetitive movement patterns in children and animals serve a dearousing function? *Developmental and Behavioral Pediatrics, 1*, 39–42.

Kircher, T., Liddle, P. F., Brammer, M. J., Williams, S. C., Murray, R. M., & McGuire, P. K. (2002). Reversed lateralization of temporal activation during speech production in thought disordered patients with schizophrenia. *Psychological Medicine, 32*, 439–449.

Kircher, T. J., Senior, C., Phillips, M. L., Rabe-Hesketh, S., Benson, P. J., Bullmore, E. T., Brammer, M., Simmons, A., Bartels, M., & David, A. S. (2001). Recognizing one's own face. *Cognition, 78*, 1–15.

Klin, A., Jones, B. A., Schultz, R., Volkmar, F., & Cohen, D. (2002). Defining and quantifying the social phenotype in autism. *American Journal of Psychiatry, 159*, 895–908.

Kling, A. S., & Brothers, L. A. (1992). The amygdala and social behavior. In J. P. Aggleton (Ed.), *The amygdala: Neurobiological aspects of emotion, memory and mental dysfunction* (pp. 353–378). New York: Wiley-Liss.

Kochetkova, V. I. (1973). *Paleoneurology*. Moscow: Moscow University Press.

Kock, E. P. (1967). *Visual agnosias*. Moscow: Meditsina.

Koestler, A. (1978). *Janus*. New York: Random House.

Kretschmer, E. (1927). *Medical psychology*. Moscow: Life and Knowledge.

Kretschmer, E. (1999). *Physique and character*. London: Routledge.

Ladavas, E., Cimatti, D., Del Pesce, M., & Tuozzi, G. (1993). Emotional evaluation with and without conscious stimulus identification: evidence from a split-brain patient. *Cognition and Emotion, 7*, 95–114.

Lam, K. S. L., Bodfish, J. W., & Piven, J. (2008). Evidence for three subtypes of repetitive behavior in autism that differ in familiality and association with other symptoms. *The Journal of Child Psychology and Psychiatry, 49*, 1193–1200.

Lehrer, J. (2008). *Proust was a neuroscientist*. Boston-New York: A Mariner Book, Houghton Mifflin Company.

Levy-Bruhl, L. (1930). *Archaic Thought* (Russian Translation from La Mentalite Primitive, Paris, 1922), eds. V.K. Nikolsky and A.V. Kissin, Moscow: "Atheist" Library.

Liegeois, F., Connely, A., Cross, J. H., Boyd, S. C., Gadian, D. G., Vargha-Khadem, F., & Baldeweg, T. (2004). Language reorganization in children with early-onset lesions of the left hemisphere: an fMRI study. *Brain, 127*, 1229–1236.

Lipska, B. K., & Weinberger, D. R. (1995). Genetic variation in vulnerability to the behavioral effects of neonatal hippocampal damage in rats. *Proceedings of the National Academy of Science United States of America, 92*, 8906–8910.

Livingstone, M., & Hubel, D. (1988). Segregation of form, color, movement, and depth: anatomy, physiology, and perception. *Science, 240*, 740–749.

Losh, M., Childress, D., Lam, K., & Piven, J. (2008). Defining key features of the broad autism phenotype: a comparison across parents of multiple- and single-incidence autism families. *American Journal of Medical Genetics B: Neuropsychiatric Genetics, 147B*, 424–433.

Lucero, A., & Glezerman, T. (2004). The right hemisphere and autistic talent: A special state of consciousness. VIII Conference of Association of Scientific Study of Consciousness (Poster). Antwerp, Belgium.

Lukianowicz, N. (1967). "Body image" disturbances in psychiatric disorders. *British Journal of Psychiatry, 113*, 31–47.
Luria, A. R. (1980). *Higher cortical functions in man*. New York: Basic Books.
Malcolm, B., Carpenter, M. D., & Sutin, J. (1983). *Human neuroanatomy*. Baltimore/London: Williams & Wilkins.
Malisza, K. L., Clancy, C., Shiloff, D., Foreman, D., Holden, J., Jones, C., Paulson, K., Summers, R., Yu, C. T., & Chudley, A. E. (2011). Functional evaluation of hidden figures object analysis in children with autistic disorder. *Journal of Autism and Developmental Disorders, 41*, 13–22.
Malkova, L., Mishkin, M., Suomi, S. J., & Bachevalier, J. (1997). Socioemotional behavior in adult rhesus monkeys after early versus late lesions of the medial temporal lobe. *Annals of New York Academy of Science, 807*, 538–540.
Mandelshtam, O. E. (1921). *Word and culture. Almanac of poet's guild*. Petrograd: Dragon.
Manes, F., Sahakian, B., Clark, L., Rogers, R., Antoun, N., Aitken, M., & Robbins, T. (2002). Decision-making processes following damage to prefrontal cortex. *Brain, 125*, 624–639.
Marindale, C., & Hasenfus, N. (1978). EEG differences as a function of creativity, stge of creative process, and effort to be original. *Biological Psychology, 6*, 157–167.
Martin, A., Wiggs, C. L., Ungerleider, L. G., & Haxby, J. V. (1996). Neural correlates of category-specific knowledge. *Nature, 379*, 649–652.
Maslov, S. Yu. (1983). *Asymmetry of cognitive mechanisms and its consequences* (Semiotics and informatics, Vol. 20, pp. 3–34). Moscow: All-Union Institute of Scientific and Technical Information (VINITI).
Maximova, E. B. (2008). *Levels of communications*. Moscow: Dialogue-MIFI.
McCrae, R. R. (1987). Creativity, divergent thinking, and openness to experience. *Journal of Personality and Social Psychology, 52*, 1258–1265.
McDougle, C. J., Kresch, L., Goodman, W. K., Naylor, S. T., Volkmar, F. R., Cohen, D. J., & Price, L. H. (1995). A case-controlled study of repetitive thoughts and behavior in adults with autistic disorder and obsessive-compulsive disorder. *American Journal of Psychiatry, 152*, 772–777.
Mednick, S. A. (1962). The associative basis of the creative process. *Psychological Review, 69*, 220–232.
Mega, M. S., & Cummings, J. L. (1994). Frontal-subcortical circuits and neuropsychiatric disorders. *Journal of Neuropsychiatry and Clinical Neurosciences, 6*, 358–370.
Merewether, F. C., & Alpert, M. (1990). The components and neuroanatomic bases of prosody. *Journal of Communication Disorders, 23*, 325–336.
Mesulam, M. (1990). Large-scale neurocognitive networks and distributed processing for attention, language and memory. *Annals of Neurology, 28*, 597–613.
Meyers, C. A., Berman, S. A., Scheibel, R. S., & Hayman, A. (1992). Case report: acquired antisocial personality disorder associated with unilateral left orbital frontal damage. *Journal of Psychiatry and Neuroscience, 17*, 121–125.
Miller, L. (1999). The Savant syndrome. Intellectual impairment and exceptional skill. *Psychological Bulletin, 125*, 31–46.
Minshew, N. J., & Goldstein, G. (1993). Is autism an amnestic disorder? Evidence from the California Verbal Learning Test. *Neuropsychology, 7*, 209–216.
Mishkin, M., Ungerleider, L. G., & Macko, K. A. (1983). Object vision and spatial vision: two cortical pathways. *Trends in Neuroscience, 6*, 414–417.
Moore, C. J., & Price, C. J. (1999). Three distinct ventral occipitotemporal regions for reading and object naming. *Neuroimage, 10*, 181–192.
Morris, J. S., Frith, S. D., Perrett, D. I., Rowland, D., Young, A. W., Calder, A. J., & Dolan, R. J. (1996). A differential neural response in the human amygdala to fearful and happy facial expressions. *Nature, 383*, 812–815.
Morris, J. S., Ohman, A., & Dolan, R. L. (1998). Conscious and unconscious emotional learning in the human amygdala. *Nature, 393*, 467–470.
Moscovitch, M. (1983). Stages of processing and hemispheric differences in language in the normal subjects. In: *Psychobiology of Language* (ed. M. Studdert-Kennedy). Cambridge, MA: MIT Press, 88–104.

Mottron, L., & Belleville, S. (1995). Perspective production in a savant autistic draughtsman. *Psychological Medicine, 25*, 639–648.

Mottron, L., Burack, J., Iarocci, G., Belleville, S., & Enns, J. (2003). Locally oriented perception with intact global processing among adolescents with high-functioning autism: evidence from multiple paradigms. *Journal of Child Psychology and Psychiatry, 44*, 904–915.

Mottron, L., Dawson, M., Soulieres, I., Hubert, B., & Burack, J. A. (2006). Enhanced perceptional functioning in autism: an update, and eight principles of autistic perception. *Journal of Autism and Developmental Disorders, 36*, 27–43.

Mullan, S., & Penfield, W. (1959). Illusions of comparative interpretation and emotion. *Archives of Neurology and Psychiatry, 81*, 269–284.

Muller, R.-A., Behen, M. E., Rothermel, R. D., Chugani, D. C., Muzik, O., Mangner, T. J., & Chugani, H. T. (1999). Brain mapping of language and auditory perception in high functioning autistic adults: a PET study. *Journal of Autism and Developmental Disorders, 29*, 19–30.

Nabokov, V. (1971). *Nikolai Gogol*. New York: A New Directions Publishing Corporation.

Nabokov, V. (1989). *Speak.* New York: Memory. Vintage International. A Division of Random House, Inc.

Nelson, B., & Rawlings, D. (2007). Its own reward: a phenomenological study of artistic creativity. *Journal of Phenomenological Psychology, 38*, 217–255.

Nelson, B., & Rawlings, D. (2010). Relating schizotypy and personality to the phenomenology of creativity. *Schizophrenia Bulletin, 36*, 388–399.

Nestadt, G., Samuels, J., Riddle, M., Bienvenu, J., Liang, K.-Y., LaBuda, M., Walkup, J., Grados, M., & Hoehn-Saric, R. (2000). A family study of obsessive-complusive disorder. *Archive of General Psychiatry, 57*, 358–363.

Newberg, A. B., & Iversen, J. (2003). The neural basis of the complex mental task of meditation: neurotransmitter and neurochemical considerations. *Medical Hypotheses, 61*, 282–291.

Nikolaenko, N. N., & Deglin, V. (1984). Semiotics of space and functional brain asymmetry. *Scientific Works of Tartu State University, 17*, 441, 448–467.

O'Leary, D. D. M., & Nakagawa, Y. (2002). Patterning centers, regulatory genes and extrinsic mechanisms controlling arealization of the neocortex. *Current Opinion in Neurobiology, 12*, 14–25.

Ozonoff, S., & Miller, J. N. (1996). An exploration of right-hemisphere contributions to the pragmatic impairments of autism. *Brain and Language, 52*, 411–434.

Palkovitz, R. J., & Wiesenfeld, A. R. (1980). Differential autonomic responses of autistic and normal children. *Journal of Autism and Developmental Disorders, 10*, 347–360.

Pandya, D. N., & Yeterian, E. H. (1985). Architecture and connections of cortical association areas. In A. Peters & E. G. Jones (Eds.), *Cerebral cortex: Association and auditory cortices* (pp. 3–60). New York: Plenum.

Pardo, J. V., Fox, P. T. & Raichle, M.E. (1991). Localization of a human system for sustained attention by positron emission tomography. *Nature*, 349, 61–64.

Pardo, J. V., Pardo, P. J., & Raichle, M. E. (1993). Neural correlates of self-induced dysphoria. *American Journal of Psychiatry, 150*, 713–719.

Park, C. C. (2001). *Exiting nirvana: A daughter life's with autism.* Boston-New York-London: Little, Brown and Company.

Park, D., & Youderian, P. (1974). Light and number: ordering principles in the world of an autistic child. *Journal of Autism and Childhood Schizophrenia, 4*, 313–323.

Passingham, R. (1997). Functional organization of motor system. In R. S. J. Frackowiak, K. J. Friston, C. D. Frith, R. J. Dolan, & J. C. Mazziottal (Eds.), *Human brain function* (pp. 243–274). San Diego: Academic Press.

Paul, R., Augustyn, A., Klin, A., & Volkmar, F. R. (2005). Perception and production of prosody by speakers with autism spectrum disorders. *Journal of Autism and Developmental Disorders, 35*, 205–220.

Paul, R., Shriberg, L. D., McSweeny, J., Cicchetti, D., Klin, A., & Volkmar, F. (2005). Brief report: relations between prosodic performance and communication and socialization rating in high functioning speakers with autism spectrum disorders. *Journal of Autism and Developmental Disorders, 35*, 861–869.

Pelphrey, K. A., Sasson, N. J., Reznick, J. S., Paul, G., Goldman, B. D., & Piven, J. (2002). Visual scanning of faces in autism. *Journal of Autism and Developmental Disorders, 32*, 249–261.

Pelphrey, K., Singerman, J. D., Allison, T., & McCarthy, G. (2003). Brain activation evoked perception of gaze shifts: the influence of context. *Neuropsychologia, 41*, 156–170.

Penfield, W. (1975). *The mystery of the mind. A critical study of consciousness and the human brain*. Princeton NJ: Princeton University Press.

Penfield, W., & Perot, P. (1963). The brain's record of auditory and visual experience. *Brain, 86*, 596–696.

Penn, H. E. (2006). Neurobiological correlates of autism: a review of recent research. *Child Neuropsychology, 12*, 57–79.

Perrett, D. I., Hietanen, J. K., Oram, M. W., & Benson, P. J. (1992). Organization and functions of cells responsive to faces in the temporal cortex. *Phylosophical Transactions of the Royal Society of London—Series B: Biological Sciences, 335*, 23–30.

Phillips, M. L. (2004). Facial processing deficits and social dysfunction: how are they related? *Brain, 127*, 1691–1692.

Pickles, A., Starr, E., Kazak, S., Bolton, P., Papanikolaou, K., & Bailey, A. (2000). Variable expression of the autism broader phenotype findings from extended pedigrees. *Journal of Child Psychology and Psychiatry, 41*, 491–502.

Pierce, K., Haist, F., Sedaghat, F., & Courchesne, E. (2004). The brain response to personally familiar faces in autism: findings of fusiform activity and beyond. *Brain, 127*, 2703–2716.

Pierce, K., Muller, R.-A., Ambrose, J., Allen, G., & Courchesne, E. (2001). Face processing occurs outside the fusiform 'face area' in autism: evidence from functional MRI. *Brain, 124*, 2059–2073.

Piven, J., Arndt, S., Bailey, J., Haveramp, S., Andreasen, N. C., & Palmer, P. (1995). An MRI study of brain size in autism. *American Journal of Psychiatry, 152*, 1145–1149.

Piven, J., Palmer, P., Jacobi, D., Childress, D., & Arndt, S. (1997). Broader autism phenotype. Evidence from a family history study of multiple-incidence autism families. *American Journal of Psychiatry, 154*, 185–190.

Porrino, L. J., Crane, A. M., & Goldman-Rakis, P. S. (1981). Direct and indirect pathways from the amygdala to the frontal lobe. *Journal of Comprehensive Neurology, 198*, 121–136.

Posner, M. I., & Peterson, S. E. (1990). The attention system of the human brain. *Annual Review of Neuroscience, 13*, 25–42.

Prior, M., & Macmillan, M. B. (1973). Maintenance of sameness in children with Kanner's syndrome. *Journal of Autism and Childhood Schizophrenia, 3*, 154–167.

Puce, A., Allison, T., Bentin, S., Gore, J., & Gregory McCarthy, G. (1998). Temporal cortex activation in humans viewing eye and mouth movements. *The Journal of Neuroscience, 18*, 2188–2199.

Randrup, A., & Munkvad, I. (1967). Stereotypical activities produced by amphetamine in several animal species and in man. *Psychopharmacologia, 11*, 300–310.

Reaven, J. A., Hepburn, S. L., & Ross, R. G. (2008). Use of the ADOS and ADI-R in children with psychosis: importance of clinical judgment. *Clinical Child Psychology and Psychiatry, 13*, 81–94.

Reisinger, L. M., Cornish, K. M., & Fombonne, E. (2011). Diagnostic differentiation of autism spectrum disorders and pragmatic language impairment. *Journal of Autism and Developmental Disorders, 41*(12), 1694–1704.

Ridley, R. M. (1994). The psychology of perseverative and stereotypical behavior. *Progress in Neurobiology, 44*, 221–231.

Ridley, R. M., & Baker, H. F. (1982). Stereotypy in monkeys and humans. *Psychological Medicine, 12*, 61–72.

Ridley, R. M., Baker, H. F., Owen, F., Cross, A. J., & Crow, T. J. (1982). Behavioral and biochemical effects of chronic amphetamine treatment on the vervet monkey. *Psychopharmacology, 78*, 245–251.

Ring, H. A., Baron-Cohen, S., Wheelwright, S., Williams, S. C. R., Brammer, M., Andrew, C., & Bullmore, E. T. (1999). Cerebral correlates of preserved cognitive skills in autism. A functional MRI study of Embedded figures task performance. *Brain, 122*, 1305–1315.

Rojas, D. S., Bawn, S. D., Benkers, T. L., Reite, M. L., & Rogers, S. J. (2002). Smaller left hemisphere planum temporal in adults with autistic disorder. *Neuroscience Letters, 328*, 237–240.

Rolls, E. T. (1992). Neurophysiology and functions of the primate amygdala. In J. P. Aggleton (Ed.), *The amygdala: Neurobiological aspects of emotion, memory, and mental dysfunction* (pp. 143–166). New York: Wiley-Liss.

Ronald, A., Happe, F., Bolton, P., Butcher, L. M., Price, T. S., Wheelright, S., et al. (2006). Genitic heritability between the three components of the autism spectrum: a twin study. *Journal of the American Academy of Child and Adolescent Psychiatry, 45*, 691–699.

Ronald, A., Happe, F., Price, T. S., Baron-Cohen, S., & Plomin, R. (2006). Phenotypic and genetic overlap between autistic traits at the extremes of the general population. *Journal of the American Academy of Child and Adolescent Psychiatry, 45*, 1206–1214.

Ropar, D., & Peebles, D. (2007). Sorting preference in children with autism: the dominance of concrete features. *Journal of Autism and Developmental Disorders, 37*, 270–280.

Rosenberg, R. E., Daniels, A. M., Law, J. K., Law, P. A., & Kaufman, W. E. (2009). Trends in autism spectrum disorder diagnoses: 1994–2007. *Journal of Autism and Developmental Disorders, 39*, 1099–1111.

Ross, E. D. (2003). Aprosodias. In T. E. Feinberg & M. J. Farath (Eds.), *Behavioral neurology and neuropsychology* (pp. 743–755). New York: McGraw-Hill.

Ross, E. D., & Mesulam, M. M. (1979). Dominant language functions of the right hemisphere? Prosody and emotional gesturing. *Archive of Neurology, 36*, 144–148.

Ross, E. D., Orbelo, D. M., Burgard, M., & Hansel, S. (1998). Functional-anatomic correlates of aprosodic deficits in patients with right brain damage. *Neurology, 50*(suppl 4), A363.

Rutter, M. L. (1970). Autistic children: infancy to adulthood. *Seminars in Psychiatry, 2*, 435–450.

Rutter, M. L. (2011). Progress in understanding autism: 2007–2010. *Journal of Autism and Developmental Disorders, 41*, 395–404.

Rylander, G. (1972). Psychosis and the punding and choreiform syndromes in addiction to central stimulant drugs. *Folia Psychiatria, Neurologia, Neurochirurgia, 75*, 203–212.

Sacks, O. (1987). The twins. In O. Sacks (Ed.), *Man who mistook his wife for a hat* (pp. 195–213). New York: Harper & Row.

Sacks, O. (1995). *An anthropologist on Mars*. New York: Alfred A. Knopf.

Saper, C. B. (2002). The central autonomic nervous system: conscious visceral perception and autonomic pattern generation. *Annual Review of Neuroscience, 25*, 433–469.

Saunders, R. C., Kolachana, B. S., Bachevalier, J., & Weinberger, D. R. (1998). Neonatal lesions of the medial temporal lobe disrupt prefrontal cortical regulation of striatal dopamine. *Nature, 393*, 169–171.

Schleicher, A., Morosan, P., Amunts, K., & Zilles, K. (2009). Quantitative architectural analysis: a new approach to cortical mapping. *Journal of Autism and Developmental Disorders, 39*, 1568–1581.

Schmaryan, A. S. (1949). *Brain pathology and psychiatry*. Moscow: Medgiz.

Schultz, R. (2005). Developmental deficits in social perception in autism: the role of the amygdala and fusiform face area. *International Journal of Developmental Neuroscience, 23*, 125–141.

Schultz, R., Gauhier, I., Klin, A., Fullbright, R., Anderson, A., Volkmar, F., Skudlarski, P., Lacadic, C., Cohen, D., & Gore, J. (2000). Abnormal ventral temporal cortical activity during face discrimination among individuals with autism and Asperger syndrome. *Archives of General Psychiatry, 57*, 331–340.

Scraggs, P. R., & Ridley, R. M. (1978). Behavioral effects of amphetamine in a small primate: relative potency of the D- and L-isomers. *Psychopharmacology, 15*, 243–245.

Selfe, L. (1977). *Nadia. A case of extraordinary drawing ability in an autistic child*. London, UK: Academic Press.

Shah, A., & Frith, U. (1983). An islet of ability in autistic children: a research note. *Journal of Child Psychology and Psychiatry, 24*, 613–620.
Shao, Y., Cuccaro, M. L., Hauser, E. R., Riford, K. L., Menold, M. M., Wolpert, C. M., Ravan, S. A., Elston, L., Decena, K., Donnelly, S. L., Abramson, R. K., Wright, H. H., DeLong, G. R., Gilbert, J. R., & Pericak-Vance, M. A. (2003). Fine mapping of autistic disorder to chromosome 15q11-q13 by use of phenotypic subtypes. *American Journal of Human Genetics, 72*, 539–548.
Sidtis, J. J., Volpe, B. T., Holtzman, J. D., Wilson, D. H., & Gazzaniga, M. S. (1981). Cognitive interaction after staged callosal section: evidence for transfer of semantic activation. *Science, 212*, 344–346.
Simmons, J. Q., & Baltaxe, C. (1975). Language patterns of adolescent autistics. *Journal of Autism and Childhood Schizophrenia, 5*, 333–351.
Skuse, D. H., Mandy, W. P. L., & Scourfield, J. (2005). Measuring autistic traits: heritability, reliability and validity of the social and communication disorders checklist. *British Journal of Psychiatry, 187*, 568–572.
Smirnov, V. (1976). *Stereotaxical neurology*. Moscow: Meditsina.
Smith, S. B. (1983). *The great mental calculators*. New York: Columbia University Press.
Spezio, M. L., Adolphs, R., Hurley, R. S. E., & Piven, J. (2007). Abnormal use of facial information in high-functioning autism. *Journal of Autism and Developmental Disorders, 37*, 929–939.
Spiker, D., Lotspeich, L. J., Dimiceli, S., Myers, R. M., & Risch, N. (2002). Behavioral phenotypic variation in autism multiplex families. Evidence for a continuous severity gradient. *American Journal of Medical Genetics, 114*, 129–136.
Spitz, H. H. (1994). Lewis Carroll's formula for calendar calculating. *American Journal on Mental Retardation, 98*, 601–606.
Stavridou, A., & Furnham, A. (1996). The relationship between psychoticism, trait-creativity and the attentional mechanism of cognitive inhibition. *Personality and Individual Differences, 21*, 143–153.
Stewart, L., Meyer, B.-U., Frith, U., & Rothwell, J. (2001). Left posterior BA37 is involved in object recognition: a TMS study. *Neuropsychologia, 39*, 1–6.
Sugihara, G., Tsuchiya, K. J., & Takei, N. (2008). Distinguishing broad autism phenotype from schizophrenia-spectrum disorders. *Journal of Autism and Developmental Disorders, 38*, 1998–1999.
Szatmari, P., Georgiades, S., Bryson, S., Zwaigenbaum, L., Roberts, W., Mahoney, W., Goldberg, J., & Tuff, L. (2006). Investigating the structure of the restricted, repetitive behaviours and interests domain of autism. *Journal of Child Psychology and Psychiatry, 47*, 582–590.
Szatmari, P., Georgiades, S., Duku, E., Zwaigenbaum, L., Glodberg, J., & Bennett, T. (2008). Alexithymia in parents of children with autism spectrum disorder. *Journal of Autism and Developmental Disorders, 38*, 1859–1865.
Tager-Flusberg, H. (1985). Basic level and superordinate level categorization by autistic, mentally retarded, and normal children. *Journal of Experimental Child Psychology, 40*, 450–469.
Teunisse, J. P., & De Gelder, B. (1994). Do autists have a generalized face processing deficit? *International Journal of Neuroscience, 77*, 1–10.
Theoharides, T. C., Kempuraj, D., & Redwood, L. (2009). Autism: an emerging 'neuroimmune disorder' in search of therapy. *Expert Opinion, 10*, 2128–2143.
Toichi, M., & Kamio, Y. (2001). Verbal associations for simple and common words in high-functioning autism. *Journal of Autism and Developmental Disorders, 31*, 483–489.
Tranel, D., Damasio, A. R., & Damasio, H. (1988). Intact recognition of facial expression, gender, and age in patients with impaired face identity. *Neurology, 38*, 690–696.
Treffert, D. A. (1988). The idiot savant: a review of the syndrome. *American Journal of Psychiatry, 145*, 563–572.
Trimble, M. R. (2007). *The soul in the brain*. Baltimore: The John Hopkins University Press.

Ullman, S. (1995). The visual analysis of shape and form. In M. S. Gazzaniga (Ed.), *The cognitive neuroscience* (pp. 339–350). Cambridge, MA: MIT Press.

Ungerer, J. A., & Sigman, M. (1987). Categorization skills and receptive language development in autistic children. *Journal of Autism and Developmental Disorders, 17*, 3–16.

Van der Geest, J. N., Kemner, C., Camfferman, G., Verbated, M. N., & van Engeland, H. (2002). Looking at images with human figures: comparison between autistic and normal children. *Journal of Autism and Developmental Disorders, 32*, 69–75.

Volkmar, F. R., & Nelson, D. S. (1990). Seizure disorder in autism. *Journal of American Academy of Child and Adolescent Psychiatry, 29*(1), 127–129.

Vuilleumier, P., Mohr, C., Valenza, N., Wetzel, C., & Landis, T. (2003). Hyperfamiliarity for unknown faces after left lateral temporo-occipital venous infarction: a double dissociation with prosopagnosia. *Brain, 126*, 889–907.

Walenski, M., Mostofsky, S. H., Gidley-Larsom, J. C., & Ullman, M. T. (2008). Brief report: enhanced picture naming in autism. *Journal of Autism and Developmental Disorders, 38*, 1395–1399.

Wang, A. T., Lee, S. S., Sigman, M., & Dapretto, M. (2006). Neural basis of irony comprehension in children with autism: the role of prosody and context. *Brain, 129*, 932–943.

Wapner, W., Hamby, S., & Gardner, H. (1981). The role of the right hemisphere in the apprehension of complex linguistic materials. *Brain and Language, 14*, 15–33.

Warrington, E. K., & Shallice, T. (1984). Category specific semantic impairments. *Brain, 107*, 829–854.

Waterhouse, L. (1988). Extraordinary visual memory and pattern perception in an autistic boy. In L. K. Obler & D. Fein (Eds.), *Exceptional brain*. New York: Guilford Press.

Weeks, S. J., & Hobson, R. P. (1987). The salience of facial expression for autistic children. *Journal of Child Psychology and Psychiatry, 28*, 137–152.

Welchew, D. E., Ashwin, C., Berkouk, K., Salvador, R., Suckling, J., Baron-Cohen, S., & Bullmore, E. (2005). Functional disconnectivity of the medial temporal lobe in Asperger's syndrome. *Biological Psychiatry, 57*, 991–998.

Willemsen-Swinkels, S. H., Bakermans-Kranenburg, M. J., Buitelaar, J. K., van Ijzedorn, M. H., Buitelaar, J. K., & van Engeland, H. (2000). Insecure and disorganized attachment in children with a pervasive developmental disorder: relationship with social interaction and heart rate. *Journal of Child Psychology and Psychiatry, 41*, 759–767.

Winner, E., & Gardner, H. (1977). The comprehension of metaphor in brain-damaged patients. *Brain, 100*, 717–729.

Witelson, S. F., Kigar, D. L., & Harvey, T. (1999). The exceptional brain of Albert Einstein. *The Lancet, 353*, 2149–2153.

Yamane, S., Kaji, S., & Kawano, K. (1988). What facial features activate face neurons in the inferotemporal cortex of the monkey? *Experimental Brain Research, 73*, 209–214.

Young, A. W., & Ellis, H. D. (1989). Childhood prosopagnosia. *Brain and Cognition, 9*, 16–47.

Young, A. W., Hay, D. C., & Ellis, A. W. (1985). The faces that launched a thousand slips: everyday difficulties and errors in recognizing people. *British Journal of Psychology, 76*, 495–523.

Young, A. W., Leafhead, K. M., & Szulecka, T. K. (1994). The Capgras and Cotard delusions. *Psychopathology, 27*, 226–231.

Young, A. W., Newcombe, F., de Haan, E. H. F., Small, M., & Hay, D. C. (1993). Face perception after brain injury. Selective impairments affecting identity and expression. *Brain, 116*, 941–959.

Zald, D. H., & Kim, S. W. (1996a). Anatomy and function of the orbital frontal cortex, I: anatomy, neurocircuitry, and obsessive-compulsive disorder. *The Journal of Neuropsychiatry and Clinical Neurosciences, 8*, 125–138.

Zald, D. H., & Kim, S. W. (1996b). Anatomy and function of the orbital frontal cortex, II: function and relevance to the obsessive-compulsive disorder. *The Journal of Neuropsychiatry and Clinical Neuroscience, 8*, 249–261.

Zeki, S., & Lamb, M. (1994). The neurology of kinetic art. *Brain, 117*, 607–636.

Zeman, A. (2001). Consciousness. *Brain, 124*, 1263–1289.

Index

T.B. Glezerman, *Autism and the Brain: Neurophenomenological Interpretation*,
DOI 10.1007/978-1-4614-4112-0, © Springer Science+Business Media New York 2013

I

L

MIX
Papier aus verantwortungsvollen Quellen
Paper from responsible sources
FSC® C105338

If you have any concerns about our products,
you can contact us on
ProductSafety@springernature.com

In case Publisher is established outside the EU,
the EU authorized representative is:
Springer Nature Customer Service Center GmbH
Europaplatz 3, 69115 Heidelberg, Germany

Printed by Libri Plureos GmbH
in Hamburg, Germany